T0305993

THE MATHEMATICAL LANGUAGE
OF QUANTUM THEORY

For almost every student of physics, their first course on quantum theory raises puzzling questions and creates an uncertain picture of the quantum world. This book presents a clear and detailed exposition of the fundamental concepts of quantum theory: states, effects, observables, channels and instruments. It introduces several up-to-date topics, such as state discrimination, quantum tomography, measurement disturbance and entanglement distillation. A separate chapter is devoted to quantum entanglement. The theory is illustrated with numerous examples, reflecting recent developments in the field. The treatment emphasizes quantum information, though its general approach makes it a useful resource for graduate students and researchers in all subfields of quantum theory. Focusing on mathematically precise formulations, the book summarizes the relevant mathematics.

TEIKO HEINOSAARI is a researcher in the Turku Centre for Quantum Physics of the Department of Physics and Astronomy at the University of Turku, Finland. His research focuses on quantum measurements and quantum information theory.

MÁRIO ZIMAN is a researcher in the Research Center for Quantum Information at the Institute of Physics at the Slovak Academy of Sciences, Bratislava, and lectures at the Faculty of Informatics at the Masaryk University in Brno. His research interests include the foundations of quantum physics, quantum estimations and quantum information theory.

The present volume is part of an informal series of books, all of which originated as review articles published in *Acta Physica Slovaca*. The journal can be accessed for free at www.physics.sk/aps.

Vladimir Buzek, editor of the journal.

THE MATHEMATICAL LANGUAGE OF QUANTUM THEORY

From Uncertainty to Entanglement

TEIKO HEINOSAARI

University of Turku, Finland

MÁRIO ZIMAN

Slovak Academy of Sciences, Slovakia

CAMBRIDGE
UNIVERSITY PRESS

Shaftesbury Road, Cambridge CB2 8EA, United Kingdom

One Liberty Plaza, 20th Floor, New York, NY 10006, USA

477 Williamstown Road, Port Melbourne, VIC 3207, Australia

314–321, 3rd Floor, Plot 3, Splendor Forum, Jasola District Centre, New Delhi – 110025, India

103 Penang Road, #05–06/07, Visioncrest Commercial, Singapore 238467

Cambridge University Press is part of Cambridge University Press & Assessment, a department of the University of Cambridge.

We share the University's mission to contribute to society through the pursuit of education, learning and research at the highest international levels of excellence.

www.cambridge.org
Information on this title: www.cambridge.org/9780521195836

First published 2012
Reprinted 2013

A catalogue record for this publication is available from the British Library

ISBN   978-0-521-19583-6   Hardback

Contents

Preface

Quantum theory is not an easy subject to master. Trained in the everyday world of macroscopic objects like locomotives, elephants and watermelons, we are insensitive to the beauty of the quantum world. Many quantum phenomena are revealed only in carefully planned experiments in a sophisticated laboratory. Some features of quantum theory may seem contradictory and inconceivable in the framework set by our experience. Rescue comes from the language of mathematics. Its mighty power extends the limits of our apprehension and gives us tools to reason systematically even if our practical knowledge fails. Mastering the relevant mathematical language helps us to avoid unnecessary quantum controversies.

Quantum theory, as we understand it in this book, is a general framework. It is not so much about *what is out there*, but, rather, determines constraints on *what is possible* and *what is impossible*. This type of constraint is familiar from the theory of relativity and from thermodynamics. We will see that quantum theory is also a framework, and one of great interest, where these kinds of question can be studied.

What are the main lessons that quantum theory has taught us? The answer, of course, depends on who you ask. Two general themes in this book reflect our answer: uncertainty and entanglement.

Uncertainty. Quantum theory is a statistical theory and there seems to be no way to escape its probabilistic nature. The intrinsic randomness of quantum events is the seed of this uncertainty. There are various different ways in which it is manifested in quantum theory. We will discuss many of these aspects, including the nonunique decomposition of a mixed state into pure states, Gleason's theorem, the no-cloning theorem, the impossibility of discriminating nonorthogonal states and the unavoidable disturbance caused by the quantum measurement process.

Entanglement. The phenomenon of entanglement provides composite quantum systems with a very puzzling and counterintuitive flavour. Many of its consequences dramatically contradict our classical experience. Measurement outcomes observed by different observers can be entangled in a curious way, allowing for the violation of local realism or the teleportation of quantum states. It is fair to say that entanglement is the key resource for quantum information-processing tasks.

It should be noted that both these themes, uncertainty and entanglement, have puzzled quantum theorists since the beginning of the quantum age. Uncertainty can be traced back to Werner Heisenberg, while the word 'entanglement' was coined by Erwin Schrödinger. After many decades uncertainty and entanglement are still under active research, probably more so than ever before.

The evolution of this book had several stages, quite similar to the life cycle of a frog. Its birth dates back to a series of lectures the authors gave in 2007–8 in the Research Center for Quantum Information, Bratislava. The audience consisted of Ph.D. students working on various subfields of quantum theory. The principal aim of the lectures was to introduce a common language for the core part of quantum theory and to formulate some fundamental theorems in a mathematically precise way.

The positive response from the students encouraged us to hatch the egg, and the tadpole stage of this book was an article based on the lecture notes. The article appeared in *Acta Physica Slovaca* in August 2008. The metamorphosis from a tadpole to an adult frog took more than two years. During that stage we benefited greatly from the comments of our friends and colleagues. We hope that this grown-up frog can serve as a guide for students wishing to get an overview of the mathematical formulation of quantum theory.

Acknowledgement. This book would have never come into existence without the guidance, encouragement and help of Vladimír Bužek. Many friends and colleagues read parts of earlier versions of the manuscript and gave us valuable comments and support. Our special thanks are addressed to Paul Busch, Andreas Doering, Viktor Eisler, Sergej Nikolajevič Filippov, Stan Gudder, Jukka Kiukas, Pekka Lahti, Leon Loveridge, Daniel McNulty, Harri Mäkelä, Marco Piani, Daniel Reitzner, Tomáš Rybár, Michal Sedlák, Peter Staňo and Kari Ylinen. Teiko Heinosaari is grateful to the Alfred Kordelin Foundation for financial support.

Introduction

What is this book about? This book is an *introduction* to the basic structure of quantum theory. Our aim is to present the most essential concepts and their properties in a pedagogical and simple way but also to keep a certain level of mathematical precision. On top of that our intention is to illustrate the formalism in examples that are closely related to current research problems. As a result, the book has a quantum information flavor although it is not a quantum information textbook. The ideas related to quantum information are presented as consequences or applications of the basic quantum formalism.

Many textbooks on quantum physics concentrate either on finite- or infinite-dimensional Hilbert spaces. In this book the idea has been to treat finite- and infinite-dimensional Hilbert-space formalisms on the same footing. To keep the book at a reasonable size and so as not to be drawn into mathematical technicalities, we have sometimes compromised and presented a theorem in a general form

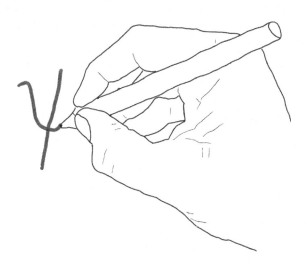

but have given its proof only in the finite-dimensional case. The last chapter, on the mathematical aspects of the phenomenon of entanglement, is written entirely in a finite-dimensional setting.

What this book is not about. We should perhaps warn the reader that this book is not about the different interpretations of quantum theory: we mostly avoid discussions of the philosophical consequences of the theory. These issues are important and interesting, but a proper understanding is easier to achieve if one knows the mathematical structure reasonably well first. Naturally, one cannot grasp the theory without some minimal interpretation. We have tried to keep the discussion away from controversial questions that do not (yet) have commonly agreed frameworks. Actually, even though two quantum theorists might disagree on some deeper foundational issues, they would agree on the basic predictions of quantum theory, such as measurement outcome probabilities. The reader will notice that many topics that could (and perhaps should) be included under the title *The Mathematical Language of Quantum Theory* are missing. For instance, the book *Mathematical Concepts of Quantum Mechanics* by Gustafson and Sigal has almost the same title as the present book but an almost disjoint content (actually, it can be recommended as a complement to this book). We believe that our book provides a compact and coherent overview of one branch of the mathematical language of quantum theory.

Prerequisites. We assume that the reader has already met quantum theory and has perhaps studied an elementary course on quantum mechanics. We also assume a basic knowledge of real analysis and linear algebra. We imagine that a typical reader is a Master's or Ph.D. student who wants to broaden his or her knowledge and learn a framework into which quantum concepts fit naturally. Clear definitions of basic concepts and their properties, together with a reasonably comprehensive index, should make the book useful as a reference work also.

Structure of the book. The book consists of 28 sections, grouped into six thematic chapters. Each section is divided into several subsections, which typically treat one question or topic. In the first chapter we recall the basics of Hilbert spaces. In each of the following five chapters we concentrate on one or two key concepts of quantum formalism. Within the body of the text there are short exercises where the reader is asked to verify a formula. These exercises are easy and straightforward, and hints are often given.

1

Hilbert space refresher

Quantum theory, in its conventional formulation, is built on the theory of Hilbert spaces and operators. In this chapter we go through this basic material, which is central for the rest of the book. Our treatment is mainly intended as a refresher and a summary of useful results. It is assumed that the reader is already familiar with some of these concepts and elementary results, at least in the case of finite-dimensional inner product spaces.

We present proofs for propositions and theorems only if the proof itself is considered to be instructive and illustrative. This gives us the freedom to present the material in a topical order rather than in the strict order of mathematical implication. Good references for this chapter are the functional analysis textbooks by Conway [45], Pedersen [113] and Reed and Simon [121]. These books also contain the proofs that we skip here.

1.1 Hilbert spaces

As an introduction, before a formal definition is given, one may think of a Hilbert space as the closest possible generalization of the inner product spaces $\mathbb{C}^d$ to infinite dimensions. Actually, there are no finite-dimensional Hilbert spaces (up to isomorphisms) other than $\mathbb{C}^d$ spaces. The crucial requirement of *completeness* becomes relevant only in infinite-dimensional spaces. This defining property of Hilbert spaces guarantees that they are well-behaved mathematical objects, and many calculations can be done almost as easily as in $\mathbb{C}^d$.

1.1.1 Finite- and infinite-dimensional Hilbert spaces

Let $\mathcal{H}$ be a complex vector space. We recall that a complex-valued function $\langle \cdot | \cdot \rangle$ on $\mathcal{H} \times \mathcal{H}$ is an *inner product* if it satisfies the following three conditions for all $\varphi, \psi, \phi \in \mathcal{H}$ and $c \in \mathbb{C}$:

- $\langle \varphi | c\psi + \phi \rangle = c \langle \varphi | \psi \rangle + \langle \varphi | \phi \rangle$ (linearity in the second argument),
- $\overline{\langle \varphi | \psi \rangle} = \langle \psi | \varphi \rangle$ (conjugate symmetry),
- $\langle \psi | \psi \rangle > 0$ if $\psi \neq 0$ (positive definiteness).

A complex vector space $\mathcal{H}$ with an inner product defined on it is an *inner product space*. An alternative name for an inner product is a *scalar product*, and then naturally one speaks of scalar product spaces.

The defining conditions for an inner product have some elementary consequences. First notice that linearity and conjugate symmetry imply that

$$\langle c\varphi + \phi | \psi \rangle = \overline{c} \langle \varphi | \psi \rangle + \langle \phi | \psi \rangle$$

and that $\langle 0 | \psi \rangle = \langle \psi | 0 \rangle = 0$.

An often-used implication, which follows from the previous equation and positive definiteness, is that

$$\langle \psi | \psi \rangle = 0 \qquad \Rightarrow \qquad \psi = 0. \tag{1.1}$$

An elementary but important result for inner product spaces is the *Cauchy–Schwarz inequality*: if $\varphi, \psi \in \mathcal{H}$ then

$$|\langle \varphi | \psi \rangle|^2 \leq \langle \varphi | \varphi \rangle \langle \psi | \psi \rangle . \tag{1.2}$$

Moreover, equality occurs if and only if φ and ψ are linearly dependent, meaning that $\varphi = c\psi$ for some complex number c. The Cauchy–Schwarz inequality will be used constantly in our calculations.

A word of warning: unlike in quantum mechanics textbooks, in most functional analysis textbooks inner products are linear in the *first* argument. This is, of course, just a harmless difference in convention, and one can define

$$\langle \varphi | \psi \rangle_{\text{quantum book}} = \langle \psi | \varphi \rangle_{\text{maths book}}$$

to obtain an inner product that is linear in the second argument.

Example 1.1 (*Inner product space* $\mathbb{C}^d$)
Let $\mathbb{C}^d$ denote the vector space of all d-tuples of complex numbers. For two vectors $\varphi = (\alpha_1, \ldots, \alpha_d)$ and $\psi = (\beta_1, \ldots, \beta_d)$, the inner product $\langle \varphi | \psi \rangle$ is defined as follows:

$$\langle \varphi | \psi \rangle = \sum_{j=1}^{d} \bar{\alpha}_j \beta_j . \tag{1.3}$$

There are also other inner products on $\mathbb{C}^d$ (try to invent one!), but when referring to $\mathbb{C}^d$ we will always assume that the inner product is that defined in (1.3). $\triangle$

An *isomorphism* is, generally speaking, a structure-preserving bijective mapping. Inner product spaces have a vector space structure and the additional structure given by the inner product. Hence, in the context of inner product spaces an isomorphism is defined as follows.

Definition 1.2 Two inner product spaces $\mathcal{H}$ and $\mathcal{H}'$ are *isomorphic* if there is a bijective linear mapping $U : \mathcal{H} \to \mathcal{H}'$ such that

$$\langle U\varphi | U\psi \rangle = \langle \varphi | \psi \rangle \tag{1.4}$$

for all $\varphi, \psi \in \mathcal{H}$. The mapping U is an *isomorphism*.

Two isomorphic spaces are completely indistinguishable as abstract inner product spaces. Isomorphisms are like costumes; two Hilbert spaces may be intrinsically the same but in the morning they have chosen different costumes and therefore they look different.

An essential concept regarding inner product spaces is orthogonality. Two vectors $\varphi, \psi \in \mathcal{H}$ are called *orthogonal* if $\langle \varphi | \psi \rangle = 0$, in which case we write $\varphi \perp \psi$. A set of vectors $X \subset \mathcal{H}$ is called *orthogonal* if any two distinct vectors belonging to X are orthogonal. Among other things, the orthogonality property can be used to define the dimension of an inner product space. First, we will draw a distinction between finite and infinite dimensions.

Definition 1.3 Let $\mathcal{H}$ be an inner product space. If for any positive integer d there exists an orthogonal set of d vectors, then $\mathcal{H}$ is *infinite dimensional*. Otherwise $\mathcal{H}$ is *finite dimensional*.

We recall the following characterization of finite-dimensional inner product spaces. This basic result is usually proved in linear algebra courses.

Proposition 1.4 If $\mathcal{H}$ is a finite-dimensional inner product space then there is a positive integer d such that:

- there are d nonzero orthogonal vectors;
- for $d' > d$, any set of d' nonzero vectors contains nonorthogonal vectors.

The number d is called the *dimension* of $\mathcal{H}$. A finite-dimensional inner product space of dimension d is isomorphic to $\mathbb{C}^d$.

Exercise 1.5 For finite-dimensional inner product spaces, the dimension can be defined equivalently as the maximal number of nonzero linearly independent vectors. Let $X = \{\varphi_1, \ldots, \varphi_n\}$ be an orthogonal set of nonzero vectors in a d-dimensional Hilbert space $\mathcal{H}$. Show that X is a linearly independent set.

[Hint: Start from the equation $c_1\varphi_1 + \cdots + c_n\varphi_n = 0$. You need to show that $c_1 = \cdots = c_n = 0$.]

Not all inner product spaces are finite dimensional. The following example, which we also use later, demonstrates this fact.

Example 1.6 (*Inner product space $\ell^2(\mathbb{N})$*)
We denote by $\mathbb{N}$ the set of natural numbers, including 0. Let $\ell^2(\mathbb{N})$ be a set of functions $f : \mathbb{N} \to \mathbb{C}$ such that the sum $\sum_{j=0}^{\infty} |f(j)|^2$ is finite. The formula

$$\langle f|g \rangle = \sum_{j=0}^{\infty} \overline{f(j)}g(j) \tag{1.5}$$

defines an inner product on $\ell^2(\mathbb{N})$. (The only nonobvious part in showing that the formula (1.5) is an inner product is verifying that the sum $\sum_{j=0}^{\infty} \overline{f(j)}g(j)$ converges whenever $f, g \in \ell^2(\mathbb{N})$. This follows from Hölder's inequality.) For each $k \in \mathbb{N}$, let δ_k be the *Kronecker function*, defined as

$$\delta_k(j) = \begin{cases} 1 & \text{if } j = k, \\ 0 & \text{if } j \neq k. \end{cases}$$

It follows that

$$\langle \delta_k|\delta_\ell \rangle = 0$$

whenever $k \neq \ell$. The inner product space $\ell^2(\mathbb{N})$ is infinite dimensional since the collection of Kronecker functions is an orthogonal set. $\triangle$

Every inner product space $\mathcal{H}$ is a *normed space*, the norm being defined as

$$\|\psi\| \equiv \sqrt{\langle \psi|\psi \rangle}. \tag{1.6}$$

The fact that the real-valued function $\psi \mapsto \|\psi\|$ defined in (1.6) is a norm means that the following three conditions are satisfied for all $\varphi, \psi \in \mathcal{H}$ and $c \in \mathbb{C}$:

- $\|\varphi\| \geq 0$ and $\|\varphi\| = 0$ if and only if $\varphi = 0$;
- $\|c\varphi\| = |c| \, \|\varphi\|$;
- $\|\varphi + \psi\| \leq \|\varphi\| + \|\psi\|$ (triangle inequality).

The first two properties follow immediately from the defining conditions of inner products. The third is easy to prove using the Cauchy–Schwarz inequality (1.2):

$$\begin{aligned} \|\varphi + \psi\|^2 &= \langle \varphi|\varphi \rangle + \langle \varphi|\psi \rangle + \langle \psi|\varphi \rangle + \langle \psi|\psi \rangle \\ &\leq \langle \varphi|\varphi \rangle + 2|\langle \varphi|\psi \rangle| + \langle \psi|\psi \rangle \\ &\leq \langle \varphi|\varphi \rangle + 2\sqrt{\langle \varphi|\varphi \rangle}\sqrt{\langle \psi|\psi \rangle} + \langle \psi|\psi \rangle \\ &= \left(\|\varphi\| + \|\psi\| \right)^2. \end{aligned}$$

Three useful formulas are stated in the following exercises.

Exercise 1.7 (*Pythagorean formula*)
Let φ and ψ be orthogonal vectors in an inner product space $\mathcal{H}$. Prove that the following equality, known as the *Pythagorean formula*, holds:

$$\|\varphi + \psi\|^2 = \|\varphi\|^2 + \|\psi\|^2 . \tag{1.7}$$

[Hint: Use (1.6) and expand the left-hand side of the equation.] In contrast with the Pythagorean formula for the real inner product space $\mathbb{R}^d$, (1.7) can hold even if φ and ψ are not orthogonal. Find two vectors which demonstrate this fact. [Hint: It is essential that $\mathcal{H}$ is a *complex* inner product space.]

Exercise 1.8 (*Bessel's inequality*)
Let $\{\psi_1, \ldots, \psi_n\}$ be an orthogonal set of n unit vectors (i.e. $\|\psi_j\| = 1$) in an inner product space $\mathcal{H}$. Show that if $\varphi \in \mathcal{H}$ then

$$\sum_{j=1}^{n} |\langle \psi_j | \varphi \rangle|^2 \le \|\varphi\|^2 . \tag{1.8}$$

[Hint: Start from the fact that

$$0 \le \left\langle \varphi - \sum_{j=1}^{n} \langle \psi_j | \varphi \rangle \psi_j \, \middle| \, \varphi - \sum_{j=1}^{n} \langle \psi_j | \varphi \rangle \psi_j \right\rangle$$

and expand the right-hand side of this inequality.]

Exercise 1.9 (*Parallelogram law*)
Let φ and ψ be vectors in an inner product space $\mathcal{H}$. Prove that the following equality, known as the *parallelogram law*, holds:

$$\|\varphi + \psi\|^2 + \|\varphi - \psi\|^2 = 2\|\varphi\|^2 + 2\|\psi\|^2 . \tag{1.9}$$

[Hint: Use (1.6) and expand the left-hand side of the equation.] It is of interest to note that the converse is also true: a normed linear space is an inner product space if the norm satisfies the parallelogram law. (A proof of this fact can be found e.g. in [62], Theorem 6.1.5.)

The norm induces a *metric* on $\mathcal{H}$. The distance between two vectors ψ and φ is given by

$$d(\psi, \varphi) \equiv \|\psi - \varphi\| . \tag{1.10}$$

With the concept of distance defined, it makes sense to speak about metric concepts in an inner product space $\mathcal{H}$. For instance, we note that, for each vector ψ, the mapping $\varphi \mapsto \langle \psi | \varphi \rangle$ from $\mathcal{H}$ to $\mathbb{C}$ is *continuous*. (This is a direct consequence of the Cauchy–Schwarz inequality.)

A metric space is called *complete* if every Cauchy sequence is convergent. (A sequence $\{\varphi_j\}$ is a Cauchy sequence if, for every $\varepsilon > 0$, there exists a positive integer N_ε such that $d(\varphi_j, \varphi_k) < \varepsilon$ whenever $j, k > N_\varepsilon$.) Loosely speaking, completeness means that every sequence which looks convergent is indeed convergent.

Since inner product spaces are not only metric spaces but also normed spaces, there is an alternative characterization of completeness. Namely, a normed space is complete if and only if every absolutely convergent series is convergent. (A series $\sum_{j=1}^{\infty} \phi_j$ is called absolutely convergent if $\sum_{j=1}^{\infty} \|\phi_j\| < \infty$.)

Every finite-dimensional inner product space is automatically complete. For infinite-dimensional inner product spaces this is not true. Furthermore there is a special name for inner product spaces that are complete.

Definition 1.10 A complete inner product space is called a *Hilbert space*.

Completeness has several useful consequences, which make Hilbert spaces much easier to deal with than general inner product spaces. One consequence is the existence of basis expansions (subsection 1.1.2).

An orthogonal set $X \subset \mathcal{H}$ is an *orthonormal set* if each vector $\psi \in X$ has unit norm. An *orthonormal basis* for a Hilbert space $\mathcal{H}$ is a maximal orthonormal set; this means that there is no other orthonormal set containing it as a proper subset. A useful criterion for the maximality of an orthonormal set $X \subset \mathcal{H}$ is the following: if ψ is orthogonal to all vectors in X then $\psi = 0$.

It can be shown that every Hilbert space has an orthonormal basis and, moreover, that all orthonormal bases of a given Hilbert space have the same cardinality. A Hilbert space $\mathcal{H}$ is called *separable* if it has a countable orthonormal basis.

Example 1.11 ($\ell^2(\mathbb{N})$ *continued*)
It can be shown that the inner product space $\ell^2(\mathbb{N})$ is complete. The set of Kronecker functions $\{\delta_0, \delta_1, \ldots\}$ is an orthonormal basis for $\ell^2(\mathbb{N})$. This can be seen by using the criterion for maximality mentioned earlier. If $f \in \ell^2(\mathbb{N})$ satisfies $\langle \delta_k | f \rangle = 0$ then $f(k) = 0$. Hence, a function f which is orthogonal to all Kronecker functions is identically zero. We conclude that $\ell^2(\mathbb{N})$ is a separable infinite-dimensional Hilbert space. $\triangle$

The following proposition should be compared with Proposition 1.4.

Proposition 1.12 Any separable infinite-dimensional Hilbert space is isomorphic to $\ell^2(\mathbb{N})$.

The idea behind Proposition 1.12 is simple, and we will give an outline of the proof without going into the details. Fix an orthonormal basis $\{\varphi_k\}_{k=0}^{\infty}$ for a separable Hilbert space $\mathcal{H}$. For each vector $\psi \in \mathcal{H}$, define a function $\tilde{\psi} : \mathbb{N} \to \mathbb{C}$ by $\tilde{\psi}(j) = \langle \varphi_j | \psi \rangle$. It then follows that $\tilde{\psi} \in \ell^2(\mathbb{N})$ and the correspondence $\psi \mapsto \tilde{\psi}$ is an isomorphism between $\mathcal{H}$ and $\ell^2(\mathbb{N})$.

> **From now on, all Hilbert spaces are assumed to be separable.**

In other words, all our Hilbert spaces are either finite dimensional or countably infinite dimensional. Typically, we denote Hilbert spaces by the letters $\mathcal{H}$ or $\mathcal{K}$. Sometimes we use $\mathcal{H}_d$ for a finite d-dimensional Hilbert space. To avoid trivial statements, we will also assume that the dimension d of our Hilbert space is at least 2.

1.1.2 Basis expansion

Let $\mathcal{H}$ be either a finite-dimensional or separable infinite-dimensional Hilbert space and let $\{\varphi_k\}_{k=1}^{d}$ be an orthonormal basis for $\mathcal{H}$. We recall that this means the following three things:

- $\langle \varphi_k | \varphi_k \rangle = 1$ for every k;
- $\langle \varphi_j | \varphi_k \rangle = 0$ for every $j \neq k$;
- if $\langle \psi | \varphi_k \rangle = 0$ for every k then $\psi = 0$.

In the case of a finite-dimensional Hilbert space, it is often useful to understand an orthonormal basis as a list of vectors rather than just as a set. In other words, we order the elements of the orthonormal basis. There is then a unique correspondence between the vectors and the d-tuples of complex numbers. This is actually just another way to express the isomorphism statement at the end of Proposition 1.4. Similarly, in the case of a separable infinite-dimensional Hilbert space we take an orthonormal basis to mean a sequence of orthogonal vectors (rather than just a set) whenever this is convenient.

Once an orthonormal basis is fixed, we can write every vector $\psi \in \mathcal{H}$ in terms of a *basis expansion*:

$$\psi = \sum_{k=1}^{d} \langle \varphi_k | \psi \rangle \, \varphi_k . \tag{1.11}$$

The exact meaning of this formula depends on whether the Hilbert space is finite or infinite dimensional.

In finite-dimensional Hilbert spaces the basis expansion is just a finite sum. The basis expansion simply expresses the fact that each vector ψ can be written as a linear combination of the basis vectors φ_k.

Exercise 1.13 Suppose that $d < \infty$ and $\psi = \sum_{k=1}^{d} c_k \varphi_k$. Prove the following: if $\{\varphi_k\}_{k=1}^{d}$ is an orthonormal basis then $c_k = \langle \varphi_k | \psi \rangle$. [Hint: Start from $\psi = \sum_{k=1}^{d} c_k \varphi_k$ and take the inner product with φ_k on both sides.]

In the case of an infinite-dimensional Hilbert space, the basis expansion (1.11) is a convergent series. To see this, first observe that

$$\sum_{k=1}^{n} \| \langle \varphi_k | \psi \rangle \, \varphi_k \|^2 = \sum_{k=1}^{n} |\langle \varphi_k | \psi \rangle|^2 \leq \|\psi\|^2$$

for any $n = 1, 2, \ldots$, where we have applied Bessel's inequality (1.8). This implies that the series

$$\sum_{k=1}^{\infty} \| \langle \varphi_k | \psi \rangle \, \varphi_k \|^2$$

converges (since the sequence of the partial sums is increasing and bounded). We then recall from subsection 1.1.1 that in a Hilbert space every absolutely convergent series is convergent. Since we have seen that the series is convergent, the basis expansion (1.11) is easy to verify. We note that the vector $\psi - \sum_{k=1}^{\infty} \langle \varphi_k | \psi \rangle \, \varphi_k$ is orthogonal to every basis vector φ_ℓ. But since $\{\varphi_k\}_{k=1}^{\infty}$ is an orthonormal basis, it follows that $\psi - \sum_{k=0}^{\infty} \langle \varphi_k | \psi \rangle \, \varphi_k = 0$. Therefore, (1.11) holds.

Let us remark that an infinite-dimensional Hilbert space does not contain a countable set of vectors that would give any other vector as a linear combination of some subset of this set. (By a linear combination we always mean a finite sum.) We can conclude from the basis expansion that any vector can be approximated arbitrarily well by a linear combination of basis vectors.

Example 1.14 (*Basis expansion in $\ell^2(\mathbb{N})$*)
We saw earlier that the Hilbert space $\ell^2(\mathbb{N})$ consists of complex functions f on $\mathbb{N}$ such that the sum $\sum_{k=0}^{\infty} |f(k)|^2$ is finite and the Kronecker functions form an orthonormal basis in $\ell^2(\mathbb{N})$. If a function f is nonzero only at a finite number of points k, then clearly we can write it as

$$f = \sum_{k:f(k)\neq 0} f(k)\delta_k . \tag{1.12}$$

This is nothing other than the basis expansion of f in the Kronecker basis. The formula (1.12) is true for all $f \in \ell^2(\mathbb{N})$, but in general it is a convergent series and need not be a finite sum.   △

The complex numbers $\langle \varphi_k | \psi \rangle$ in (1.11) are called the *Fourier coefficients* of ψ with respect to the orthonormal basis $\{\varphi_k\}_{k=1}^{d}$. They give not only the basis expansion of ψ but also the best approximation of ψ if one is allowed to use only some fixed subset of the basis vectors. Namely, fix a positive integer $m \leq d$. For any choice of m complex numbers $c_1, \ldots, c_m$, we have

$$\left\| \psi - \sum_{k=1}^{m} \langle \varphi_k | \psi \rangle \varphi_k \right\| \leq \left\| \psi - \sum_{k=1}^{m} c_k \varphi_k \right\|. \tag{1.13}$$

To see that this inequality holds, we write the square of the right-hand side in the form

$$\left\| \psi - \sum_{k=1}^{m} c_k \varphi_k \right\|^2 = \| \psi \|^2 - \sum_{k=1}^{m} |\langle \varphi_k | \psi \rangle|^2 + \sum_{k=1}^{m} |c_k - \langle \varphi_k | \psi \rangle|^2$$

and then observe that the last sum is positive unless $c_k = \langle \varphi_k | \psi \rangle$ for every $k = 1, \ldots, m$.

Another useful thing about Fourier coefficients is that one can use them to calculate the norm of a vector. The norm of ψ in (1.11) is given by *Parseval's formula*:

$$\| \psi \|^2 = \sum_{k=1}^{d} |\langle \varphi_k | \psi \rangle|^2. \tag{1.14}$$

This is obtained from the following chain of equalities:

$$\| \psi \|^2 = \langle \psi | \psi \rangle = \left\langle \sum_{k=1}^{d} \langle \varphi_k | \psi \rangle \varphi_k \middle| \sum_{j=1}^{d} \langle \varphi_j | \psi \rangle \varphi_j \right\rangle$$

$$= \sum_{k=1}^{d} \sum_{j=1}^{d} \overline{\langle \varphi_k | \psi \rangle} \langle \varphi_j | \psi \rangle \langle \varphi_k | \varphi_j \rangle = \sum_{k=1}^{d} |\langle \varphi_k | \psi \rangle|^2.$$

In the infinite-dimensional case, the fact that the sums can be taken out of the inner product is justified by the continuity of the latter.

1.1.3 Example: $L^2(\Omega)$

Up to now, we have encountered only one particular infinite-dimensional Hilbert space, namely $\ell^2(\mathbb{N})$. At the abstract level, there are no other separable infinite-dimensional Hilbert spaces as they are all isomorphic (recall Proposition 1.12). However, in applications Hilbert spaces typically have some concrete form. The benefit of using one particular concrete form rather than another is that certain operators may be easier to handle or a calculation may be easier to perform.

One very useful class of concrete Hilbert spaces is that consisting of so-called *square integrable functions*. Actually, $\ell^2(\mathbb{N})$ also belongs to this class but then the integration is just a sum on $\mathbb{N}$ and square integrability means that the sum

$\sum_{j=0}^{\infty} |f(j)|^2$ is finite. (In technical terms, the integration is with respect to the counting measure on $\mathbb{N}$.)

To introduce other examples of this type of Hilbert space, let Ω be the real line $\mathbb{R}$ or an interval on $\mathbb{R}$. We denote by $L^2(\Omega)$ the set of complex-valued functions on Ω for which the integral $\int_\Omega |f(x)|^2 dx$ is finite. To be more precise, functions are required to be measurable and two functions are identified if they are equal almost everywhere. With these conventions, $L^2(\Omega)$ is a separable infinite-dimensional Hilbert space equipped with an inner product

$$\langle f|g \rangle = \int_\Omega \overline{f(x)} g(x) \, dx . \tag{1.15}$$

There is, of course, some work to do to show that (1.15) is an inner product and that it makes $L^2(\Omega)$ a separable Hilbert space. These details can be found, in for instance, the textbooks by Rudin [122] or Folland [61].

We conclude that a unit vector in $L^2(\Omega)$ is a function $f : \Omega \to \mathbb{C}$ satisfying $\|f\|^2 = \int_\Omega |f(x)|^2 dx = 1$. It is possible (and convenient) to choose an orthonormal basis for $L^2(\Omega)$ consisting of continuous functions. For a nice example, let $\Omega = [0, 2\pi)$. A function f on $[0, 2\pi)$ can be alternatively thought of as a periodic function on $\mathbb{R}$ with period 2π. Clearly, for each $n \in \mathbb{Z}$, the function $e_n(x) := e^{inx}$ belongs to $L^2([0, 2\pi))$. Two functions e_n and e_m with $n \neq m$ are orthogonal, since

$$\langle e_n|e_m \rangle = \int_0^{2\pi} e^{i(m-n)x} \, dx = 0 \tag{1.16}$$

and we have $\|e_n\| = \sqrt{2\pi}$. One can actually prove that the set

$$\{e_n/\sqrt{2\pi} : n \in \mathbb{Z}\}$$

is an orthonormal basis in $L^2([0, 2\pi))$. For any $f \in L^2([0, 2\pi))$, one obtains

$$\langle e_n|f \rangle = \frac{1}{\sqrt{2\pi}} \int_0^{2\pi} e^{-inx} f(x) \, dx .$$

These numbers are (up to a constant factor) just the usual Fourier coefficients of f.

This example of square integrable functions has a natural extension from complex-valued functions to vector-valued functions. We denote by $L^2(\Omega; \mathbb{C}^d)$ the set of functions from Ω to $\mathbb{C}^d$ for which the integral $\int_\Omega \|f(x)\|^2 \, dx$ is finite. The norm inside the integral is the norm in $\mathbb{C}^d$. The inner product in $L^2(\Omega; \mathbb{C}^d)$ is defined as

$$\langle f|g \rangle = \int_\Omega \langle f(x)|g(x) \rangle \, dx ,$$

where the inner product under the integral sign is the inner product of $\mathbb{C}^d$. In a similar way we can start from any Hilbert space $\mathcal{H}$ and define $L^2(\Omega; \mathcal{H})$.

1.2 Operators on Hilbert spaces

In this section we introduce some elementary properties of linear mappings acting on Hilbert spaces. For the sake of simplicity we assume that the domain and codomain Hilbert spaces of such linear mappings are identical. Especially in the context of this convention, 'operator' is used as a synonym for 'linear mapping'.

There are several important classes of operators and each has a very specific structure. In this section we will familiarize ourselves with five operator classes, each presented in its own subsection.

1.2.1 The C-algebra of bounded operators*

A mapping $T : \mathcal{H} \to \mathcal{H}$ is *linear* if it preserves linear combinations:

$$T(c\psi + \phi) = cT(\psi) + T(\phi) \qquad \text{for every } \psi, \phi \in \mathcal{H} \text{ and } c \in \mathbb{C}.$$

We typically omit the parentheses and write $T\psi$ instead of $T(\psi)$.

Definition 1.15 A linear mapping $T : \mathcal{H} \to \mathcal{H}$ is a *bounded operator* if there exists a number $t \geq 0$ such that

$$\|T\psi\| \leq t\|\psi\| \qquad \text{for all } \psi \in \mathcal{H}. \tag{1.17}$$

We denote by $\mathcal{L}(\mathcal{H})$ the set of bounded operators on $\mathcal{H}$.

It is a basic result in functional analysis that the bounded operators are exactly the continuous linear mappings. In a finite-dimensional Hilbert space $\mathcal{H}$ every linear mapping is continuous, hence a bounded operator. Even more, every linear mapping $T : X \to \mathcal{H}$ defined on a linear subspace $X \subset \mathcal{H}$ of a finite dimensional Hilbert space $\mathcal{H}$ has an extension to a bounded operator $\widetilde{T} : \mathcal{H} \to \mathcal{H}$. (If $\widetilde{T}$ is an extension then $\widetilde{T}(\psi) = T(\psi)$ for every $\psi \in X$.) In an infinite-dimensional Hilbert space this fact is no longer true. In the following example we demonstrate the concept of an unbounded operator.

Example 1.16 (*Unbounded operator*)
For each $f \in \ell^2(\mathbb{N})$, define a function $Nf : \mathbb{N} \to \mathbb{C}$ by the formula

$$(Nf)(n) = nf(n).$$

It may happen that Nf is not a vector in $\ell^2(\mathbb{N})$. For instance, let

$$f(n) = \begin{cases} 0 & \text{if } n = 0, \\ 1/n & \text{if } n > 0. \end{cases}$$

Then $f \in \ell^2(\mathbb{N})$ but $Nf \notin \ell^2(\mathbb{N})$. The set

$$\mathcal{D}(N) = \{f \in \ell^2(\mathbb{N}) : Nf \in \ell^2(\mathbb{N})\}$$

is a linear subspace of $\ell^2(\mathbb{N})$ and N is a linear mapping from $\mathcal{D}(N)$ into $\ell^2(\mathbb{N})$. For each $k \in \mathbb{N}$, we have $N\delta_k = k\delta_k$ and hence $\|N\delta_k\| = k\,\|\delta_k\|$. This shows that there cannot be a bounded operator $\widetilde{N}$ on $\ell^2(\mathbb{N})$ which would be an extension of N. $\triangle$

For a bounded operator T, we use the following notation:

- $\ker(T) := \{\psi \in \mathcal{H} : T\psi = 0\}$ (kernel);
- $\mathrm{ran}(T) := \{\psi \in \mathcal{H} : \psi = T\varphi \text{ for some } \varphi \in \mathcal{H}\}$ (range);
- $\mathrm{supp}(T) := \{\psi \in \mathcal{H} : \psi \perp \varphi \text{ for all } \varphi \in \ker(T)\} \equiv \ker(T)^{\perp}$ (support).

It is easy to verify that all these subsets correspond to linear subspaces of $\mathcal{H}$. Since $\ker(T)$ is a preimage of the closed set $\{0\}$ in a continuous mapping, it is closed; $\mathrm{supp}(T)$ is also a closed set but $\mathrm{ran}(T)$ need not be closed. The dimension of $\mathrm{supp}(T)$ is called the *rank* of T.

The set of bounded operators is a vector space. Two bounded operators can be added in the usual way and scalar multiplication with a complex number c is also defined in an unsurprising manner:

$$(S + T)\psi = S\psi + T\psi\,,$$
$$(cT)\psi = c(T\psi)\,.$$

We denote by O and I the *null operator* and the *identity operator*, respectively; they are defined by $O\psi = 0$ and $I\psi = \psi$ for every $\psi \in \mathcal{H}$. The null operator plays the role of the null vector in the vector space of bounded operators.

The vector space $\mathcal{L}(\mathcal{H})$ is a normed space with a norm defined via the formula

$$\|T\| := \sup_{\|\psi\|=1} \|T\psi\|\,. \tag{1.18}$$

In other words, $\|T\|$ is the least number t that satisfies (1.17). This norm on $\mathcal{L}(\mathcal{H})$ is called the *operator norm*. One can show that $\mathcal{L}(\mathcal{H})$ is complete in the operator norm topology. Complete normed vector spaces are called *Banach spaces*.

It follows directly from the definition of the operator norm that if $T \in \mathcal{L}(\mathcal{H})$ then for every $\psi \in \mathcal{H}$ we have

$$\|T\psi\| \leq \|T\|\,\|\psi\|\,. \tag{1.19}$$

Together with the Cauchy–Schwarz inequality this implies that, for every $\varphi, \psi \in \mathcal{H}$,

$$|\langle\varphi|T\psi\rangle| \leq \|\varphi\|\,\|\psi\|\,\|T\|\,. \tag{1.20}$$

Both (1.19) and (1.20) are very useful inequalities.

We can multiply two operators by forming their composition. If $S, T \in \mathcal{L}(\mathcal{H})$ then by using inequality (1.19) twice we get

$$\|ST\psi\| \leq \|S\| \|T\psi\| = \|S\| \|T\| \|\psi\|.$$

Thus, we conclude that

$$\|ST\| \leq \|S\| \|T\|. \tag{1.21}$$

This shows, in particular, that the product operator ST is a bounded operator and hence that $ST \in \mathcal{L}(\mathcal{H})$. In other words, $\mathcal{L}(\mathcal{H})$ is an *algebra*. Moreover, inequality (1.21) implies that multiplication in $\mathcal{L}(\mathcal{H})$ is continuous in each variable.

For each bounded operator T, we can define the *adjoint operator* T^* by the formula

$$\langle \varphi | T^* \psi \rangle = \langle T\varphi | \psi \rangle, \tag{1.22}$$

which is required to hold for all $\varphi, \psi \in \mathcal{H}$. (One has to prove that the adjoint operator exists. This follows from the Fréchet–Riesz theorem, which we state in subsection 1.3.4.) The mapping $T \mapsto T^*$ is *conjugate linear*, meaning that

$$(cT + S)^* = \bar{c}T^* + S^*.$$

Moreover, if $S, T \in \mathcal{L}(\mathcal{H})$ then

$$(ST)^* = T^*S^*, \tag{1.23}$$

$$(T^*)^* = T. \tag{1.24}$$

Exercise 1.17 Prove the properties (1.23) and (1.24) directly from the defining condition (1.22).

The operator norm and the adjoint mapping are like a couple who are 'meant' to be together – their properties are nicely linked. The following equality is a key feature of $\mathcal{L}(\mathcal{H})$.

Proposition 1.18 A bounded operator T and its adjoint T^* satisfy

$$\|T\| = \|T^*\| = \|T^*T\|^{1/2}. \tag{1.25}$$

Proof For $\psi \in \mathcal{H}$, $\|\psi\| = 1$, we obtain

$$\|T\psi\|^2 = \left| \|T\psi\|^2 \right| = |\langle T\psi | T\psi \rangle| = \left| \langle \psi | T^*T\psi \rangle \right| \overset{(1.20)}{\leq} \|\psi\|^2 \|T^*T\|$$

$$= \|T^*T\| \overset{(1.21)}{\leq} \|T^*\| \|T\|,$$

which implies that

$$\|T\|^2 \leq \|T^*T\| \leq \|T^*\| \|T\|. \tag{1.26}$$

This shows that $\|T\| \leq \|T^*\|$. Replacing T by T^* into this inequality and using the identity $(T^*)^* = T$ we also get $\|T^*\| \leq \|T\|$. Therefore, $\|T\| = \|T^*\|$. Using this fact in (1.26) proves the claim. $\square$

We can summarize the previous discussion by saying that $\mathcal{L}(\mathcal{H})$ is an example of a C^*-*algebra*. This means that:

- $\mathcal{L}(\mathcal{H})$ is an algebra;
- $\mathcal{L}(\mathcal{H})$ is a complete normed space (i.e. a Banach space);
- the adjoint mapping $T \mapsto T^*$ on $\mathcal{L}(\mathcal{H})$ is conjugate linear and satisfies (1.23) and (1.24);
- the operator norm on $\mathcal{L}(\mathcal{H})$ satisfies (1.21) and (1.25).

These are the basic properties of $\mathcal{L}(\mathcal{H})$ and also the facts most often used for proving other relevant properties.

After this intensive burst of information, it is now time for an example.

Example 1.19 (*Shift operators*)
Let us continue the discussion on our favourite Hilbert space $\ell^2(\mathbb{N})$. It is often convenient to write an element $\zeta \in \ell^2(\mathbb{N})$ as

$$\zeta = (\zeta_0, \zeta_1, \ldots),$$

where $\zeta_i \equiv \zeta(i)$. Let $A : \ell^2(\mathbb{N}) \to \ell^2(\mathbb{N})$ be the *(forward) shift operator*, defined by

$$A(\zeta_0, \zeta_1, \ldots) = (0, \zeta_0, \zeta_1, \ldots).$$

We have

$$\|A\zeta\| = \sum_{j=0}^{\infty} |\zeta_j|^2 = \|\zeta\|,$$

and therefore A is bounded and $\|A\| = 1$. To calculate the adjoint operator A^*, let $\zeta = (\zeta_0, \zeta_1, \ldots), \eta = (\eta_0, \eta_1, \ldots) \in \ell^2(\mathbb{N})$. The defining condition (1.22) for A^* gives

$$\langle \eta | A^* \zeta \rangle = \langle A\eta | \zeta \rangle = \sum_{j=0}^{\infty} \bar{\eta}_j \zeta_{j+1}.$$

As the vector η is arbitrary, we conclude that

$$A^*(\zeta_0, \zeta_1, \zeta_2, \ldots) = (\zeta_1, \zeta_2, \ldots).$$

The operator A^* is called the *backward shift operator*. $\triangle$

In a finite d-dimensional Hilbert space $\mathcal{H}$, the set $\mathcal{L}(\mathcal{H})$ consists of all linear mappings. We can hence identify $\mathcal{L}(\mathcal{H})$ with the set $\mathcal{M}_d(\mathbb{C})$ of $d \times d$ complex matrices. We simply fix an orthonormal basis $\{\varphi_j\}_{j=1}^d$ for $\mathcal{H}$, and for each $T \in \mathcal{L}(\mathcal{H})$ we define a matrix $[T_{jk}]$ by

$$[T_{jk}] = \langle \varphi_j | T \varphi_k \rangle . \tag{1.27}$$

Conversely, starting with a matrix $[T_{jk}]$ we recover the action of T on a vector $\psi \in \mathcal{H}$ via the formula

$$T\psi = \sum_{j,k=1}^d [T_{jk}] \langle \varphi_k | \psi \rangle \varphi_j. \tag{1.28}$$

This is just the ordinary rule from matrix calculus when ψ is written in the orthonormal basis $\{\varphi_j\}_{j=1}^d$.

A comparison of (1.27) and (1.22) shows that the adjoint operator T^* corresponds to the transposed and complex conjugated matrix, i.e.,

$$[T_{jk}^*] = \overline{[T_{kj}]} .$$

This is sometimes called the *conjugate transpose* matrix, which is evidently a very descriptive name.

If $\mathcal{H}$ is infinite dimensional, we can still fix an orthonormal basis for $\mathcal{H}$ and define a matrix $[T_{jk}]$ by (1.27) for each bounded operator T. Using a basis expansion (subsection 1.1.2) we can recover the action of T on every vector ψ from the matrix $[T_{jk}]$, and the formula is still (1.28) except that the sum is infinite. But we now have the problem that a given infinite matrix may not correspond to any bounded operator, and it may not be very easy to see when this happens. Sometimes one has to fix an orthonormal basis for $\mathcal{H}$ in order to perform calculations. However, as a starting point we want to think of $\mathcal{L}(\mathcal{H})$ as consisting of linear mappings and not of matrices, even if the dimension of $\mathcal{H}$ is finite.

It is clear that the diagonal elements $[T_{jj}]$ of a matrix $[T_{jk}]$ do not determine it completely. However, if the diagonal elements of an operator T are known in *all* orthonormal bases then T is determined. This is the content of Proposition 1.21 below. The auxiliary equation presented in Exercise 1.20 will be useful also later.

Exercise 1.20 (*Polarization identity*)
Let $T \in \mathcal{L}(\mathcal{H})$ and $\phi, \psi \in \mathcal{H}$. Verify the following identity, known as the *polarization identity*:

$$\langle \phi | T \psi \rangle = \tfrac{1}{4} \sum_{k=0}^3 i^k \langle \psi + i^k \phi | T (\psi + i^k \phi) \rangle . \tag{1.29}$$

[Hint: Expand the right-hand side of (1.29).] Notice that with $T = I$ this becomes

$$\langle \phi | \psi \rangle = \tfrac{1}{4} \sum_{k=0}^{3} i^k \, \| \psi + i^k \phi \|^2 . \tag{1.30}$$

Hence, if we know the norms of all vectors in $\mathcal{H}$, we can recover all inner products.

Proposition 1.21 Let $S, T \in \mathcal{L}(\mathcal{H})$. If $\langle \psi | S \psi \rangle = \langle \psi | T \psi \rangle$ for every $\psi \in \mathcal{H}$ then $S = T$.

Proof Assume that $\langle \psi | S \psi \rangle = \langle \psi | T \psi \rangle$ for every $\psi \in \mathcal{H}$. By the polarization identity (1.29), we have $\langle \phi | S \psi \rangle = \langle \phi | T \psi \rangle$ for every $\psi, \phi \in \mathcal{H}$. In particular, choosing $\phi = (S - T)\psi$ we conclude that $\langle (S - T)\psi | (S - T)\psi \rangle = 0$ for every $\psi \in \mathcal{H}$. Therefore, $S\psi = T\psi$ for every $\psi \in \mathcal{H}$. $\square$

One of the important concepts in linear algebra is that of the eigenvalues of a matrix. It plays a central role in functional analysis also, but one needs to generalize it and introduce the concept of a spectrum.

Definition 1.22 Let T be a bounded operator. A number $\lambda \in \mathbb{C}$ is

- an *eigenvalue* of T if there exists a nonzero vector $\psi \in \mathcal{H}$ such that $T\psi = \lambda\psi$. The vector ψ is an *eigenvector* of T associated with the eigenvalue λ.
- in the *spectrum* of T if the inverse mapping of the bounded operator $\lambda I - T$ does not exist.

It is clear that all the eigenvalues of T are in the spectrum of T since eigenvectors belong to the kernel of $\lambda I - T$ and an invertible mapping cannot have any nonzero vector in its kernel. However, the spectrum may also contain numbers other than just the eigenvalues. In a finite-dimensional Hilbert space the eigenvalues of T are the solutions of the equation $\det(T - \lambda I) = 0$. In particular, every operator has eigenvalues. This is no longer true in infinite-dimensional Hilbert space, as we demonstrate in the following example.

Example 1.23 (*Operator without eigenvalues*)
The shift operator A defined in Example 1.19 does not have any eigenvalues. This is easy to see by making the assumption that ψ is an eigenvector of A with eigenvalue λ and writing ψ in the basis expansion $\psi = \sum_k c_k \delta_k$. Since $A\delta_k = \delta_{k+1}$, we get the following set of equations:

$$\lambda c_0 = 0 ,$$
$$\lambda c_1 = c_0 ,$$
$$\lambda c_2 = c_1 ,$$
$$\vdots$$

This can happen only if $c_0 = c_1 = \cdots = 0$, hence $\psi = 0$. Let us note that the backward shift operator A^*, however, does have eigenvalues. For instance, the vector δ_0 is its eigenvector associated with the eigenvalue 0. △

Even though a bounded operator need not have any eigenvalues, every nonzero bounded operator has a nonempty spectrum. It can be shown that the spectrum of a bounded operator T is a compact subset of $\mathbb{C}$ and is bounded by the operator norm $\|T\|$.

Exercise 1.24 Let λ be an eigenvalue of $T \in \mathcal{L}(\mathcal{H})$. Show that $|\lambda| \leq \|T\|$. [Hint: Start from $T\psi = \lambda\psi$ and use (1.19).]

Exercise 1.25 Prove the following: if ψ_1 and ψ_2 are eigenvectors of a bounded operator T associated with the same eigenvalue λ then every linear combination of ψ_1 and ψ_2 is also an eigenvector associated with λ.

1.2.2 Partially ordered vector space of selfadjoint operators

Definition 1.26 A bounded operator $T \in \mathcal{L}(\mathcal{H})$ is *selfadjoint* if $T = T^*$. We denote by $\mathcal{L}_s(\mathcal{H})$ the set of bounded selfadjoint operators on $\mathcal{H}$.

Since the adjoint mapping $T \mapsto T^*$ is linear with respect to real linear combinations, the real linear combinations of selfadjoint operators remain selfadjoint; hence the set $\mathcal{L}_s(\mathcal{H})$ is a real vector space.

Exercise 1.27 Unlike the set of bounded operators $\mathcal{L}(\mathcal{H})$, the set of selfadjoint operators $\mathcal{L}_s(\mathcal{H})$ is not an algebra; the product of two selfadjoint operators is not always selfadjoint. Prove the following: the product ST of two selfadjoint operators S, T is selfadjoint if and only if S and T commute, i.e., $ST = TS$. [Hint: Remember that $(ST)^* = T^*S^*$.]

Even though the condition for self-adjointness seems quite restrictive, the complex linear combinations of selfadjoint operators actually fill up $\mathcal{L}(\mathcal{H})$. To see this, let $T \in \mathcal{L}(\mathcal{H})$. We denote $T_R = \frac{1}{2}(T + T^*)$ and $T_I = -\frac{1}{2}i(T - T^*)$. These operators are called the *real* and the *imaginary parts* of T, respectively. It is straightforward to verify that T_R and T_I are selfadjoint and that $T = T_R + iT_I$.

With these terms defined we can say that a bounded operator T is selfadjoint if and only if the imaginary part T_I of T vanishes (and hence $T = T_R$). The following characterization of selfadjoint operators corresponds to this expression.

Proposition 1.28 A bounded operator $T \in \mathcal{L}(\mathcal{H})$ is selfadjoint if and only if $\langle\psi|T\psi\rangle$ is a real number for every $\psi \in \mathcal{H}$.

Proof By Proposition 1.21, a bounded operator T is selfadjoint if and only if $\langle \psi | T\psi \rangle = \langle \psi | T^*\psi \rangle$ for every $\psi \in \mathcal{H}$. But $\langle \psi | T^*\psi \rangle = \overline{\langle T\psi | \psi \rangle}$, so the previous condition is the same as requiring that $\langle \psi | T\psi \rangle = \overline{\langle \psi | T\psi \rangle}$ for every $\psi \in \mathcal{H}$. □

The previous observation motivates the following definition.

Definition 1.29 A bounded operator $T \in \mathcal{L}(\mathcal{H})$ is *positive* if $\langle \psi | T\psi \rangle \geq 0$ for every $\psi \in \mathcal{H}$.

A comparison of Definition 1.29 and Proposition 1.28 shows that positive operators are selfadjoint. Notice that there is a $\geq$ symbol instead of a $>$ symbol in the definition of positive operators. This is commonly the way in which positivity is defined in functional analysis. The easiest example of a positive operator is tI, where t is a nonnegative real number and I is the identity operator.

Exercise 1.30 Let T be a positive operator and λ its eigenvalue. Show that $\lambda \geq 0$.

The concept of positivity defines a partial ordering in $\mathcal{L}_s(\mathcal{H})$ in a natural way – this is spelled out in the following definition.

Definition 1.31 Let $S, T \in \mathcal{L}_s(\mathcal{H})$. We write $S \geq T$ if the operator $S - T$ is positive.

Clearly, a selfadjoint operator T is positive exactly when $T \geq O$. It is straightforward to verify that the relation $\geq$ in Definition 1.31 is a partial ordering. This relation has some further properties, which connect the order structure and the vector space structure of $\mathcal{L}_s(\mathcal{H})$. Namely, let $T_1, T_2, T_3 \in \mathcal{L}_s(\mathcal{H})$ and $\alpha \in \mathbb{R}, \alpha \geq 0$. It follows directly from the definition of positivity that:

- if $T_1 \geq T_2$ then $T_1 + T_3 \geq T_2 + T_3$;
- if $T_1 \geq T_2$ then $\alpha T_1 \geq \alpha T_2$.

These properties mean that the relation $\geq$ makes $\mathcal{L}_s(\mathcal{H})$ a *partially ordered vector space*.

The following results illustrate the partial ordering relation of selfadjoint operators. They will be useful later.

Exercise 1.32 Show that

$$- \|T\| I \leq T \leq \|T\| I$$

for all $T \in \mathcal{L}_s(\mathcal{H})$. [Hint: Use (1.20).]

It is useful to notice that the product T^*T is positive for every $T \in \mathcal{L}(\mathcal{H})$. Namely, for every $\psi \in \mathcal{H}$ we have

$$\langle \psi | T^*T\psi \rangle = \langle T\psi | T\psi \rangle = \|T\psi\|^2 \geq 0.$$

The converse holds also: any positive operator S can be written as $S = T^*T$ for some $T \in \mathcal{L}(\mathcal{H})$. This fact follows from the next result, also known as the *square root lemma*.

Theorem 1.33 (*Square root lemma*)
Let $T \in \mathcal{L}_s(\mathcal{H})$ be a positive operator. There is a unique positive operator, denoted by $T^{1/2}$, satisfying $(T^{1/2})^2 = T$. The operator $T^{1/2}$ (also denoted by $\sqrt{T}$) is called the *square root* of T. It has the following properties.

(a) If $S \in \mathcal{L}(\mathcal{H})$ and $ST = TS$, then $ST^{1/2} = T^{1/2}S$.
(b) If T is invertible then $T^{1/2}$ is invertible also and $(T^{1/2})^{-1} = (T^{-1})^{1/2}$.

The square root lemma is important in many situations. In the following we derive some useful consequences.

Proposition 1.34 Let $T \in \mathcal{L}_s(\mathcal{H})$. If $O \leq T \leq I$ then $O \leq T^2 \leq T$.

Proof Let $\psi \in \mathcal{H}$. Then on the one hand

$$\langle \psi | T^2 \psi \rangle = \langle T^{1/2}\psi | T T^{1/2}\psi \rangle \leq \langle T^{1/2}\psi | T^{1/2}\psi \rangle = \langle \psi | T\psi \rangle \ ;$$

hence $T^2 \leq T$. On the other hand,

$$\langle \psi | T^2 \psi \rangle = \langle T\psi | T\psi \rangle \geq 0 \, ,$$

and thus $T^2 \geq O$. $\square$

Proposition 1.35 Let $T \in \mathcal{L}_s(\mathcal{H})$ be a positive operator and $\psi \in \mathcal{H}$. If $\langle \psi | T\psi \rangle = 0$ then $T\psi = 0$.

Proof Suppose that $\langle \psi | T\psi \rangle = 0$. Then

$$0 = \langle \psi | T\psi \rangle = \langle T^{1/2}\psi | T^{1/2}\psi \rangle \, ,$$

and hence $T^{1/2}\psi = 0$. Applying the operator $T^{1/2}$ on both sides of this equality gives $T\psi = 0$. $\square$

Armed with the square root lemma, we are ready for the following definition, which leads to a very useful decomposition theorem.

Definition 1.36 Let $T \in \mathcal{L}(\mathcal{H})$. We write $|T| := (T^*T)^{1/2}$ and call this operator the *absolute value* of T.

Theorem 1.37 (*Polar decomposition*)
Let $T \in \mathcal{L}(\mathcal{H})$. There exists a bounded operator $V \in \mathcal{L}(\mathcal{H})$ such that:

- $T = V|T|$;
- $\|V\psi\| = \|\psi\|$ for every $\psi \in \text{supp}(V)$.

The operator V can be chosen to be an isomorphism on $\mathcal{H}$ if $\dim \mathcal{H} < \infty$, or T is invertible, or T is selfadjoint.

The second condition in Theorem 1.37 means that V is a *partial isometry*. An operator U is called an *isometry* if it satisfies $\|U\psi\| = \|\psi\|$ for every $\psi \in \mathcal{H}$. This implies that $U\psi = 0$ only if $\psi = 0$; hence $\ker(U) = \{0\}$ for all isometries. Thus we see that a partial isometry V is an isometry if and only if $\ker(V) = \{0\}$.

There is one more operator decomposition to be introduced. At the beginning of this subsection we saw that a bounded operator T can be written as a linear combination $T = T_\mathrm{R} + iT_\mathrm{I}$ of its real and imaginary parts, which are selfadjoint operators. This splitting can be continued; every selfadjoint operator can be written as a difference of two positive operators. For every $T \in \mathcal{L}_\mathrm{s}(\mathcal{H})$, the operators $\|T\| I \pm T$ are positive (see Exercise 1.32) and

$$T = \tfrac{1}{2}\big(\|T\|I + T\big) - \tfrac{1}{2}\big(\|T\|I - T\big).$$

Sometimes it is more convenient to use two different positive operators,

$$T^+ := \tfrac{1}{2}\big(|T| + T\big), \qquad T^- := \tfrac{1}{2}\big(|T| - T\big)$$

and these are called the *positive part* and the *negative part* of T, respectively. Obviously, $T = T^+ - T^-$. It can be shown that T^+ and T^- are positive operators.

1.2.3 *Orthocomplemented lattice of projections*

Definition 1.38 A selfadjoint operator $P \in \mathcal{L}_\mathrm{s}(\mathcal{H})$ is a *projection* if $P^2 = P$. We denote by $\mathcal{P}(\mathcal{H})$ the set of projections.

Projections are positive operators; if P is a projection and $\psi \in \mathcal{H}$ then

$$\langle\psi|P\psi\rangle = \langle\psi|P^2\psi\rangle = \langle P\psi|P\psi\rangle \geq 0.$$

We now introduce the most important example of a projection.

Example 1.39 (*One-dimensional projection*)
Let $\eta \in \mathcal{H}$ be a unit vector. We define an operator P_η on $\mathcal{H}$ as

$$P_\eta\psi = \langle\eta|\psi\rangle\,\eta. \tag{1.31}$$

It is straightforward to check that $P_\eta^* = P_\eta$ and $P_\eta^2 = P_\eta$. Therefore, P_η is a projection. It is clear from (1.31) that the range of P_η is the one-dimensional subspace $\mathbb{C}\eta = \{c\eta \mid c \in \mathbb{C}\}$. For this reason P_η is called a one-dimensional projection. $\quad\triangle$

In the following two propositions we will describe the elementary properties of projections.

Proposition 1.40 Let P be a projection and $O \neq P \neq I$. Then:

(a) $\|P\| = 1$;
(b) the eigenvalues of P are 0 and 1;
(c) for every $\psi \in \mathcal{H}$, there are orthogonal vectors $\psi_0, \psi_1 \in \mathcal{H}$ such that

$$P\psi_0 = 0, \qquad P\psi_1 = \psi_1 \qquad \text{and} \qquad \psi = \psi_0 + \psi_1 .$$

Proof (a): An application of (1.25) gives

$$\|P\| = \|P^2\| = \|P^*P\| = \|P\|^2 .$$

Since $\|P\| \neq 0$ by the assumption, we conclude that $\|P\| = 1$.

(b): First suppose that $\psi \in \mathcal{H}$ is an eigenvector of P with eigenvalue λ. Then $P\psi = \lambda\psi$ and

$$P\psi = P^2\psi = \lambda P\psi = \lambda^2\psi .$$

Hence, $\lambda^2 = \lambda$, which means that either $\lambda = 0$ or $\lambda = 1$. To prove that P has both these eigenvalues, choose nonzero vectors $\psi, \phi \in \mathcal{H}$ such that $P\psi \neq 0$ and $P\phi \neq \phi$ (this can be done since $O \neq P \neq I$). Then, using $P^2 = P$, one can verify that $P\psi$ is an eigenvector of P with eigenvalue 1 and $(I - P)\phi$ is an eigenvector of P with eigenvalue 0.

(c): Write $\psi_0 = (I - P)\psi$ and $\psi_1 = P\psi$. These vectors have the required properties. □

Proposition 1.41 Let P be a projection and $\psi \in \mathcal{H}$. The following conditions are equivalent:

(i) $\psi \in \mathrm{ran}(P)$;
(ii) $P\psi = \psi$;
(iii) $\|P\psi\| = \|\psi\|$.

Proof It is trivial that (ii)⟹(iii). In the following we prove that (i)⟹(ii) and (iii)⟹(i).

Assume that (i) holds. This means that there is a vector $\varphi \in \mathcal{H}$ such that $P\varphi = \psi$. But then

$$P\psi = PP\varphi = P\varphi = \psi$$

and, hence, (ii) holds. We conclude that (i)⟹(ii).

Assume that (iii) holds. We write $\psi = P\psi + (I - P)\psi$ as in the proof of Proposition 1.40(c). Since $P\psi$ and $(I - P)\psi$ are orthogonal vectors, we can apply the Pythagorean formula (see Exercise 1.7) and obtain

$$\|\psi\|^2 = \|P\psi\|^2 + \|(I - P)\psi\|^2 .$$

Assumption (iii) now implies that $\|(I - P)\psi\| = 0$, whence it follows that $(I - P)\psi = 0$. Thus $P\psi = \psi$. $\qquad\qquad\qquad\qquad\qquad\qquad\qquad\qquad\square$

Let us then return to the structure of the set of projections. This set is a subset of the set of selfadjoint operators and therefore inherits the order structure defined in subsection 1.2.2. The partial ordering of two projections can be specified as follows.

Proposition 1.42 Let P and Q be projections. The following conditions are equivalent:

(i) $P \geq Q$;
(ii) $PQ = Q$;
(iii) $QP = Q$;
(iv) $QP = PQ = Q$;
(v) $P - Q$ is a projection.

Proof We prove this by showing that (i)$\Rightarrow$(ii)$\Rightarrow$(iii)$\Rightarrow$(iv)$\Rightarrow$(v)$\Rightarrow$(i).

Suppose that (i) holds and fix $\psi \in \mathcal{H}$. Let us first assume that $P \neq O$. Then by Proposition 1.40 we have $\|P\| = 1$, and an application of (1.19) gives

$$\|PQ\psi\| \leq \|P\| \|Q\psi\| = \|Q\psi\| .$$

However, using (i) we get

$$\|PQ\psi\|^2 = \langle PQ\psi | PQ\psi \rangle = \langle Q\psi | PQ\psi \rangle \overset{(i)}{\geq} \langle Q\psi | QQ\psi \rangle$$
$$= \langle Q\psi | Q\psi \rangle = \|Q\psi\|^2$$

and hence $\|PQ\psi\| \geq \|Q\psi\|$. Therefore $\|PQ\psi\| = \|Q\psi\|$. This, together with Proposition 1.41, implies that $PQ\psi = Q\psi$. As this is true for any $\psi \in \mathcal{H}$, we conclude that $PQ = Q$ and that (ii) holds. If $P = O$ then (i) implies that $O \leq Q \leq O$, which means that $Q = O$. Hence, in this case also, (i) implies (ii).

Suppose that (ii) holds. We get

$$Q = Q^* = (PQ)^* = Q^*P^* = QP ,$$

and therefore (iii) holds. In the same way, (iii) implies (ii) and they are thus equivalent. This means that (iii) implies (iv).

Assume that (iv) holds. The operator $P - Q$ is selfadjoint, as are both P and Q. Using (iv) we get

$$(P - Q)^2 = P - PQ - QP + Q = P - Q .$$

Hence $P - Q$ is a projection, and we conclude that (iv) implies (v).

Finally, assume that (v) holds. As a projection is a positive operator, we have $\langle\psi|(P-Q)\psi\rangle \geq 0$ for every $\psi \in \mathcal{H}$. This means that $\langle\psi|P\psi\rangle \geq \langle\psi|Q\psi\rangle$ for every $\psi \in \mathcal{H}$ and hence (i) holds. Thus (v) implies (i). □

The order structure of $\mathcal{P}(\mathcal{H})$ is not only easy to handle (as we have seen in Proposition 1.42) but also pleasant in another way – it can be shown that $\mathcal{P}(\mathcal{H})$ is a *lattice*. This means that any two projections P and Q have a greatest lower bound (infimum) and least upper bound (supremum).

There are even more enrichments relating to the order structure of $\mathcal{P}(\mathcal{H})$. If P is a projection, we write $P^{\perp} := I - P$ and call this operator the *complement* of P. The operator $P^{\perp}$ is clearly selfadjoint and is a projection since

$$P^{\perp}P^{\perp} = (I-P)(I-P) = I - P - P + P^2 = I - P = P^{\perp}.$$

Exercise 1.43 Verify the following properties of $P^{\perp}$:

- $(P^{\perp})^{\perp} = P$;
- if $Q \leq P$ then $P^{\perp} \leq Q^{\perp}$.

There is another specific property of the complement mapping. For two projections P and Q we have that

- if $Q \leq P$ and $Q \leq P^{\perp}$ then $Q = O$.

This property follows from Proposition 1.42 since $Q \leq P$ and $Q \leq P^{\perp}$ give $PQ = Q$ and $P^{\perp}Q = Q$. Adding these two equations together implies that $Q = O$. In other words, the infimum of P and $P^{\perp}$ is O.

The three properties listed above mean that the mapping $P \mapsto P^{\perp}$ on $\mathcal{P}(\mathcal{H})$ is an *orthocomplementation*. We thus conclude that $\mathcal{P}(\mathcal{H})$ is an *orthocomplemented lattice*.

Unlike the set of selfadjoint operators $\mathcal{L}_s(\mathcal{H})$, the set of projections $\mathcal{P}(\mathcal{H})$ is not a vector space; the sum of two projections is a projection only under some extra conditions. These are listed below.

Proposition 1.44 Let P and Q be projections. Then the following conditions are equivalent:

(i) $P^{\perp} \geq Q$;
(ii) $Q^{\perp} \geq P$;
(iii) $PQ = O$;
(iv) $QP = O$;
(v) $PQ = QP = O$;
(vi) $P + Q$ is a projection.

Proof We prove the proposition by showing that (i)$\Rightarrow$(ii)$\Rightarrow$(iii)$\Rightarrow$(iv)$\Rightarrow$(v)$\Rightarrow$(vi)$\Rightarrow$(i). The fact that (i) implies (ii) is a direct consequence of Exercise 1.43.

Assume then that (ii) holds and fix $\psi \in \mathcal{H}$. Denote $\psi_1 = Q\psi$. We then have

$$\begin{aligned} \|P\psi_1\|^2 = \langle P\psi_1|P\psi_1\rangle &= \langle \psi_1|P\psi_1\rangle \\ &\leq \langle \psi_1|Q^\perp \psi_1\rangle = \langle \psi_1|Q^\perp Q\psi\rangle = 0\,. \end{aligned}$$

This implies that $P\psi_1 = 0$ and hence $PQ\psi = 0$. As ψ was an arbitrary vector we conclude that $PQ = O$. Therefore (ii) implies (iii).

Suppose that (iii) holds. We obtain

$$O = O^* = (PQ)^* = Q^*P^* = QP\,,$$

and therefore (iv) holds. In the same way, (iv) implies (iii) and they are thus equivalent. This means that (iv) implies (v).

Assume that (v) holds. Then

$$(P + Q)^2 = P^2 + PQ + QP + Q^2 = P + Q\,.$$

Thus, $P + Q$ is a projection. Therefore, (v) implies (vi).

Finally, assume that (vi) holds. This implies that $P + Q \leq I$. Hence, (i) holds. $\qquad\square$

Two projections P and Q satisfying one (and hence all) of the conditions in Proposition 1.44 are called *orthogonal*.

Example 1.45 (*Orthogonal one-dimensional projections*)
Let us continue the discussion started in Example 1.39. Let $\eta, \phi \in \mathcal{H}$ be two unit vectors and P_η, P_ϕ the corresponding one-dimensional projections. For a vector $\psi \in \mathcal{H}$, we get

$$P_\phi P_\eta \psi = \langle \eta|\psi\rangle \langle \phi|\eta\rangle \phi\,. \tag{1.32}$$

This shows that on the one hand $P_\phi P_\eta = O$ if $\langle \phi|\eta\rangle = 0$. On the other hand, choosing $\psi = \eta$ in (1.32) we see that $P_\phi P_\eta = O$ only if $\langle \phi|\eta\rangle = 0$. We conclude that two one-dimensional projections are orthogonal if and only if the unit vectors defining them are orthogonal. $\qquad\triangle$

As can be seen from Proposition 1.44, we can combine two orthogonal projections to get a third projection. In this way, one-dimensional projections can be used as building blocks to obtain other projections. Actually, every projection is either a finite or countably infinite sum of one-dimensional projections. We will explain this construction but will not prove the details.

Let P be a projection. As shown in Proposition 1.41, the range of P consists of its eigenvectors with eigenvalue 1. Hence, ran(P) is a linear subspace of $\mathcal{H}$.

Let us first assume that ran(P) is finite dimensional with dimension r. We choose an orthonormal basis $\{\eta_k\}_{k=1}^r$ for ran(P). For every $k = 1, \ldots, r$, we then have a one-dimensional projection $P_k \equiv P_{\eta_k}$. For $k \neq l$ the projections P_k and P_l are orthogonal, as explained in Example 1.45. By Proposition 1.44 the sum $\sum_{k=1}^r P_k$ is a projection, and it can be shown that $\sum_{k=1}^r P_k = P$. If ran(P) is infinite dimensional, the same procedure still works. In this case we first note that the set ran(P) is closed. Indeed, it follows from Proposition 1.41 that $\psi \in$ ran(P) exactly when $P^\perp \psi = 0$. Hence, ran(P) is the preimage of the closed set $\{0\}$ in the continuous mapping $P^\perp$, which implies that ran(P) is closed. We conclude that ran(P) is a closed linear subspace of $\mathcal{H}$ and therefore that it has an orthonormal basis $\{\eta_k\}_{k=1}^\infty$. The infinite sum $\sum_{k=1}^\infty P_k$ converges in the weak operator topology (see subsection 1.3.1 below) and we have again $\sum_{k=1}^\infty P_k = P$.

We end this subsection by presenting another useful property of projections. Earlier, in Proposition 1.42, we saw that two projections P and Q satisfy $Q \leq P$ if and only if $QP = PQ = Q$. The 'only if' part is true even when Q is a positive operator but not necessarily a projection, as we prove below.

Proposition 1.46 Let P be a projection and T a positive operator. If $T \leq P$ then $TP = PT = T$.

Proof If $T \leq P$ then $P - T$ is a positive operator. Thus, for every $\varphi \in \mathcal{H}$ satisfying $\langle \varphi | P \varphi \rangle = 0$, we have also $\langle \varphi | T \varphi \rangle = 0$. By Proposition 1.35 we conclude that $T\varphi = 0$ whenever $P\varphi = 0$.

For each vector $\psi \in \mathcal{H}$, we use Proposition 1.40 and write ψ as a sum $\psi = \psi_0 + \psi_1$ with $P\psi_0 = 0$ and $P\psi_1 = \psi_1$. By the above paragraph $T\psi_0 = 0$. Therefore,

$$T\psi = T(\psi_0 + \psi_1) = T\psi_1 = TP\psi_0 + TP\psi_1 = TP\psi .$$

We conclude that $T = TP$, and this further implies that

$$T = T^* = (TP)^* = P^*T^* = PT .$$

Collecting these equations we get $TP = PT = T$. $\qquad\square$

1.2.4 Group of unitary operators

In subsection 1.1.1 we defined the concepts of isomorphic inner product spaces and isomorphisms. Any Hilbert space $\mathcal{H}$ is, trivially, isomorphic with itself; one isomorphism is the identity mapping $I : \psi \mapsto \psi$. There are also other isomorphisms on a given Hilbert space and they play important roles in various different situations.

Proposition 1.47 Let U be a linear mapping on $\mathcal{H}$. The following conditions are equivalent:

  (i) U is an isomorphism;
 (ii) U is a surjective isometry;
(iii) U is bounded and $UU^* = U^*U = I$.

Proof It is clear that (i)$\Rightarrow$(ii). In the following we prove that (ii)$\Rightarrow$(iii) and (iii)$\Rightarrow$(i).

Assume that U is a surjective isometry. Since $\|U\psi\| = \|\psi\|$ for every $\psi \in \mathcal{H}$, it follows that U is a bounded operator with $\|U\| = 1$. It also follows that U is injective; for $\psi, \varphi \in \mathcal{H}$ we have $\|U\psi - U\varphi\| = \|\psi - \varphi\|$ and hence $U\psi = U\varphi$ only if $\psi = \varphi$. Therefore, U is bijective and has an inverse U^{-1}. We have

$$\langle \varphi | \varphi \rangle = \langle U\varphi | U\varphi \rangle = \langle \varphi | U^*U\varphi \rangle \tag{1.33}$$

for all $\varphi \in \mathcal{H}$. By Proposition 1.21 we obtain $U^*U = I$, implying that $U^{-1} = U^*$. Hence U satisfies $UU^* = U^*U = I$. We conclude that (ii)$\Rightarrow$(iii).

Suppose then that U is a bounded operator and $UU^* = U^*U = I$. Then U is bijective with $U^{-1} = U^*$. For all $\varphi, \psi \in \mathcal{H}$, we have

$$\langle \varphi | \psi \rangle = \langle \varphi | U^*U\psi \rangle = \langle U\varphi | U\psi \rangle , \tag{1.34}$$

hence U is an isomorphism. $\square$

It is common to rename the isomorphisms on $\mathcal{H}$ in the following way.

Definition 1.48 A bounded operator $U \in \mathcal{L}(\mathcal{H})$ is *unitary* if it satisfies $UU^* = U^*U = I$. We denote by $\mathcal{U}(\mathcal{H})$ the set of unitary operators on $\mathcal{H}$.

In a finite-dimensional Hilbert space the condition $UU^* = I$ (or similarly $U^*U = I$) implies that $\det(U) \neq 0$; hence U is invertible and $U^{-1} = U^*$. Therefore, to check whether an operator U is unitary it is enough to verify that either $UU^* = I$ or $U^*U = I$. This is again one small difference between finite- and infinite-dimensional Hilbert spaces. Namely, in an infinite-dimensional Hilbert space both the conditions $UU^* = I$ and $U^*U = I$ are needed to guarantee that a bounded operator U is unitary. For instance, the forward and backward shift operators, defined in Example 1.19, satisfy $A^*A = I$, but $AA^* \neq I$ since $AA^*\delta_0 = 0$ and hence they are not unitary.

In the proof of Proposition 1.47 we saw that $\|U\| = 1$ for a unitary operator U. Hence, an eigenvalue λ of U satisfies $|\lambda| \leq 1$ (see Exercise 1.24). It is actually easy to see that all eigenvalues of U must satisfy $|\lambda| = 1$. Namely, suppose that $U\psi = \lambda\psi$ for some $\lambda \in \mathbb{C}$ and some nonzero vector $\psi \in \mathcal{H}$. Since

$$|\lambda|^2 \langle \psi | \psi \rangle = \langle U\psi | U\psi \rangle = \langle \psi | \psi \rangle ,$$

we conclude that $|\lambda| = 1$. Thus, all the eigenvalues of U are of the form $\lambda = e^{ia}$, $a \in \mathbb{R}$. Note, however, that a unitary operator in an infinite-dimensional Hilbert space need not have eigenvalues at all. One can show that $|\lambda| = 1$ holds for any λ in the spectrum of U.

In linear algebra courses, unitary operators are usually introduced as mappings that transform one orthonormal basis to another. This characteristic property is valid also in the infinite-dimensional case, and we have the following simple but useful result.

Proposition 1.49 Let $\mathcal{H}$ be a Hilbert space and $U \in \mathcal{L}(\mathcal{H})$. The following are equivalent:

(i) U is unitary;
(ii) for every orthonormal basis $\{\varphi_j\}$, $\{U\varphi_j\}$ is also an orthonormal basis.
(iii) there exists an orthonormal basis $\{\varphi_j\}$ such that $\{U\varphi_j\}$ is also an orthonormal basis.

Proof Suppose that (i) holds and let $\{\varphi_j\}$ be an orthonormal basis for $\mathcal{H}$. It is clear from (1.4) that $\{U\varphi_j\}$ is an orthonormal set. To prove that it is a maximal orthonormal set (and hence an orthonormal basis), suppose that $\psi \in \mathcal{H}$ is such that $\langle \psi | U\varphi_j \rangle = 0$ for every j. This implies that $\langle U^*\psi | \varphi_j \rangle = 0$ for every j, which means that $U^*\psi = 0$ as $\{\varphi_j\}$ is an orthonormal basis. Since U^* is bijective we conclude that $\psi = 0$. This means that $\{U\varphi_j\}$ is maximal. Therefore (i) implies (ii).

It is clear that (ii) implies (iii). To complete the proof, we therefore need to show that (iii) implies (i).

Suppose that (iii) holds. For a vector $\psi \in \mathcal{H}$, the basis expansion with respect to $\{U\varphi_j\}$ gives

$$\psi = \sum_j \langle U\varphi_j | \psi \rangle U\varphi_j = U\left(\sum_j \langle \varphi_j | U^*\psi \rangle \varphi_j\right).$$

We interpret the last expression as a basis expansion with respect to $\{\varphi_j\}$ and hence obtain $\psi = UU^*\psi$. As this is true for every ψ, we conclude that $UU^* = I$.

To see that $U^*U = I$, let us first write the vector ψ in the orthonormal basis $\{\varphi_j\}$ and then act with U, obtaining

$$U\psi = U\left(\sum_j \langle \varphi_j | \psi \rangle \varphi_j\right) = \sum_j \langle \varphi_j | \psi \rangle U\varphi_j.$$

However, the basis expansion of $U\psi$ with respect to $\{U\varphi_j\}$ gives

$$U\psi = \sum_j \langle U\varphi_j | U\psi \rangle U\varphi_j = \sum_j \langle \varphi_j | U^*U\psi \rangle U\varphi_j.$$

Since the Fourier coefficients are unique, we conclude that $\langle \varphi_j | U^*U\psi \rangle = \langle \varphi_j | \psi \rangle$ for every j. This implies that $U^*U\psi = \psi$. As this is true for every ψ, we conclude that $U^*U = I$. $\qquad\square$

Unitary operators form a group with respect to operator multiplication. In particular, the identity operator I is the unit element of the group $\mathcal{U}(\mathcal{H})$. Let U and V be two unitary operators. Using (1.23) we get

$$(UV)(UV)^* = UVV^*U^* = I$$

and

$$(UV)^*(UV) = V^*U^*UV = I,$$

which shows that the product UV is a unitary operator. Moreover, we have $U^{-1} = U^*$ and, applying (1.24), we get that U^{-1} is unitary too. Thus all the group axioms are satisfied.

There is an intimate connection between unitary and selfadjoint operators. In order to see this we need to introduce the exponential map on $\mathcal{L}(\mathcal{H})$. Let T be a bounded operator. For each $k = 0, 1, 2, \ldots$, we make the following definitions:

$$F_k(T) := \sum_{n=0}^{k} \frac{T^n}{n!}, \qquad f_k(T) := \sum_{n=0}^{k} \frac{\|T^n\|}{n!},$$

where $T^0 = I$. We have

$$f_k(T) \leq \sum_{n=0}^{k} \frac{\|T\|^n}{n!} \leq \sum_{n=0}^{\infty} \frac{\|T\|^n}{n!} = e^{\|T\|},$$

which shows that the increasing sequence $f_0(T), f_1(T), \ldots$ of real numbers has an upper bound and thus converges. This means that the series $\sum_{n=0}^{\infty} T^n/n!$ is absolutely convergent. Since $\mathcal{L}(\mathcal{H})$ is a complete normed space, an absolutely convergent series converges (see the discussion before Definition 1.10). We denote by e^T the limit of the series, that is,

$$e^T := \sum_{n=0}^{\infty} \frac{T^n}{n!} = \lim_{k\to\infty} F_k(T). \qquad (1.35)$$

The mapping $T \mapsto e^T$ is called the *exponential map* on $\mathcal{L}(\mathcal{H})$.

Product and adjoint operators are continuous mappings on $\mathcal{L}(\mathcal{H})$. Using these facts it is straightforward to verify the following formulas for $a, b \in \mathbb{C}$ and $T \in \mathcal{L}(\mathcal{H})$:

$$e^{aT}e^{bT} = e^{(a+b)T},$$
$$(e^{aT})^* = e^{\bar{a}T^*}.$$

In particular, if T is selfadjoint then $(e^{iT})^* = e^{-iT}$ and

$$e^{iT}e^{-iT} = e^{-iT}e^{iT} = e^0 = I\,,$$

showing that e^{iT} is a unitary operator. In summary, every selfadjoint operator defines a unitary operator via the exponential map.

The group of unitary operators has a half brother, the set of antiunitary operators. As we will see later, antiunitary operators turn out to be of less physical relevance than unitary operators. However, they play their own role in some special situations.

Definition 1.50 A mapping $A : \mathcal{H} \to \mathcal{H}$ is an *antiunitary operator* if

$$A(\varphi + c\psi) = A\varphi + \bar{c}A\psi$$

and

$$\langle A\varphi | A\psi \rangle = \langle \psi | \varphi \rangle$$

for all $\varphi, \psi \in \mathcal{H}$ and $c \in \mathbb{C}$. We denote by $\overline{\mathcal{U}}(\mathcal{H})$ the set of antiunitary operators on $\mathcal{H}$.

Notice that $\|A\psi\| = \|\psi\|$ for every $\psi \in \mathcal{H}$ since

$$\|A\psi\|^2 = \langle A\psi | A\psi \rangle = \langle \psi | \psi \rangle = \|\psi\|^2\,.$$

Another useful remark is that a composition of two antiunitary operators A_1 and A_2 is a unitary operator. Namely, for all $\varphi, \psi \in \mathcal{H}$ we get

$$\langle (A_1A_2)\varphi | (A_1A_2)\psi \rangle = \langle A_2\psi | A_2\varphi \rangle = \langle \varphi | \psi \rangle\,,$$

and the linearity of A_1A_2 can be seen in a similar direct way.

Exercise 1.51 Show that the composition of a unitary operator and an antiunitary operator is an antiunitary operator.

There is a prototypical class of antiunitary operators. Let us fix an orthonormal basis $\{\varphi_j\}$ for $\mathcal{H}$. Every vector $\psi \in \mathcal{H}$ has a unique basis expansion,

$$\psi = \sum_j c_j\varphi_j\,,$$

and we can therefore define a mapping J by

$$J\psi = \sum_j \bar{c}_j\varphi_j\,.$$

Since $\sum_j |\bar{c}_j|^2 = \sum_j |c_j|^2$, the sum on the right-hand side converges, hence defining a vector in $\mathcal{H}$. It is straightforward to verify that J is an antiunitary operator, and it clearly satisfies $J^2 = I$. We call it the *complex conjugate operator* corresponding to the orthonormal basis $\{\varphi_j\}$.

Proposition 1.52 Let J be the complex conjugate operator corresponding to an orthonormal basis $\{\varphi_j\}$. Each antiunitary operator A can be written in the form $A = JU$, where U is a unitary operator uniquely determined by A.

Proof Let A be an antiunitary operator. Write $U = JA$. Since U is a composition of two antiunitary operators, it is unitary. We now have $JU = J^2A = A$. If $A = JU = JU'$ then $J^2U = J^2U'$. Therefore $U = U'$ and U is uniquely determined by A. □

We conclude from Proposition 1.52 that an antiunitary operator $A = JU$ is a bijection and that its inverse is $A^{-1} = U^*J$. The inverse A^{-1} is a product of a unitary operator and an antiunitary operator, therefore antiunitary itself. Together with our earlier observations, this implies that the set $\mathcal{U}(\mathcal{H}) \cup \overline{\mathcal{U}}(\mathcal{H})$ of all unitary and antiunitary operators is a group.

For each antiunitary operator A, we write $A^* \equiv A^{-1}$. This notation is reasonable since, for every $\varphi, \psi \in \mathcal{H}$, we have

$$\langle\varphi|A\psi\rangle = \langle I\varphi|A\psi\rangle = \langle AA^{-1}\varphi|A\psi\rangle = \langle\psi|A^{-1}\varphi\rangle .$$

This we can write as

$$\langle\psi|A^*\varphi\rangle = \overline{\langle A\psi|\varphi\rangle} ,$$

which agrees with the usual definition of the adjoint operator (formula (1.22) in subsection 1.2.2) except for the additional complex conjugation.

1.2.5 Ideal of trace class operators

In the finite-dimensional Hilbert space $\mathbb{C}^d$, the *trace* of an operator T can be obtained by writing T as a matrix in some orthonormal basis and then summing the diagonal entries of the matrix. This number, denoted by tr[T], does not depend on the chosen orthonormal basis, and we thus have

$$\text{tr}[T] = \sum_{j=1}^d \langle\varphi_j|T\varphi_j\rangle \tag{1.36}$$

for any orthonormal basis $\{\varphi_j\}_{j=1}^d$ of $\mathbb{C}^d$. As one knows from linear algebra, tr[T] equals the sum of the eigenvalues of T, counting multiplicity (in other words, if an eigenvalue λ occurs n times in the spectrum of T then it occurs in the sum of eigenvalues as $n\lambda$).

In an infinite-dimensional Hilbert space the trace is still a useful concept but things are not as straightforward as in $\mathbb{C}^d$. We can define the trace in a meaningful way only in a proper subset of bounded operators. 'Meaningful' here means two things: the trace has to be finite and independent of the chosen orthonormal basis.

Let $\mathcal{H}$ be a separable infinite-dimensional Hilbert space and $\{\varphi_j\}_{j=1}^{\infty}$ an orthonormal basis for $\mathcal{H}$. For any positive operator $T \in \mathcal{L}(\mathcal{H})$, we write

$$\text{tr}[T] = \sum_{j=1}^{\infty} \langle \varphi_j | T \varphi_j \rangle . \tag{1.37}$$

On the right-hand side we then have a sum of nonnegative numbers. It may happen that the sum on the right-hand side does not converge. In this case we write $\text{tr}[T] = \infty$. Now we will show that $\text{tr}[T]$ does not depend on the orthonormal basis chosen. Let $\{\psi_j\}_{j=1}^{\infty}$ be another orthonormal basis for $\mathcal{H}$. Then, using Parseval's formula (1.14) twice we get

$$\sum_{j=1}^{\infty} \langle \psi_j | T \psi_j \rangle = \sum_{j=1}^{\infty} \left\| T^{1/2} \psi_j \right\|^2 = \sum_{j=1}^{\infty} \left(\sum_{k=1}^{\infty} |\langle \varphi_k | T^{1/2} \psi_j \rangle|^2 \right)$$

$$= \sum_{k=1}^{\infty} \left(\sum_{j=1}^{\infty} |\langle \psi_j | T^{1/2} \varphi_k \rangle|^2 \right) = \sum_{k=1}^{\infty} \left\| T^{1/2} \varphi_k \right\|^2$$

$$= \sum_{k=1}^{\infty} \langle \varphi_k | T \varphi_k \rangle .$$

Interchanging the order of the sums is allowed since all the terms are nonnegative.

Example 1.53 (*Trace of a one-dimensional projection*)
Let $\eta \in \mathcal{H}$ be a unit vector and P_η the corresponding one-dimensional projection, as defined in Example 1.39. Let us choose an orthonormal basis $\{\varphi_j\}_{j=1}^{\infty}$ for $\mathcal{H}$ such that $\varphi_1 = \eta$. Then

$$\text{tr}[P_\eta] = \sum_{j=1}^{\infty} \langle \varphi_j | P_\eta \varphi_j \rangle = \langle \varphi_1 | P_\eta \varphi_1 \rangle = \langle \varphi_1 | \varphi_1 \rangle = 1 .$$

We conclude that $\text{tr}[P_\eta] = 1$. △

Exercise 1.54 Let $S, T \in \mathcal{L}(\mathcal{H})$ be positive operators. Prove the following properties of the trace:

(a) $\text{tr}[S + T] = \text{tr}[S] + \text{tr}[T]$;
(b) $\text{tr}[\alpha T] = \alpha \, \text{tr}[T]$ for all $\alpha \geq 0$;
(c) $\text{tr}[UTU^*] = \text{tr}[T]$ for all unitary operators U.

[Hint: For (a) and (b) use definition (1.37) directly. For (c), Proposition 1.49 is useful.]

Let us remind ourselves that in the discussion so far we have considered only positive operators. In order to extend the trace map to a larger set of operators, we need the following definition. Recall from Definition 1.36 in subsection 1.2.2 that $|T| = (T^*T)^{1/2}$.

Definition 1.55 A bounded operator T is a *trace class operator* if $\text{tr}[|T|] < \infty$. We denote by $\mathcal{T}(\mathcal{H})$ the set of trace class operators.

If $\mathcal{H}$ is infinite dimensional, the set $\mathcal{T}(\mathcal{H})$ is a proper subset of $\mathcal{L}(\mathcal{H})$. For instance, the identity operator I gives $\text{tr}[|I|] = \text{tr}[I] = \infty$, so I is therefore not a trace class operator. For all trace class operators we have the following result (see e.g. [121]).

Proposition 1.56 If $T \in \mathcal{T}(\mathcal{H})$ and $\{\varphi_j\}_{j=1}^{\infty}$ is an orthonormal basis for $\mathcal{H}$ then $\sum_{j=1}^{\infty} |\langle \varphi_j | T \varphi_j \rangle| < \infty$. The number

$$\text{tr}[T] := \sum_{j=1}^{\infty} \langle \varphi_j | T \varphi_j \rangle \qquad (1.38)$$

is called the *trace* of T and is independent of which orthonormal basis is chosen.

All the effort we have made may seem superfluous since, in the end, we have defined the trace of any trace class operator with the same formula as we used earlier for positive trace class operators. The point is that if an operator T is not a trace class operator then its trace can be finite in some orthonormal basis but infinite in another. Hence we need to know that an operator belongs to the trace class before blindly applying the formula (1.38).

In the following we list some useful facts about trace class operators but do not provide any proofs. The first piece of information is that the set of trace class operators $\mathcal{T}(\mathcal{H})$ is a vector space and the mapping

$$T \mapsto \text{tr}\big[|T|\big] =: \|T\|_{\text{tr}}$$

is a norm on $\mathcal{T}(\mathcal{H})$. In particular, the triangle inequality

$$\|S + T\|_{\text{tr}} \leq \|S\|_{\text{tr}} + \|T\|_{\text{tr}} \qquad (1.39)$$

holds for all $S, T \in \mathcal{T}(\mathcal{H})$. The norm $\|\cdot\|_{\text{tr}}$ is called the *trace norm*. A useful property is that the operator norm is bounded by the trace norm,

$$\|T\| \leq \|T\|_{\text{tr}}$$

for every $T \in \mathcal{T}(\mathcal{H})$.

The set $T(\mathcal{H})$ of trace class operators is not only a linear subspace of $\mathcal{L}(\mathcal{H})$, it is an *ideal*. This means that even though tr$[S]$ is not defined for all bounded operators S, it is always defined for the products TS and ST whenever T is a trace class operator. Moreover,

$$\text{tr}[TS] = \text{tr}[ST] \qquad \text{and} \qquad \left|\text{tr}[TS]\right| \leq \|T\|_{\text{tr}} \|S\| . \qquad (1.40)$$

There is still more structure to explore. For two trace class operators S, T, we write

$$\langle T | S \rangle_{\text{HS}} := \text{tr}\left[T^* S\right] .$$

This is an inner product on $T(\mathcal{H})$ and is called the *Hilbert–Schmidt inner product*. The induced norm

$$\|T\|_{\text{HS}} := \left(\text{tr}\left[T^* T\right]\right)^{1/2}$$

is called the *Hilbert–Schmidt norm* and the associated Cauchy–Schwarz inequality takes the form

$$\left|\text{tr}[ST]\right|^2 \leq \text{tr}\left[S^* S\right] \text{tr}\left[T^* T\right] . \qquad (1.41)$$

It turns out that the Hilbert–Schmidt inner product makes sense for a larger class of operators than the trace class operators. The operators satisfying $\|T\|_{\text{HS}} < \infty$ are called *Hilbert–Schmidt operators* and form a separable Hilbert space. A more detailed discussion of this class of operators is not needed for our purposes.

Example 1.57 (*Orthogonal projections revisited*)
In subsection 1.2.3 we defined orthogonality for projections; two projections P and Q are orthogonal if $PQ = O$. Has this any connection with orthogonality in the Hilbert–Schmidt sense?

We saw in Example 1.53 that one-dimensional projections have trace equal to 1. Therefore, a projection P is a trace class operator if and only if it can be written as finite sum of orthogonal one-dimensional projections (recall the discussion at the end of subsection 1.2.3). Let us suppose that two projections P and Q are trace class; then it makes sense to consider their Hilbert–Schmidt inner product. Clearly, if $PQ = O$ then $\langle P | Q \rangle_{\text{HS}} = \text{tr}[PQ] = 0$ and the projections are thus orthogonal in the Hilbert–Schmidt inner product.

Suppose that $\langle P | Q \rangle_{\text{HS}} = 0$. It follows that

$$0 = \text{tr}[PQ] = \text{tr}\left[P^2 Q^2\right] = \text{tr}\left[Q P^2 Q\right] .$$

Since $Q P^2 Q$ is a positive operator, we conclude that $\langle \psi | Q P^2 Q \psi \rangle = 0$ for every $\psi \in \mathcal{H}$. This implies that $\|P Q \psi\| = 0$ for every $\psi \in \mathcal{H}$, hence $PQ = O$. △

All the three different norms are easy to calculate for a selfadjoint trace class operator T if we know its eigenvalues. In this case we have

$$\|T\| = \max_j\{|\lambda_j|\}, \quad \|T\|_{\text{tr}} = \sum_j |\lambda_j|, \quad \|T\|_{\text{HS}} = \sqrt{\sum_j |\lambda_j|^2}, \qquad (1.42)$$

where $\{\lambda_j\}$ are the eigenvalues of T (including multiplicities). It is now easy to verify that

$$\|T\| \le \|T\|_{\text{HS}} \le \|T\|_{\text{tr}} .$$

This chain of inequalities is true also for a non-selfadjoint trace class operator.

Exercise 1.58 The simple expressions in (1.42) are not true in general. For instance, if $\mathcal{H} = \mathbb{C}^2$ and

$$T = \begin{pmatrix} 0 & 0 \\ 1 & 0 \end{pmatrix},$$

then the only eigenvalue of T is 0. However, T is not the zero operator and therefore its norm cannot be 0. Calculate $\|T\|_{\text{HS}}$.

1.3 Additional useful mathematics

In this section we will go through some miscellaneous topics needed in later chapters.

1.3.1 Weak operator topology

A separable infinite-dimensional Hilbert space is the closest infinite-dimensional analogue of $\mathbb{C}^d$. However, the infinite dimensionality makes the treatment a bit more delicate. The discussion in this section is redundant for a finite-dimensional Hilbert space. Hence, we will assume here that $\mathcal{H}$ is a separable infinite-dimensional Hilbert space.

The topology in $\mathcal{L}(\mathcal{H})$ determined by the operator norm, defined in (1.18), is too strong for many purposes. This means, roughly speaking, that not all operator sequences that we would like to consider as convergent do actually converge. Several different topologies appear naturally for $\mathcal{L}(\mathcal{H})$. In the perspective of quantum mechanics, the weak operator topology is usually the most relevant. This topology does not come from a norm and the open sets in it are somewhat lengthy to define. However, here we do not need an explicit description of them; it is enough for our purposes to specify when a sequence converges in the weak operator topology.

Definition 1.59 A sequence $\{T_j\} \subset \mathcal{L}(\mathcal{H})$ converges to a bounded operator T *in the weak operator topology*, or *weakly*, if

$$\lim_j |\langle \varphi | T \psi \rangle - \langle \varphi | T_j \psi \rangle| = 0 \qquad \text{for every } \varphi, \psi \in \mathcal{H}. \tag{1.43}$$

One should compare this definition with the fact that a sequence $\{T_j\} \subset \mathcal{L}(\mathcal{H})$ converges to a bounded operator T in the operator norm topology if

$$\lim_j \|T - T_j\| = \lim_j \sup_{\|\psi\|=1} \|(T - T_j)\psi\| = 0. \tag{1.44}$$

Proposition 1.60 If a sequence $\{T_j\} \subset \mathcal{L}(\mathcal{H})$ converges to a bounded operator T in the operator norm topology then it also converges to T in the weak operator topology.

Proof For every $\varphi, \psi \in \mathcal{H}$, we get

$$|\langle \varphi | T \psi \rangle - \langle \varphi | T_j \psi \rangle| = |\langle \varphi | (T - T_j) \psi \rangle| \le \|\varphi\| \, \|\psi\| \, \|T - T_j\| \, .$$

Here we have applied (1.20). $\qquad \square$

The converse implication in Proposition 1.60 is not valid. To demonstrate this, let us write $T_j = A^j$ for each $j = 1, 2, \ldots$, where A is the shift operator defined in Example 1.19. Hence, T_j acts in the following way:

$$T_j(\zeta_0, \zeta_1, \ldots) = (0, \ldots, 0, \zeta_0, \zeta_1, \ldots) \, .$$

Then $\lim_j T_j = O$ in the weak operator topology. Indeed, if $\eta \in \ell^2(\mathbb{N})$ then on the one hand

$$\lim_{j \to \infty} |\langle \eta | T_j \zeta \rangle| = \lim_{j \to \infty} \left| \sum_{k=0}^{\infty} \bar{\eta}_{k+j} \zeta_k \right| \le \|\zeta\|^2 \lim_{j \to \infty} \sum_{k=j}^{\infty} |\eta_k|^2 = 0 \, .$$

On the other hand we have $\|T_j \zeta\| = \|\zeta\|$, and this implies that the sequence $\{T_j\}$ cannot converge to O in the operator norm topology. It follows from Proposition 1.60 that $\{T_j\}$ does not converge in the operator norm topology, as the only option would be that $\{T_j\}$ converges to O.

We conclude that the operator norm topology and the weak operator topology are different, and thus one has to specify the topology when the convergence of operators is discussed. If not otherwise stated, we will understand all formulas in the weak sense, i.e. if a sequence or a sum is said to converge then it converges with respect to the weak operator topology. In practice this means that strange-looking operator formulas like

$$\flat \heartsuit = \diamond \emptyset$$

should be interpreted as

$$\langle \varphi | \flat \heartsuit \psi \rangle = \langle \varphi | \diamond \, \text{\textrlhookdbar} \, \psi \rangle \qquad \forall \varphi, \psi \in \mathcal{H}.$$

The rule of thumb is: an operator formula need to be 'sandwiched' between vectors if its meaning is otherwise unclear.

1.3.2 Dirac notation and rank-1 operators

In subsection 1.2.3 we defined the one-dimensional projection P_η for each unit vector $\eta \in \mathcal{H}$. In the so-called *Dirac notation* this projection is written as

$$P_\eta \equiv |\eta\rangle\langle\eta|.$$

Generally, if $\eta, \phi \in \mathcal{H}$, we define a linear mapping $|\eta\rangle\langle\phi|$ on $\mathcal{H}$ by

$$|\eta\rangle\langle\phi| \, \psi = \langle\phi|\psi\rangle \, \eta. \tag{1.45}$$

Using the Cauchy–Schwarz inequality we get

$$\left\| |\eta\rangle\langle\phi|\psi \right\| = \left\| \langle\phi|\psi\rangle \, \eta \right\| = \|\eta\| \, |\langle\phi|\psi\rangle| \leq \|\eta\| \, \|\phi\| \, \|\psi\| \, ,$$

showing that $|\eta\rangle\langle\phi|$ is a bounded operator. If η and ϕ are nonzero vectors then the range of $|\eta\rangle\langle\phi|$ is the one-dimensional subspace $\mathbb{C}\eta = \{c\eta \mid c \in \mathbb{C}\}$ and $|\eta\rangle\langle\phi|$ is called a *rank-1 operator*.

A moment's thought shows that $|\eta\rangle\langle\phi|$ behaves as a 'rearranged' inner product. For instance, the following rules apply for every $c \in \mathbb{C}$ and $\eta, \phi \in \mathcal{H}$:

$$c|\eta\rangle\langle\phi| = |c\eta\rangle\langle\phi| = |\eta\rangle\langle\bar{c}\phi|,$$
$$(|\eta\rangle\langle\phi|)^* = |\phi\rangle\langle\eta|.$$

From these equations we conclude that $|\eta\rangle\langle\phi|$ is selfadjoint if and only if $\phi = r\eta$ for some $r \in \mathbb{R}$. Therefore, a selfadjoint rank-1 operator is of the form rP_η for some $r \in \mathbb{R}$ and a unit vector $\eta \in \mathcal{H}$.

It is also straightforward to verify that the product of two rank-1 operators $|\eta\rangle\langle\phi|$ and $|\xi\rangle\langle\psi|$ is given by the formula

$$|\eta\rangle\langle\phi||\xi\rangle\langle\psi| = \langle\phi|\xi\rangle|\eta\rangle\langle\psi|. \tag{1.46}$$

Sometimes even a single vector η is written in the form $|\eta\rangle$. With this convention, (1.45) and (1.46) become simply reordering rules for the vectors. Moreover, $\langle\phi|$ is taken to denote the linear functional

$$\psi \mapsto \langle\phi|\psi\rangle \, .$$

Again, this is consistent with the other definitions. In his famous textbook *The Principles of Quantum Mechanics* [54], Dirac gave attractive and intuitive names to his notation. He called an inner product $\langle \eta | \phi \rangle$ a *bracket*. Since we can decompose it as a 'product' of $\langle \eta |$ and $| \phi \rangle$, Dirac called $\langle \eta |$ a *bra vector* and $| \phi \rangle$ a *ket vector*.

The convenience of Dirac's notation starts to become clear in the following formulas, which we will find useful later.

Example 1.61 It is sometimes convenient to write the identity operator I on $\mathcal{H}$ as a sum:

$$I = \sum_{k=1}^{d} |\varphi_k\rangle\langle\varphi_k|, \tag{1.47}$$

where $\{\varphi_k\}_{k=1}^{d}$ is an orthonormal basis for $\mathcal{H}$. If $\mathcal{H}$ is infinite dimensional then this type of expression should be understood in the weak sense (see subsection 1.3.1). Hence it is then just a shorthand notation for the following formula, true for all $\psi, \xi \in \mathcal{H}$,

$$\langle\psi|\xi\rangle = \sum_{k=1}^{d} \langle\psi|\varphi_k\rangle \langle\varphi_k|\xi\rangle. \tag{1.48}$$

This formula follows by writing the basis expansion of ξ in the basis $\{\varphi_k\}_{k=1}^{d}$. △

Exercise 1.62 Sometimes a formula is easier to derive by first passing to Dirac notation. Let P_η and P_ϕ be two one-dimensional projections. Show that

$$P_\eta P_\phi P_\eta = \mathrm{tr}\big[P_\eta P_\phi\big] P_\eta.$$

(Hint: As we have already hinted, write the projections in Dirac notation.)

Positive rank-1 operators are the smallest among all positive operators in the sense that a positive operator below a positive rank-1 operator has to be a rank-1 operator itself. The following simple result is sometimes useful.

Proposition 1.63 Let R be a positive rank-1 operator and T a positive operator. If $T \leq R$, then $T = tR$ for some number $0 \leq t \leq 1$.

Proof If $R = O$, the claim is trivial. Hence, assume that $R \neq O$. We can write R in the form $R = r|\eta\rangle\langle\eta|$ for some number $r > 0$ and unit vector $\eta \in \mathcal{H}$. If $T \leq R$ then $T \leq r|\eta\rangle\langle\eta|$. From Proposition 1.46 follows that

$$T = T|\eta\rangle\langle\eta| = |\eta\rangle\langle\eta|T. \tag{1.49}$$

First, note that $T\eta = \langle\eta|T\eta\rangle \eta$. Inserting this into (1.49) shows that $T = \langle\eta|T\eta\rangle |\eta\rangle\langle\eta|$. Since $\langle\eta|T\eta\rangle \leq \langle\eta|R\eta\rangle = r$, we conclude that $T = tR$ for some number $0 \leq t \leq 1$. □

Sometimes it may be convenient to use even more compressed Dirac notation. We can denote the vectors of a fixed orthonormal basis by kets with numerical labels. Hence, $|n\rangle$, $n = 1, \ldots, d$, can denote an orthonormal basis for a d-dimensional Hilbert space.

Example 1.64 (*Existence of an orthogonal unitary basis*)
If $\mathcal{H} = \mathbb{C}^d$ then the dimension of the inner product space $\mathcal{T}(\mathcal{H})$ is d^2. It is sometimes useful to choose an orthogonal basis for $\mathcal{T}(\mathcal{H})$ consisting of unitary operators only. To demonstrate that this kind of basis exists, fix an orthonormal basis $|l\rangle$, $l = 0, \ldots, d - 1$, for $\mathbb{C}^d$. We then define d^2 operators

$$U_{rs} = \sum_{l=0}^{d-1} e^{-i2\pi sl/d} |l \ominus_d r\rangle \langle l|,$$

where $r, s = 0, \ldots, d - 1$ and $\ominus_d$ denotes the difference modulo d. We have

$$\mathrm{tr}\left[U_{rs}^* U_{r's'}\right] = \sum_{l,l'=0}^{d-1} e^{i2\pi(sl-s'l')/d} \,\mathrm{tr}\left[|l\rangle \langle l \ominus_d r |l' \ominus_d r'\rangle \langle l'|\right]$$

$$= \sum_{l=0}^{d-1} e^{i2\pi(s-s')l)/d} \langle l \ominus_d r |l \ominus_d r'\rangle = d\delta_{rr'}\delta_{ss'}.$$

Here we have used the fact that for $s \neq s'$ the geometric sum formula gives

$$\sum_{l=0}^{d-1} e^{i2\pi(s-s')l/d} = \frac{1 - e^{i2\pi(s-s')}}{1 - e^{i2\pi(s-s')/d}} = 0$$

and for $s = s'$ we have

$$\sum_{l=0}^{d-1} e^{i2\pi(s-s')l/d} = \sum_{l=0}^{d-1} 1 = d.$$

Thus, the operators U_{rs} form an orthogonal basis of $\mathcal{T}(\mathcal{H})$. It is straightforward to verify that $U_{rs}U_{rs}^* = U_{rs}^* U_{rs} = I$, hence each U_{rs} is unitary. Notice that $U_{00} = I$. $\triangle$

1.3.3 Spectral and singular-value decompositions

An operator T is called *normal* if $T^*T = TT^*$. In particular, all selfadjoint and unitary operators are normal. Normal operators have a very useful description via the spectral decomposition theorem.

Consider complex numbers $\lambda_1, \ldots, \lambda_k$ and an orthogonal set of unit vectors $\varphi_1, \ldots, \varphi_k$. The operator

$$T = \lambda_1 |\varphi_1\rangle\langle\varphi_1| + \cdots + \lambda_k |\varphi_k\rangle\langle\varphi_k|$$

is normal and the absolute value $|T|$ of T is given by

$$|T| = |\lambda_1||\varphi_1\rangle\langle\varphi_1| + \cdots + |\lambda_k||\varphi_k\rangle\langle\varphi_k|.$$

From this we conclude that T is a trace class operator and $\|T\|_{\text{tr}} = \sum_{j=1}^{k} |\lambda_j|$.

More generally, if $\lambda_1, \lambda_2, \ldots$ is a sequence of complex numbers and $\varphi_1, \varphi_2, \ldots$ is an orthonormal set of vectors then the sum

$$T = \sum_j \lambda_j |\varphi_j\rangle\langle\varphi_j|$$

defines a normal trace class operator, provided that $\sum_j |\lambda_j| < \infty$.

Theorem 1.65 (*Spectral decomposition*)
Let T be a normal trace class operator. There exists a sequence $\{\lambda_j\}$ of complex numbers and an orthonormal basis $\{\varphi_j\}$ such that

$$T = \sum_j \lambda_j |\varphi_j\rangle\langle\varphi_j|. \tag{1.50}$$

If an operator T is normal but not trace class, it still has a spectral decomposition. But, instead of a finite or infinite sum, one may have to write its spectral decomposition as an integral. We will not discuss this topic further, owing to its more involved technicalities.

Nonnormal trace class operators look almost the same as normal trace class operators. To give a warm-up example, let $\lambda_1, \ldots, \lambda_k$ be complex numbers and let $\{\varphi_j\}_{j=1}^{k}$, $\{\phi_j\}_{j=1}^{k}$ be two orthonormal sets. The operator

$$T = \sum_{j=1}^{k} \lambda_j |\varphi_j\rangle\langle\phi_j|$$

is trace class. Writing $\lambda_j = r_j e^{i\theta_j}$, we get

$$T = \sum_{j=1}^{k} r_j e^{i\theta_j} |\varphi_j\rangle\langle\phi_j| = \sum_{j=1}^{k} r_j |\varphi_j'\rangle\langle\phi_j|,$$

where $\{\varphi_j'\}$ is another orthonormal set. Thus, by slightly modifying the orthonormal set we can pass from complex numbers λ_j to positive real numbers r_j.

Let us now consider an arbitrary trace class operator T. The absolute value $|T|$ of T is a positive trace class operator. The eigenvalues of $|T|$, counted with their multiplicity, are called the *singular values of* T. We denote them by $s_j(T)$. From their definition it follows that $s_j(T) \geq 0$ (since $|T|$ is positive) and $\sum_j s_j(T) < \infty$ (since $|T|$ is trace class).

Theorem 1.66 (*Singular-value decomposition*)
Let T be a trace class operator. There exist orthonormal sets $\{\varphi_j\}$ and $\{\phi_j\}$ in $\mathcal{H}$ such that

$$T = \sum_j s_j(T)|\varphi_j\rangle\langle\phi_j|. \tag{1.51}$$

All trace class operators are special cases of *compact operators*. In mathematics books, Theorems 1.65 and 1.66 are typically developed in a more general form and proved for all compact operators.

1.3.4 Linear functionals and dual spaces

A linear mapping f from a complex vector space V into the field of complex numbers is called a *linear functional*. If V is a normed space then we denote by V^* the set of all continuous linear functionals. It is called the *dual space* of V. The dual space V^* is a vector space itself when the linear structure is defined pointwise, i.e. $(f_1 + cf_2)(v) = f_1(v) + cf_2(v)$ for all $v \in V$ and $c \in \mathbb{C}$. We can also define a norm on V^* by setting

$$\|f\| := \sup_{\|v\|=1} |f(v)|.$$

In this way, the dual space V^* becomes a normed space.

Continuous linear functionals arise in many situations, and it is convenient to characterize the various dual spaces. Let us first investigate the dual spaces of Hilbert spaces. Let $\mathcal{H}$ be a Hilbert space. Each vector $\phi \in \mathcal{H}$ defines a linear functional f_ϕ on $\mathcal{H}$ by the formula

$$f_\phi(\psi) = \langle\phi|\psi\rangle .$$

An application of the Cauchy–Schwarz inequality shows that f_ϕ is continuous, hence an element of the dual space $\mathcal{H}^*$. The content of the following theorem is that all continuous linear functionals on $\mathcal{H}$ have this form.

Theorem 1.67 (*Fréchet–Riesz theorem*)
Let $f \in \mathcal{H}^*$. There exists a unique vector $\phi \in \mathcal{H}$ such that $f = f_\phi$. Moreover, $\|f_\phi\| = \|\phi\|$.

The name of the previous theorem varies between mathematics books: it is sometimes called the Riesz lemma or the Riesz representation theorem. Notice that if a vector $\phi \in \mathcal{H}$ corresponds to f_ϕ then the scalar multiple $c\phi$ corresponds to

$f_{c\phi} = \bar{c} f_\phi$. For this reason, the one-to-one correspondence $\phi \leftrightarrow f_\phi$ is not linear but conjugate linear.

Let us then investigate the dual space $\mathcal{T}(\mathcal{H})^*$ of trace class operators. Suppose for a moment that $\mathcal{H}$ is a finite-dimensional Hilbert space and $\mathcal{T}(\mathcal{H})$ is hence a finite-dimensional normed space. By fixing an orthonormal basis in $\mathcal{H}$ we can identify $\mathcal{H}$ with $\mathbb{C}^d$ and $\mathcal{T}(\mathcal{H})$ with the set of $d \times d$ complex matrices $\mathcal{M}_d(\mathbb{C})$. We denote by $H^{(j,k)} \in \mathcal{M}_d(\mathbb{C})$ the matrix having (j, k)th entry equal to 1 and all other entries equal to 0. Let f be a linear functional on $\mathcal{M}_d(\mathbb{C})$. We define a matrix $S = [s_{jk}]$ by setting $s_{jk} = f(H^{(k,j)})$. For any matrix $T = [t_{jk}] \in \mathcal{M}_d(\mathbb{C})$, we then obtain

$$f(T) = f\left(\sum_{j,k=1}^{d} t_{jk} H^{(j,k)}\right) = \sum_{j,k=1}^{d} t_{jk} f(H^{(j,k)}) = \sum_{j,k=1}^{d} t_{jk} s_{kj} = \operatorname{tr}[ST] .$$

We conclude that all linear functionals on $\mathcal{M}_d(\mathbb{C})$ can be represented by matrices through the above trace formula. This is consistent with Theorem 1.67 since we can equip $\mathcal{M}_d(\mathbb{C})$ with the Hilbert–Schmidt inner product and then $\operatorname{tr}[ST] = \langle S^* | T \rangle_{\mathrm{HS}}$.

Let us then assume that $\mathcal{H}$ is a separable infinite-dimensional Hilbert space. As we saw in subsection 1.2.5, the set $\mathcal{T}(\mathcal{H})$ of trace class operators is a vector space and the trace norm makes it a normed space. For each bounded operator $S \in \mathcal{L}(\mathcal{H})$, we define a linear functional f_S on $\mathcal{T}(\mathcal{H})$ by the formula

$$f_S(T) = \operatorname{tr}[ST] . \tag{1.52}$$

The crucial thing here is that $\mathcal{T}(\mathcal{H})$ is an ideal in $\mathcal{L}(\mathcal{H})$, therefore ST is a trace class operator. Using (1.40) we see that f_S is continuous and thus $f_S \in \mathcal{T}(\mathcal{H})^*$. Notice also that

$$f_{S_1} + f_{S_2} = f_{S_1+S_2}$$

for two bounded operators $S_1, S_2 \in \mathcal{L}(\mathcal{H})$.

Each bounded operator S determines a vector f_S in the dual space $\mathcal{T}(\mathcal{H})^*$ of $\mathcal{T}(\mathcal{H})$, and the converse is true also. Namely, each continuous linear functional on $\mathcal{T}(\mathcal{H})$ is of the form f_S for some $S \in \mathcal{L}(\mathcal{H})$. For a proof of the following result, see e.g. [46].

Theorem 1.68 The mapping $S \mapsto f_S$ is a linear bijection from $\mathcal{L}(\mathcal{H})$ to $\mathcal{T}(\mathcal{H})^*$, and $\|S\| = \|f_S\|$ for every $S \in \mathcal{L}(\mathcal{H})$.

In other words, Theorem 1.68 states that the dual space $\mathcal{T}(\mathcal{H})^*$ of $\mathcal{T}(\mathcal{H})$ can be identified with $\mathcal{L}(\mathcal{H})$ and the identification is given by formula (1.52). Some

additional features make this identification even more useful. In the following we
list two.

Proposition 1.69 Let $S \in \mathcal{L}(\mathcal{H})$ and f_S be as in (1.52). Then:

(a) $S \geq O$ if and only if $f_S(T) \geq 0$ whenever $T \geq O$;
(b) $S = S^*$ if and only if $f_S(T) \in \mathbb{R}$ whenever $T = T^*$.

Generally, a linear functional f defined on $\mathcal{L}(\mathcal{H})$ or some of its subset is called
positive if $f(T) \geq 0$ whenever $T \geq O$. Item (a) therefore states that f_S is positive
if and only if S is positive.

Exercise 1.70 Prove one half of Proposition 1.69, i.e. that if f_S is positive then S
in positive. [Hint: Calculate $f_S(P_\psi)$ for one-dimensional projections P_ψ and recall
that projections are positive.]

If $\dim \mathcal{H} = d < \infty$ then $\mathcal{L}(\mathcal{H}) = \mathcal{T}(\mathcal{H}) = M_d(\mathbb{C})$ and our previous discussion
on $\mathcal{T}(\mathcal{H})^*$ applies equally well to $\mathcal{L}(\mathcal{H})^*$. This is no longer true if $\mathcal{H}$ is infinite
dimensional. The following useful result is proved in [51].

Theorem 1.71 For a positive linear functional f on $\mathcal{L}_s(\mathcal{H})$, the following conditions are equivalent:

(i) f is *normal*, i.e. for each norm-bounded increasing sequence $\{T_j\}_{j=1}^\infty \subset \mathcal{L}_s(\mathcal{H})$
    with limit T, the sequence $\{f(T_j)\}_{j=1}^\infty$ converges to $f(T)$.
(ii) There exists a positive trace class operator S such that $f(T) = \text{tr}[ST]$ for all
     $T \in \mathcal{L}_s(\mathcal{H})$.

1.3.5 Tensor product

Forming the tensor product is a way to create a new Hilbert space out of two (or
more) Hilbert spaces. Several different constructions lead to the same thing, and
for everyday calculations it is not necessary to remember all the details, just some
simple computational rules. First, therefore we will describe tensor product spaces
in an informal way before sketching a precise construction.

Let $\mathcal{H}$ and $\mathcal{K}$ be two finite-dimensional inner product spaces. We can form a new
inner product space $\mathcal{H} \otimes \mathcal{K}$, called the *tensor product* of $\mathcal{H}$ and $\mathcal{K}$, in the following
way. The elements of $\mathcal{H} \otimes \mathcal{K}$ are expressions of the form

$$\sum_{j=1}^n \psi_j \otimes \zeta_j, \tag{1.53}$$

where $\psi_j \in \mathcal{H}, \zeta_j \in \mathcal{K}$ and $n \in \mathbb{N}$. The quantity $\psi \otimes \zeta$ is assumed to be linear with
respect to both arguments, so that we have

$$c(\psi \otimes \zeta) = (c\psi) \otimes \zeta = \psi \otimes (c\zeta),$$
$$(\psi_1 + \psi_2) \otimes \zeta = \psi_1 \otimes \zeta + \psi_2 \otimes \zeta,$$
$$\psi \otimes (\zeta_1 + \zeta_2) = \psi \otimes \zeta_1 + \psi \otimes \zeta_2,$$

for every $\psi_1, \psi_2, \psi \in \mathcal{H}$, $\zeta_1, \zeta_2, \zeta \in \mathcal{K}$ and $c \in \mathbb{C}$. The addition of two elements is defined in a natural way:

$$\sum_{j=1}^{n} \psi_j \otimes \zeta_j + \sum_{j=n+1}^{m} \psi_j \otimes \zeta_j = \sum_{j=1}^{m} \psi_j \otimes \zeta_j.$$

Finally, an inner product on $\mathcal{H} \otimes \mathcal{K}$ is defined by

$$\langle \psi_1 \otimes \zeta_1 | \psi_2 \otimes \zeta_2 \rangle = \langle \psi_1 | \psi_2 \rangle \langle \zeta_1 | \zeta_2 \rangle \tag{1.54}$$

and then extending by linearity to all elements. In this way, $\mathcal{H} \otimes \mathcal{K}$ becomes an inner product space. If $\{\varphi_l\}$ is an orthonormal basis for $\mathcal{H}$ and $\{\phi_j\}$ an orthonormal basis for $\mathcal{K}$ then $\{\varphi_l \otimes \phi_j\}$ is an orthonormal basis for $\mathcal{H} \otimes \mathcal{K}$. In particular, the dimension of $\mathcal{H} \otimes \mathcal{K}$ is the product of the dimensions of $\mathcal{H}$ and $\mathcal{K}$.

In the same way as for vectors, two operators $S \in \mathcal{L}(\mathcal{H})$ and $T \in \mathcal{L}(\mathcal{K})$ determine an operator $S \otimes T$ acting in the tensor product space $\mathcal{H} \otimes \mathcal{K}$. If $\psi \in \mathcal{H}$ and $\zeta \in \mathcal{K}$ then

$$S \otimes T \, \psi \otimes \zeta = S\psi \otimes T\zeta.$$

This action can then be extended to all vectors in $\mathcal{H} \otimes \mathcal{K}$ by linearity.

Exercise 1.72 Let $\mathcal{H}$ be a finite-dimensional Hilbert space with an orthonormal basis $\{\varphi_j\}_{j=1}^{d}$. Let us fix the orthonormal basis $\{\varphi_j \otimes \varphi_k\}_{j,k=1}^{d}$ for the tensor product space $\mathcal{H} \otimes \mathcal{H}$. The ordering of the basis vectors is taken to be $\varphi_1 \otimes \varphi_1, \varphi_1 \otimes \varphi_2, \ldots,$ $\varphi_2 \otimes \varphi_1, \ldots$ If S and T are two operators in $\mathcal{H}$ and their matrices in the basis $\{\varphi_j\}_{j=1}^{d}$ are $[s]$ and $[t]$, respectively, then we claim that the $d^2 \times d^2$ matrix corresponding to $S \otimes T$ is

$$\begin{bmatrix} s_{11}[t] & s_{12}[t] & \cdots & s_{1d}[t] \\ s_{21}[t] & s_{22}[t] & \cdots & s_{2d}[t] \\ \vdots & \vdots & \ddots & \vdots \\ s_{d1}[t] & s_{d2}[t] & \cdots & s_{dd}[t] \end{bmatrix}.$$

Verify this claim!

If $\mathcal{H}$ and $\mathcal{K}$ are two Hilbert spaces (not necessarily finite dimensional) then the above construction still leads to an inner product space. However, if $\mathcal{H}$ and $\mathcal{K}$ are not both finite dimensional, we need to take the completion of this inner product space to obtain a Hilbert space. This Hilbert space is then the tensor product space and denoted by $\mathcal{H} \otimes \mathcal{K}$. In this case not all vectors in $\mathcal{H} \otimes \mathcal{K}$ are finite sums of the form (1.53). However, every vector in $\mathcal{H} \otimes \mathcal{K}$ can be approximated arbitrarily well with vectors of the form (1.53).

Let us then take a brief look at an explicit construction of a tensor product space. This is required when one needs to prove something concerning the structure of the tensor product space. Let $\mathcal{H}$ and $\mathcal{K}$ be two Hilbert spaces. For each $\psi \in \mathcal{H}$ and $\zeta \in \mathcal{K}$, we define a mapping $\psi \otimes \zeta$ from $\mathcal{H} \times \mathcal{K}$ to $\mathbb{C}$ by the formula

$$\psi \otimes \zeta(\varphi, \xi) = \langle \varphi | \psi \rangle \langle \xi | \zeta \rangle .$$

The set V of all finite linear combinations of such mappings is a linear space, and it becomes an inner product space when we make the definition

$$\langle \psi_1 \otimes \zeta_1 | \psi_2 \otimes \zeta_2 \rangle = \langle \psi_1 | \psi_2 \rangle \langle \zeta_1 | \zeta_2 \rangle \tag{1.55}$$

and extend this by linearity to all elements in V. The tensor product of $\mathcal{H} \otimes \mathcal{K}$ is the completion of V under the inner product (1.55). For more details see e.g. [121]. There are also other ways to construct $\mathcal{H} \otimes \mathcal{K}$. One possibility is to use bounded antilinear mappings; see e.g. [60].

Exercise 1.73 In many cases a tensor product space can be given a concrete equivalent form. In subsection 1.1.1 we encountered two Hilbert spaces, $\mathbb{C}^d$ and $\ell^2(\mathbb{N})$. Their tensor product space $\ell^2(\mathbb{N}) \otimes \mathbb{C}^d$ is isomorphic with $\ell^2(\mathbb{N}; \mathbb{C}^d)$. This latter space is quite like $\ell^2(\mathbb{N})$ but its elements are functions $f : \mathbb{N} \rightarrow \mathbb{C}^d$ satisfying $\sum_k \| f(k) \|^2 < \infty$ (recall the end of subsection 1.1.3). Choose the usual orthonormal bases for $\ell^2(\mathbb{N})$ and $\mathbb{C}^d$ and hence obtain an orthonormal basis for $\ell^2(\mathbb{N}) \otimes \mathbb{C}^d$. Then find an orthonormal basis for $\ell^2(\mathbb{N}; \mathbb{C}^d)$ and an isomorphism $U : \ell^2(\mathbb{N}) \otimes \mathbb{C}^d \rightarrow \ell^2(\mathbb{N}; \mathbb{C}^d)$ mapping the orthonormal basis of $\ell^2(\mathbb{N}) \otimes \mathbb{C}^d$ into the orthonormal basis of $\ell^2(\mathbb{N}; \mathbb{C}^d)$.

2

States and effects

One of the main purposes of a physical theory is to describe and predict events observed in physical experiments. In this chapter we introduce the key physical concepts used in the description of quantum experiments and present their mathematical formalization as Hilbert space operators.

2.1 Duality of states and effects

Fundamental in quantum theory is the concept of a state, which is usually understood as a description of an ensemble of similarly prepared systems. An *effect*, on the other hand, is a measurement apparatus that produces either 'yes' or 'no' as an outcome. The concept of an effect was coined by Ludwig [96] and made popular by Kraus [89]. The duality of states and effects means, essentially, that when a state and an effect are specified we can calculate a probability distribution. This can be compared with the outcomes of real experiments, and it gives a physical meaning to the mathematical Hilbert space machinery.

2.1.1 Basic statistical framework

A basic situation in physics is the following: we have an object system under investigation, and we are trying to learn something about it by doing an experiment. As a result, a measurement outcome is registered. A statistical theory, such as quantum theory, does not in general predict which individual outcomes will occur in any particular measurement; it merely predicts their probabilities of occurrence. Hence, we take the output of an experiment (which consists of many measurement runs) to be a probability distribution on a set Ω of the possible measurement outcomes. In order to verify these probabilistic predictions, an experiment should thus be understood as an action which is (or can be) repeated in similar conditions many times.

Example 2.1 (*Polarizer*)

Let us imagine a simple optical experiment in which photons emitted from a light source are directed towards a piece of material called a polarizer (see Figure 2.1). We expect that the polarizer, like any other object, will absorb and thus filter out some incoming photons. However, it possesses a special feature. By setting two identical polarizers in a sequence one can achieve any filtering rate, depending on their mutual orientation. The point is that the polarizers filter the photons according to their quantum state of polarization. As a result of an individual experimental run we either register a photon passage or not. However, we cannot predict the behavior of an individual photon. What we observe in this experiment is that the ratio of the outgoing and incoming light intensities (the numbers of photons) depends on the adjustment of the polarizer and on the light source. This fraction is predicted by quantum theory. We denote by N_p, N_i the number of passing and incoming photons, respectively. Then the ratio N_p/N_i is an approximation to the probability that an incoming photon will pass the polarizer. △

Commonly, it is useful to divide an experiment into preparation and measurement phases. In a given experiment the division may be quite arbitrary, but this does not matter. We simply assume that there is a collection of possible preparations and a collection of possible measurements and that any preparation and any measurement can be combined to form an experiment leading to a probability distribution of measurement outcomes. Therefore, the preparations specify probability distributions for every possible measurement on the system. Alternatively, any fixed measurement will specify a probability distribution for every preparation. This short explanation of what we call the *basic statistical framework* is depicted in Figure 2.2. We will now introduce some further terminology and add some essential assumptions.

Figure 2.1 A polarizer filters incoming photons according to their polarization. We cannot predict the behavior of an individual photon, but we can predict the ratio of passing and incoming photons for an ensemble of similarly prepared photons.

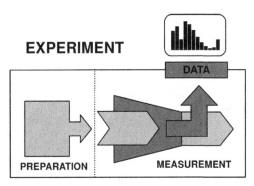

Figure 2.2 Basic statistical framework. An experiment is divided into preparation and measurement. The output of an experiment is a probability distribution of measurement outcomes.

Example 2.2 (*Polarizing beamsplitter*)

How can we divide the simple polarization experiment described in Example 2.1 into preparation and measurement parts? Naturally, the light source is a preparation device and the remaining part of the setup is the measurement part, leading to the observation of one of two outcomes: a photon is either detected or not. Let us stress that we are assuming implicitly an important property of the light source. For the statistics of any experiment it is important that we are able to calculate the total number of individual runs. This is achieved if we have evidence of each individual run of the experiment. However, in practice it can happen that a light source emits photons but we do not know when this will occur. In such cases the evidence must be provided by the observation of an outcome in the measurement part of the experiment. This issue is serious only if such an 'evidence-free' preparation is combined with a measurement in which one outcome is identified as the absence of the observation of any other outcome. The 'no count' event on a photodetector placed in the beam is an example of such outcome. If the photon source is evidence-free, the simple polarization experiment described in Example 2.1 cannot be accommodated within the basic statistical framework unless we are able to detect the absorption of photons inside the polarizer.

There exists a simple solution for evidence-free photon sources also. A polarizing beamsplitter is a special polarization sensitive device which splits an incoming beam of light into two beams, depending on the polarization properties of the photons (see Figure 2.3). Its action is very similar to that of a polarizer, but for our purposes its importance is that no photon is lost. Those photons that would be absorbed inside a polarizer are instead transmitted along a different optical path and hence can be detected by an additional photodetector. Using such an experimental

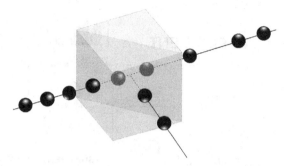

Figure 2.3 A polarizing beamsplitter splits an incoming beam of light into two beams with different (mutually orthogonal) polarizations.

setup the detection of a photon by either detector provides us with evidence of the successful preparation of the photon. The photon source can be considered as constituting the preparation procedure and the polarizing beamsplitter and detectors represent the measurement part of the experiment.   △

Two preparation procedures can be superficially quite different and yet lead to the same probability distribution in any chosen measurement. If this is the case, we say that the two preparations are equivalent. We define a *state* of the system to be a collection of equivalent preparation procedures. We denote by S the set of all such states. Quite often it is easier to think of a state as describing an ensemble of similarly prepared systems, as already mentioned, and this should not cause any confusion.

An *elementary event* in any physical experiment is of the form

'The recorded measurement outcome is x.'

or slightly more generally

'The recorded measurement outcome belongs to a subset $X \subset \Omega$.'

Here, by '*elementary*' we mean that this type of claim can be evaluated by two options 'yes' or 'no'. One can collect all the elementary events observed in various experiments that occur with the same probability for all states; this collection of equivalent events is called an *effect*. In other words, an effect determines a function $\varrho \mapsto E(\varrho)$ from the set of states S into the interval $[0, 1]$. We identify effects with this kind of function, and we agree that the number $E(\varrho)$ is the probability of a 'yes' answer.

Example 2.3 (*Identity and zero effects*)
The *identity effect* I assigns a probability 1 for every state ϱ. It is hence a constant mapping $I(\varrho) = 1$. Physically it corresponds to a measurement with a single

outcome that will always be registered. Analogously, the *zero effect O* assigns a probability 0 for every state ϱ. It corresponds to an event that never happens. $\triangle$

Suppose that we perform an experiment in which the measurement is fixed but we alternate randomly between two different preparation procedures (states), so that they occur with probabilities λ and $1 - \lambda$. This mixing gives us a new preparation procedure. As a result we have a way to combine two states into a third. Namely, for states ϱ_1, ϱ_2 and a number $0 \leq \lambda \leq 1$, there exists a state, denoted by $\lambda \varrho_1 + (1 - \lambda)\varrho_2$, which is a *mixture* of ϱ_1 and ϱ_2 with weights λ and $1 - \lambda$. Any measurement performed on this new state $\lambda \varrho_1 + (1 - \lambda)\varrho_2$ should give results that are consistent with the results obtained when measuring ϱ_1 and ϱ_2.

Consequently, we come to a *basic assumption* demanding that the statistical correspondence between states and effects is consistent with the mixing property of states: if ϱ_1 and ϱ_2 are states and $0 \leq \lambda \leq 1$ then

$$E(\lambda \varrho_1 + (1 - \lambda)\varrho_2) = \lambda E(\varrho_1) + (1 - \lambda)E(\varrho_2) \qquad (2.1)$$

for any effect E. We also make the simplifying assumption that all mappings from the set of states to the interval $[0, 1]$ satisfying (2.1) are effects. In the end, one has to test experimentally whether a given effect can be actually realized.

Example 2.4 Consider the detection of photons passing through a horizontally oriented linear polarizer. Suppose that we are randomly switching between two different preparation procedures (states), ϱ_1 and ϱ_2. In the first procedure, photons emitted from the light source pass through a polarizing beamsplitter and only a horizontally polarized path enters the measurement process. In the second procedure this beam passes through a quarter-wave plate, changing the linear polarization into a circular polarization. Hence, randomly (with probability $\frac{1}{2}$) the quarter-wave plate is present in the optical path of the photon. What is the probability that a photon will be detected?

We denote by E the effect corresponding to a detection event. We know the detection probabilities for the first and second preparation procedures individually. In the first case the photon will be detected with certainty, whereas the detection probability is $\frac{1}{2}$ if the quarter-wave plate is present. The statistics of the sub-ensembles of outcomes for the particular preparation procedures must be in accordance with these probabilities, i.e. $E(\varrho_1) = 1$ and $E(\varrho_2) = \frac{1}{2}$. As an example, let us consider the experimental record

$$+_2, +_1, -_2, +_1, -_2, +_1, +_1, +_2, -_2, +_2, +_2, +_1, +_1, -_2, +_1, +_1, \ldots,$$

where the number denotes the choice of preparation procedure and $\pm$ indicates whether the photon was detected. Collecting all the outcomes associated with the first preparation procedure we find $E(\varrho_1) = 1$ and for the second preparation procedure we have $E(\varrho_2) = \frac{1}{2}$. For the whole sequence we get $E(\varrho) = \frac{3}{4}$. Since both preparation procedures were chosen with equal probabilities (i.e. $\lambda = \frac{1}{2}$), this data is consistent with the equation $E(\varrho) = \lambda E(\varrho_1) + (1 - \lambda)E(\varrho_2)$. △

We can turn things upside down and consider states ϱ to be functions on the set of effects:

$$f_\varrho(E) := E(\varrho). \qquad (2.2)$$

Our definitions imply that two states ϱ_1 and ϱ_2 are the same if and only if the associated functions f_{ϱ_1} and f_{ϱ_2} coincide. Therefore, we can think the set of states as a subset of all real-valued functions defined on the set of effects. A convex combination of two functions f and f' is defined in the usual way:

$$(\lambda f + (1 - \lambda)f')(E) = \lambda f(E) + (1 - \lambda)f'(E).$$

It follows from the basic assumption (2.1) that the mixture of states defined earlier agrees with the convex combination of their related functions and hence that the set of states can be identified with a convex subset of the real vector space of real-valued functions on effects. The basic assumption (2.1) can be rephrased by saying that effects are *affine* mappings.

To summarize, the basic statistical framework consists of two ingredients, states and effects, assumed to have the following properties:

- the states form a convex set;
- the effects are affine mappings from the set of states to the interval [0, 1].

2.1.2 State space

As noted in the conclusion of the previous subsection, the states of a quantum system are assumed to form a convex set. Since the basic statistical framework is very general and covers all kinds of experiment, it is clear that the convex set corresponding to quantum states must be of some specific type.

There are several equivalent ways of representing mathematically the state space of a quantum system. In the most common Hilbert space formulation, states are described by positive trace class operators of trace 1, also called *density matrices*. We write

$$S(\mathcal{H}) := \{\varrho \in T(\mathcal{H}) \mid \varrho \geq 0, \operatorname{tr}[\varrho] = 1\},$$

and from now on we identify the set of states with $S(\mathcal{H})$. Let us note that the set $S(\mathcal{H})$ is not only convex but actually σ-*convex*, which means that if $\{\varrho_j\}$ is a sequence in $S(\mathcal{H})$ and $\{\lambda_j\}$ is a sequence of positive numbers summing to 1 then the sum $\sum_j \lambda_j \varrho_j$ converges in $\mathcal{T}(\mathcal{H})$ with respect to the trace norm and the limit belongs to $\mathcal{S}(\mathcal{H})$.

The dimension of the Hilbert space $\mathcal{H}$ is a specific property of the quantum system under consideration. As two Hilbert spaces with the same dimension are isomorphic, the dimension is, from the mathematical point of view, the *only* characteristic feature of the state space of the system. As we will see later, in Chapter 3, the dimension represents the maximal number of perfectly distinguishable states in a single experimental run.

An example of a state is a one-dimensional projection. Indeed, we saw in Section 1.2 that one-dimensional projections are positive and have trace 1. Also, recall that any one-dimensional projection is of the form P_φ for some unit vector $\varphi \in \mathcal{H}$. In Dirac notation we write $P_\varphi = |\varphi\rangle\langle\varphi|$. Sometimes even a unit vector $\varphi \in \mathcal{H}$ is called a state, and then one is actually referring to P_φ. We use the term *vector state* (or *state vector*) in this kind of context. Remember, however, that two different unit vectors φ and ψ define the same one-dimensional projection if φ and ψ differ by only a constant factor.

By forming convex combinations of one-dimensional projections we get more states. Actually, a basic fact about $\mathcal{S}(\mathcal{H})$ is that all elements in this set are (either finite or infinite) convex combinations of one-dimensional projections. The following theorem is a direct consequence of the spectral decomposition, Theorem 1.65.

Theorem 2.5 A state $\varrho \in \mathcal{S}(\mathcal{H})$ has a *canonical convex decomposition* of the form

$$\varrho = \sum_j \lambda_j P_j, \tag{2.3}$$

where $\{\lambda_j\}$ is a finite or infinite sequence of (not necessarily different) positive numbers summing to 1 and $\{P_j\}$ is a sequence of associated orthogonal one-dimensional projections, i.e. $P_j P_k = \delta_{jk} P_j$.

The word 'canonical' refers to the property that the projections $\{P_j\}$ are orthogonal. However, a state ϱ can have several different canonical convex decompositions. There is a unique canonical convex decomposition if and only if all the numbers λ_j are different. Let us stress that in a canonical decomposition we assume that $\lambda_j \neq 0$.

Example 2.6 (*Kernel and support of a state*)
Many mathematical properties of a state ϱ can be read from its canonical convex decomposition (2.3). For a vector $\psi \in \mathcal{H}$, the condition $\varrho\psi = 0$ implies that $\langle\psi|\varrho\psi\rangle = 0$. This requires that, for every j, $\langle\psi|P_j\psi\rangle = 0$, which we can also write as $\langle\psi|(\sum_j P_j)\psi\rangle = 0$. Using Proposition 1.35 we conclude that

$$\psi \in \ker(\varrho) \quad \text{if and only if} \quad \left(\sum_j P_j\right)\psi = 0. \qquad (2.4)$$

The sum $\sum_j P_j$ is a projection (since the P_j are orthogonal). Therefore, on the one hand each $\varphi \in \mathcal{H}$ can be written in the form $\varphi = \varphi_0 + \varphi_1$, where $(\sum_j P_j)\varphi_0 = 0$ and $(\sum_j P_j)\varphi_1 = \varphi_1$ (see Proposition 1.40). From (2.4) we see that $\varphi_0 \in \ker(\varrho)$. On the other hand, for every $\psi \in \ker(\varrho)$ we get $\langle \varphi_1 | \psi \rangle = 0$, hence $\varphi_1 \in \mathrm{supp}(\varrho)$. We thus conclude that

$$\varphi \in \mathrm{supp}(\varrho) \quad \text{if and only if} \quad \left(\sum_j P_j\right)\varphi = \varphi. \qquad (2.5)$$

We also see that the rank of ϱ is the number of one-dimensional projections P_j.

It is worth noticing that the kernel and support of ϱ do not depend on its eigenvalues λ_j, but only on the sum of the projections P_j. △

Let us recall some general terminology related to convex sets. An element ϱ is called *extremal* if it cannot be written as a proper convex combination of some other elements. As a mathematical condition, extremality means that if $\varrho = \lambda \varrho_1 + (1-\lambda)\varrho_2$ for some number $0 < \lambda < 1$ then this implies that $\varrho_1 = \varrho_2 = \varrho$. Convex decompositions with $\varrho_1 = \varrho_2$ are called *trivial*.

Exercise 2.7 Prove the following: if an element ϱ can be written as a convex combination $\varrho = \lambda_1\varrho_1 + \cdots + \lambda_n\varrho_n$ of n elements then it can also be written as a convex combination of two elements $\varrho = \lambda\varrho_1 + (1-\lambda)\varrho_2$. Conclude that to decide whether an element ϱ is extremal, it is enough to consider convex decompositions into two elements. [Hint: $\varrho = \lambda_1\varrho_1 + (1-\lambda_1)(\lambda_2\varrho_2/(1-\lambda_1) + \cdots + \lambda_n\varrho_n/(1-\lambda_1))$.]

An element that is not extremal has uncountably many (nontrivial) convex decompositions. Namely, suppose that $\varrho = \lambda\varrho_1 + (1-\lambda)\varrho_2$ for some $0 < \lambda < 1$ and $\varrho_1 \neq \varrho_2$. Then we can define $\varrho'_\mu = \mu\varrho_1 + (1-\mu)\varrho_2$ for any number μ satisfying $\lambda < \mu < 1$ and write

$$\varrho = \frac{\lambda}{\mu}\varrho'_\mu + \left(1 - \frac{\lambda}{\mu}\right)\varrho_2.$$

This is a nontrivial convex decomposition, as $\varrho'_\mu \neq \varrho_2$. Since μ can be any number from the interval $(\lambda, 1)$, the element ϱ has uncountably many different convex decompositions.

Exercise 2.8 Prove the following: if an element ϱ is not extremal then there are elements $\varrho_1 \neq \varrho_2$ such that $\varrho = \frac{1}{2}\varrho_1 + \frac{1}{2}\varrho_2$.

Example 2.9 (*Extremal probability distributions*)
Let us consider the set of probability distributions on a finite set $\{1, \ldots, n\}$, i.e. vectors $\vec{p} = (p_1, \ldots, p_n) \in \mathbb{R}^n$ such that $p_j \geq 0$ and $\sum_j p_j = 1$. It is straightforward to see that this set is convex, meaning that $\lambda\vec{p}_1 + (1-\lambda)\vec{p}_2$ is a valid

probability distribution if $\lambda \in [0, 1]$ and $\vec{p}_1, \vec{p}_2$ are probability distributions. What are the extremal elements? Any probability vector $\vec{p}$ can be written as

$$\vec{p} = \sum_j p_j \vec{\delta}_j ,\qquad (2.6)$$

where $\vec{\delta}_j$ is the probability vector with vanishing entries except the jth, which is equal to 1. Obviously (2.6) is a convex decomposition of the vector $\vec{p}$. The decomposition is trivial only if $\vec{p}$ is equal to $\vec{\delta}_j$ for some j. One can directly verify the defining property of the extremal elements and so conclude that the $\vec{\delta}_j$ are the only extremal elements of the convex set of probability distributions. Thus, the number of extremal points equals the dimension (the number of entries) of the probability vectors. Moreover, let us note that, since the extremal elements $\{\vec{\delta}_j\}$ form a basis of the real vector space $\mathbb{R}^n$, the decomposition of $\vec{p}$ into extremal elements is unique. $\qquad\qquad\qquad\qquad\qquad\qquad\qquad\qquad\qquad\qquad\qquad\qquad\qquad\qquad \triangle$

The convex structure of $\mathcal{S}(\mathcal{H})$ leads to the following twofold classification.

Definition 2.10 An extremal element of the convex set $\mathcal{S}(\mathcal{H})$ is called a *pure state*. Any other element of $\mathcal{S}(\mathcal{H})$ is called a *mixed state*.

As a preparation for the following mathematical characterization of pure states, we make a short observation. If ϱ is a state then the operator ϱ^2 is a positive trace class operator, and $\text{tr}[\varrho^2] > 0$. By definition, states satisfy the normalization $\text{tr}[\varrho] = 1$. This leads to a bound for the trace of ϱ^2. Namely, using the inequalities presented in subsection 1.2.5 we get

$$\text{tr}[\varrho^2] \leq \|\varrho\| \, \text{tr}[\varrho] = \|\varrho\| \leq \text{tr}[\varrho] = 1 .$$

We conclude that to each state ϱ we can assign a number $0 < \text{tr}[\varrho^2] \leq 1$. The following proposition utilizes this number for a purity criterion and also identifies one-dimensional projections as the only pure states.

Proposition 2.11 Let $\varrho \in \mathcal{S}(\mathcal{H})$. The following conditions are equivalent:

(i) ϱ is a pure state;
(ii) ϱ is a one-dimensional projection;
(iii) $\text{tr}[\varrho^2] = 1$.

Proof We prove the proposition by showing that (i)$\Rightarrow$(ii)$\Rightarrow$(iii)$\Rightarrow$(i).

Suppose that ϱ is a pure state. Then its canonical decomposition (2.3) can contain only one term, $\lambda_1 P_1$. Since $\text{tr}[\varrho] = 1$ it follows that $\lambda_1 = 1$ and hence that $\varrho = P_1$. Thus (i) implies (ii).

Let us now assume that (ii) holds. A projection satisfies $\varrho^2 = \varrho$. Therefore $\text{tr}[\varrho^2] = \text{tr}[\varrho] = 1$. Thus (iii) follows.

Finally, we assume that (iii) holds. Let us write ϱ in the form $\varrho = \lambda\varrho_1 + (1-\lambda)\varrho_2$ for some $0 < \lambda < 1$. We want to prove that this is necessarily a trivial convex decomposition, i.e. $\varrho_1 = \varrho_2$. We obtain

$$1 = \mathrm{tr}[\varrho^2] = \lambda^2 \, \mathrm{tr}[\varrho_1^2] + (1-\lambda)^2 \, \mathrm{tr}[\varrho_2^2] + 2\lambda(1-\lambda)\,\mathrm{tr}[\varrho_1\varrho_2]$$
$$\leq \lambda^2 + (1-\lambda)^2 + 2\lambda(1-\lambda)|\mathrm{tr}[\varrho_1\varrho_2]| \leq 1 \,.$$

In the last inequality we used the Cauchy–Schwarz inequality $|\mathrm{tr}[\varrho_1\varrho_2]| \leq \sqrt{\mathrm{tr}[\varrho_1^2]}\sqrt{\mathrm{tr}[\varrho_2^2]}$. Since 1 is the first and also the last number, we must have $|\mathrm{tr}[\varrho_1\varrho_2]| = 1$, which implies the saturation of the Cauchy–Schwarz inequality. Consequently, $\varrho_1 = c\varrho_2$ for some complex number c. As $\mathrm{tr}[\varrho_1] = \mathrm{tr}[\varrho_2] = 1$ we have $c = 1$ and therefore $\varrho_1 = \varrho_2$. In conclusion, if $\mathrm{tr}[\varrho^2] = 1$ then the state ϱ admits only trivial convex decompositions, which means that it is pure. $\square$

The distinction between pure and mixed states is based respective on the nonexistence or existence of a nontrivial convex decomposition. Is it possible to make a finer classification of states?

Item (iii) in Proposition 2.11 provides us with a function on the state space that has the value 1 if and only if the state is pure, thus enabling us to decide on the mentioned binary question: extremal versus mixed. However, this function is not constant for mixed states, which indicates that it could be used to quantify mixedness. We need to take a closer look at its mathematical properties.

Definition 2.12 The *purity* $\mathcal{P}(\varrho)$ of a state ϱ is defined as

$$\mathcal{P}(\varrho) := \mathrm{tr}[\varrho^2] = \sum_j \lambda_j^2 \,,$$

where the λ_j are the eigenvalues of ϱ (including multiplicities).

Proposition 2.13 The purity $\mathcal{P}$ satisfies the following properties.

(a) $\mathcal{P}$ is a convex map: $\mathcal{P}(\lambda\varrho_1 + (1-\lambda)\varrho_2) \leq \lambda\mathcal{P}(\varrho_1) + (1-\lambda)\mathcal{P}(\varrho_2)$.
(b) $\mathcal{P}$ is invariant under unitary conjugation: $\mathcal{P}(U\varrho U^*) = \mathcal{P}(\varrho)$.
(c) $\mathcal{P}(\varrho) = 1$ if and only if ϱ is a pure state.

Proof (a): We prove the claim by showing that the difference

$$\Delta \equiv \lambda\mathcal{P}(\varrho_1) + (1-\lambda)\mathcal{P}(\varrho_2) - \mathcal{P}(\lambda\varrho_1 + (1-\lambda)\varrho_2)$$

is positive. A direct calculation gives

$$\Delta = (\lambda - \lambda^2)\mathrm{tr}[\varrho_1^2] + [(1-\lambda) - (1-\lambda)^2]\mathrm{tr}[\varrho_2^2] - 2\lambda(1-\lambda)\mathrm{tr}[\varrho_1\varrho_2]$$
$$= \lambda(1-\lambda)\mathrm{tr}[\varrho_1^2 + \varrho_2^2 - \varrho_1\varrho_2 - \varrho_2\varrho_1] = \lambda(1-\lambda)\mathrm{tr}[(\varrho_1 - \varrho_2)^2] \,.$$

Since $0 \leq \lambda \leq 1$ and $(\varrho_1 - \varrho_2)^2 \geq O$ it follows that Δ is positive, thus $\mathcal{P}$ is convex.

(b): If ϱ is a state and U is a unitary operator, we get

$$P(U\varrho U^*) = \text{tr}\left[U\varrho U^* U\varrho U^*\right] = \text{tr}\left[\varrho^2\right] = P(\varrho).$$

(c): This is proved in Proposition 2.11. □

These properties, especially convexity, reflect the heuristic expectations that any measure of mixedness should satisfy. In particular, we expect that a mixture of states cannot be less mixed (or more pure) than any of its constituents. Indeed, from convexity it follows that

$$P(\lambda\varrho_1 + (1 - \lambda)\varrho_2) \leq \max\{P(\varrho_1), P(\varrho_2)\}.$$

Similarly, since there is no preferred orthonormal basis in the Hilbert space, a change of basis cannot affect the quality of mixedness; thus it is perfectly reasonable that $U\varrho U^*$ and ϱ are equally mixed. (This unitary freedom is further discussed in section 2.3.)

Exercise 2.14 Show that the purity $P(\varrho)$ of a state ϱ can be increased if it is mixed with another, appropriate, state. In particular, find $\varrho_1, \varrho_2 \in S(\mathcal{H})$ and $\lambda \in [0, 1]$ such that

$$P(\lambda\varrho_1 + (1 - \lambda)\varrho_2) > \min\{P(\varrho_1), P(\varrho_2)\}.$$

[Hint: Pure states are not the best try. If you need more help, wait for Example 2.28.]

In Example 2.9 we saw that the convex set of probability distributions on a finite set contains a finite number of extremal elements and, moreover, that each probability distribution can be expressed as a unique convex combination of these extremal elements. The following example presents a different convex set (an n-dimensional sphere) with uncountably many extremal elements and with the property that every mixed element has a convex decomposition into two extremal elements.

Example 2.15 (*Extremal elements of the sphere*)
Let $n \geq 2$. We want to demonstrate that an element x is an extremal point of the sphere

$$\mathbb{S}^{n-1} = \{x \in \mathbb{R}^n : \|x\| \leq 1\}$$

if and only if $\|x\| = 1$ and that any (nonextremal) element $x \in \mathbb{S}^{n-1}$ can be written as a convex combination of two extremal points.

We will prove something slightly more general. Consider the unit sphere (i.e. the set of vectors x satisfying $\|x\| \leq 1$) in an inner product space $\mathcal{V}$. First we will show that the unit norm elements (i.e. those with $\|x\| = 1$) are extremal points of this sphere. Assume that $x = \frac{1}{2}(x_1 + x_2)$. The triangle inequality implies that

$$1 = \|x\| = \tfrac{1}{2}\|x_1 + x_2\| \le \tfrac{1}{2}(\|x_1\| + \|x_2\|) \le 1 . \tag{2.7}$$

The last inequality is saturated if and only if $\|x_1\| = \|x_2\| = 1$. In order to prove that $x_1 = x_2 = x$ we use the parallelogram law for the points $\frac{1}{2}x_1$ and $\frac{1}{2}x_2$. We obtain

$$\tfrac{1}{4}\|x_1 - x_2\|^2 + \tfrac{1}{4}\|(x_1 + x_2)\|^2 = \tfrac{1}{2}(\|x_1\|^2 + \|x_2\|^2) ;$$

hence $\|x_1 - x_2\| = 0$ and therefore $x_1 = x_2$.

In order to prove that the boundary points are the only extremal points we will show that an arbitrary point can be expressed as a convex combination of two points on the boundary. Consider $x \ne 0$ and define $x_1 = x / \|x\|$ and $x_2 = -x/ \|x\|$. These points belong to the boundary, hence they are extremal. We have

$$x = \tfrac{1}{2}(1 + \|x\|)x_1 + \tfrac{1}{2}(1 - \|x\|)x_2 ,$$

which is a nontrivial convex decomposition if $\|x\| < 1$.

Let us note that the last argument is valid in an arbitrary normed space (and also when the norm is not induced by an inner product). However, the converse is not necessarily true unless the normed space is actually an inner product space. Consider for example the norm in the vector space $\mathbb{R}^n$ given by the maximal component, i.e. $\|x\| = \max_j |x_j|$. This norm is not generated by an inner product, hence the parallelogram law does not hold. The unit sphere in this norm forms a cube in $\mathbb{R}^n$ and the extremal points of the cube are exactly its corners. The other unit vectors are not extremal.   △

It is now time to get back to the convex set of quantum states and analyze the properties of their convex decompositions. Now that we have identified the pure-states, we can study the decomposability of mixed states into pure states. How many pure-state decompositions does a given mixed state have? And what is the minimal number of pure states in such a decomposition?

A mixed state ϱ has a canonical decomposition (2.3) into a convex mixture of orthogonal pure states. The number of pure states equals the number of nonzero eigenvalues (including their multiplicities), which equals the rank r of the operator ϱ. In fact, this is the minimal number of pure states that one needs for any pure-state decomposition of ϱ. This follows from the facts that pure states are rank-1 operators and that the rank of a sum of operators is at most the sum of the individual ranks.

To illustrate the richness of the convex decompositions of a mixed state into pure states we describe a method presented by Cassinelli *et al.* [39]. Let $\varrho \in \mathcal{S}(\mathcal{H})$ be a mixed state. Choose an orthonormal basis $\{\varphi_j\}_{j=1}^d$ for $\mathcal{H}$ (here either $d < \infty$ or $d = \infty$). For each j, define $\lambda_j := \langle \varphi_j | \varrho \varphi_j \rangle$. These numbers are nonnegative and they satisfy the identity

$$\sum_{j=1}^d \lambda_j = \sum_{j=1}^d \langle \varphi_j | \varrho \varphi_j \rangle = \text{tr}[\varrho] = 1.$$

For each j such that $\lambda_j \neq 0$, we write

$$\phi_j := \lambda_j^{-1/2} \varrho^{1/2} \varphi_j. \tag{2.8}$$

The vectors ϕ_j are unit vectors and

$$\varrho = \sum_j \lambda_j |\phi_j\rangle\langle\phi_j|, \tag{2.9}$$

where the sum contains those terms with $\lambda_j \neq 0$. Indeed, for any $\psi \in \mathcal{H}$ we get

$$\left\langle \psi \Big| \Big(\sum_j \lambda_j |\phi_j\rangle\langle\phi_j| \Big) \psi \right\rangle = \sum_{j=1}^d \left| \langle \psi | \varrho^{1/2} \varphi_j \rangle \right|^2 = \sum_{j=1}^d \left| \langle \varrho^{1/2} \psi | \varphi_j \rangle \right|^2$$

$$= \sum_{j=1}^d \langle \varrho^{1/2} \psi | \varphi_j \rangle \langle \varphi_j | \varrho^{1/2} \psi \rangle = \langle \psi | \varrho \psi \rangle,$$

which shows that (2.9) holds. The decomposition in (2.9) is not yet exactly what we want, as it may happen that $|\phi_i\rangle\langle\phi_i| = |\phi_j\rangle\langle\phi_j|$ for two different indices i and j. However, we can simply sum up these kinds of terms and we obtain a mixture consisting of (at most d) different pure states.

As a result we find that any mixed state can be written in many different ways as a convex combination of pure states, where at most d of these are needed but more are allowed. In particular, the decomposition defined in (2.9) shows that a pure state P_ϕ can be in some decomposition of a state ϱ if the vector ϕ is in the range of the operator $\varrho^{1/2}$. It was shown by Hadjisavvas [71] that these two conditions are equivalent.

Exercise 2.16 Deduce from the previous discussion that every mixed quantum state has uncountably many different convex decompositions into pure states.

It is possible to characterize all the decompositions of a mixed state into pure states. In the case of finite convex mixtures this was done by Hughston *et al.* [81]. A complete characterization, taking into account σ-convex (infinite but countable) mixtures, was given by Cassinelli *et al.* [39]. We will not discuss these results here, but in the following proposition we characterize the ambiguity of finite pure-state decompositions of mixed states; this proposition was derived by Jaynes [84].

Proposition 2.17 If a mixed state ϱ has a convex decomposition

$$\varrho = \sum_{j=1}^{n} p_j |\phi_j\rangle\langle\phi_j| \tag{2.10}$$

into pure states then all its convex decompositions into pure states have the form

$$\varrho = \sum_{k=1}^{m} q_k |\varphi_k\rangle\langle\varphi_k|, \tag{2.11}$$

where the vectors $\varphi_1, \ldots, \varphi_m \in \mathcal{H}$ and the numbers $q_1, \ldots, q_m$ satisfy the system of equations

$$\sqrt{p_j}\,\phi_j = \sum_{k=1}^{m} u_{jk}\sqrt{q_k}\,\varphi_k; \tag{2.12}$$

the complex numbers u_{jk} satisfy $\sum_j u_{jk}\bar{u}_{jk'} = \delta_{kk'}$.

Proof Inserting (2.12) into (2.10) gives

$$\sum_j p_j |\phi_j\rangle\langle\phi_j| = \sum_{j,k,k'} \sqrt{q_k q_{k'}}\, u_{jk}\bar{u}_{jk'} |\varphi_k\rangle\langle\varphi_{k'}| = \sum_k q_k |\varphi_k\rangle\langle\varphi_k|,$$

hence (2.11) is a convex decomposition of ϱ.

Conversely, suppose that both (2.10) and (2.11) are valid convex decompositions. We have to show that (2.12) holds. Let $\varrho = \sum_{l=1}^{r} \lambda_l |\psi_l\rangle\langle\psi_l|$ be a canonical decomposition. If necessary, we add terms with $\lambda_i = 0$ so that the vectors ψ_l can be taken to form an orthonormal basis of $\mathcal{H}$. We then write basis expansions as follows:

$$\sqrt{p_j}\phi_j = \sum_l c_{jl}\psi_l, \qquad \sqrt{q_k}\varphi_k = \sum_l d_{kl}\psi_l,$$

where c_{jl}, d_{jl} are complex numbers. The identity

$$\varrho = \sum_l \lambda_l |\psi_l\rangle\langle\psi_l| = \sum_j p_j |\phi_j\rangle\langle\phi_j| = \sum_j \sum_{l,l'} c_{jl}\bar{c}_{jl'} |\psi_l\rangle\langle\psi_{l'}|$$

requires that $\sum_j c_{jl}\bar{c}_{jl'} = \lambda_l \delta_{ll'}$. Similarly, the d_{kl} satisfy $\sum_k d_{kl}\bar{d}_{kl'} = \lambda_l \delta_{ll'}$. We write $u_{jk} = \sum_l c_{jl}\bar{d}_{kl}/\lambda_l$ for every $j = 1, \ldots, n$ and $k = 1, \ldots, m$. (This is possible since $\lambda_l \neq 0$ whenever $c_{jl}\bar{d}_{kl} \neq 0$.) Then a direct calculation gives $\sum_k u_{jk}\sqrt{q_k}\varphi_k = \sqrt{p_j}\phi_j$ and $\sum_j u_{jk}\bar{u}_{jk'} = \delta_{kk'}$. $\qquad\square$

Let us now switch our attention to the topological aspect of the set of quantum states. The set $\mathcal{S}(\mathcal{H})$ (see the start of subsection 2.1.2) is a convex subset of the real normed space $\mathcal{T}_s(\mathcal{H})$ of selfadjoint trace class operators. We say that a state ϱ belongs to the *boundary* of $\mathcal{S}(\mathcal{H})$ if, for each $\epsilon > 0$, there exists an operator

$\xi_\epsilon \in \mathcal{T}_s(\mathcal{H})$ such that $\|\varrho - \xi_\epsilon\|_{\mathrm{tr}} < \epsilon$ but $\xi_\epsilon \notin \mathcal{S}(\mathcal{H})$. What is the relation between the set of extremal elements and the boundary?

Proposition 2.18 If a state ϱ has eigenvalue 0 then it belongs to the boundary of $\mathcal{S}(\mathcal{H})$. In particular, all pure states are boundary points of $\mathcal{S}(\mathcal{H})$.

Proof Assume that ϱ has eigenvalue 0 and let $\varphi \in \mathcal{H}$ be a corresponding normalized eigenvector. Fix $\epsilon > 0$. On the one hand, the operator $\xi_\epsilon := \varrho - \frac{1}{2}\epsilon|\varphi\rangle\langle\varphi|$ is trace class and selfadjoint but not positive, since $\langle\varphi|\xi_\epsilon\varphi\rangle = -\frac{1}{2}\epsilon < 0$. In particular, $\xi_\epsilon \notin \mathcal{S}(\mathcal{H})$. On the other hand,

$$\|\xi_\epsilon - \varrho\|_{\mathrm{tr}} = \tfrac{1}{2}\epsilon \, \||\varphi\rangle\langle\varphi|\|_{\mathrm{tr}} = \tfrac{1}{2}\epsilon < \epsilon.$$

Consequently, ϱ belongs to the boundary of $\mathcal{S}(\mathcal{H})$.

Any projection P with $O \neq P \neq I$ has eigenvalues 0 and 1 (see Proposition 1.40). Hence, all pure states (being one-dimensional projections) are boundary points of $\mathcal{S}(\mathcal{H})$. $\qquad\square$

For $\dim \mathcal{H} \geq 3$ we can easily define mixed states which are on the boundary of $\mathcal{S}(\mathcal{H})$. For instance, let $\varphi_1, \varphi_2, \varphi_3$ be three orthogonal unit vectors. Fix a number $0 < \lambda < 1$ and define

$$\varrho = \lambda|\varphi_1\rangle\langle\varphi_1| + (1 - \lambda)|\varphi_2\rangle\langle\varphi_2|.$$

It is then clear that ϱ is a mixed state. However,

$$\varrho\varphi_3 = \lambda\langle\varphi_1|\varphi_3\rangle\,\varphi_1 + (1 - \lambda)\langle\varphi_2|\varphi_3\rangle\,\varphi_2 = 0,$$

and hence ϱ has eigenvalue 0. Proposition 2.18 then implies that ϱ is on the boundary of $\mathcal{S}(\mathcal{H})$. The case $\dim \mathcal{H} = 2$ is an exception, because for this dimension the boundary points of $\mathcal{S}(\mathcal{H})$ coincide with the set of pure states.

We now summarize the main facts about quantum state space.

- Pure states are one-dimensional projections.
- Mixed states are convex combinations of pure states.
- There are infinitely many different ways in which a mixed state can be expressed in terms of pure states.
- The minimal number of pure states in a convex decomposition of a given mixed state equals the rank of the mixed state. In particular, each mixed state can be expressed as a convex combination of d (or fewer) pure states.
- The pure states belong to the boundary of the state space, but if $d \geq 3$ then the boundary does not coincide with the set of pure states.

The quantum state space $\mathcal{S}(\mathcal{H})$ is defined as a certain subset of Hilbert space operators. One may wonder whether the quantum state space could have an equivalent but more down-to-earth description. The properties listed above can be used to argue that the state space is neither equivalent to a set of finite probability distributions nor to a sphere in $\mathbb{R}^n$ (except in the case $d = 2$). For example, the fact that for a sphere any mixed state can be expressed as a convex combination of two pure elements does not hold for mixed quantum states unless their rank is 2; the two-dimensional state space does not possess all the generic properties of larger-dimensional Hilbert spaces. This will become clear in the next subsection.

2.1.3 State space for a finite-dimensional system

In this subsection we will illustrate the concepts of subsection 2.1.2 in the case of a finite-dimensional system and also discuss some further topics specific to the finite-dimensional case. An extensive and entertaining presentation of finite-dimensional state spaces is given in the book by Bengtsson and Życzkowski [14].

A convenient way to illustrate the state space for a finite-dimensional Hilbert space $\mathcal{H}$ is to adopt the so-called *Bloch representation*. As explained in subsection 1.2.5, the vector space $T(\mathcal{H})$ is an inner product space endowed with the Hilbert–Schmidt inner product $\langle A|B \rangle_{HS} = \text{tr}[A^*B]$. Since we are assuming that $\mathcal{H}$ is finite dimensional, $T(\mathcal{H})$ contains all linear mappings on $\mathcal{H}$.

Let $\{\varphi_j\}_{j=1}^d$ be an orthonormal basis for $\mathcal{H}$. The operators $e_{jk} = |\varphi_j\rangle\langle\varphi_k|$, when expressed as matrices in the orthonormal basis $\{\varphi_j\}$, contain only one nonzero entry, at the jth row and kth column. We can write any operator A on $\mathcal{H}$ as a linear combination $A = \sum_{jk} a_{jk} e_{jk}$, where $a_{jk} = \langle\varphi_j|A\varphi_k\rangle$ are the entries of the matrix representation of A. Moreover, we have

$$\text{tr}\left[e_{jk}^* e_{rs}\right] = \text{tr}\left[|\varphi_k\rangle\langle\varphi_j||\varphi_r\rangle\langle\varphi_s|\right] = \langle\varphi_j|\varphi_r\rangle\,\text{tr}\left[|\varphi_k\rangle\langle\varphi_s|\right] = \delta_{jr}\delta_{ks}\,,$$

and the operators e_{jk} thus form an orthonormal basis for $T(\mathcal{H})$. We conclude that for a d-dimensional Hilbert space $\mathcal{H}$ the inner product space $T(\mathcal{H})$ has dimension d^2.

For selfadjoint operators the matrix elements satisfy the relation

$$a_{jk} = \langle\varphi_j|A\varphi_k\rangle = \langle A\varphi_j|\varphi_k\rangle = \overline{\langle\varphi_k|A\varphi_j\rangle} = \bar{a}_{kj}\,.$$

These conditions reduce the number of independent entries of A. For diagonal elements ($j = k$) they imply that the a_{jj} are real. For off-diagonal entries they imply that the upper diagonal part determines the lower diagonal part (or vice versa). There are $\frac{1}{2}d(d-1)$ off-diagonal complex elements in the upper diagonal part. Consequently, the dimension of the real vector space $T_s(\mathcal{H})$ equals $d + d(d-1) = d^2$. Therefore, the real dimension of $T_s(\mathcal{H})$ is the same as the complex dimension of $T(\mathcal{H})$. Moreover, we can choose an orthonormal basis of $T(\mathcal{H})$ in which the selfadjoint operators coincide with real linear combinations of the basis elements.

Exercise 2.19 (*Orthonormal basis of selfadjoint operators*)
Verify that the selfadjoint operators

$$\Lambda_{jj} = e_{jj}, \qquad \Lambda_{jk}^{+} = \tfrac{1}{\sqrt{2}}(e_{jk} + e_{kj}), \qquad \Lambda_{jk}^{-} = -\tfrac{1}{\sqrt{2}}i(e_{jk} - e_{kj}),$$

where $j = 1, \ldots, d$ and $k > j$, form an orthonormal basis for $\mathcal{T}_s(\mathcal{H})$.

Example 2.20 (*An orthonormal basis consisting of density matrices?*)
Can we choose an orthonormal basis for $\mathcal{T}_s(\mathcal{H})$ consisting of density matrices only? Since the orthogonality of pure states is equivalent to the orthogonality of the associated unit vectors in $\mathcal{H}$ (recall Examples 1.45 and 1.57) it follows that pure states cannot form an orthonormal operator basis. Namely, there are at most d mutually orthogonal pure states whereas the dimension of $\mathcal{T}_s(\mathcal{H})$ is d^2.

This fact does not change if we use mixed states also: there are still at most d mutually orthogonal states. This follows from the fact that two states ϱ_1, ϱ_2 are orthogonal in their Hilbert–Schmidt inner product only if they have orthogonal supports, i.e. supp$(\varrho_1) \perp$ supp(ϱ_2). Again, there can be at most d mutually orthogonal subspaces while we need d^2 mutually orthogonal operators to form an orthonormal basis.

To see that $\langle \varrho_1 | \varrho_2 \rangle_{\mathrm{HS}} = 0$ implies that supp$(\varrho_1) \perp$ supp(ϱ_2), we write down canonical convex decompositions $\varrho_1 = \sum_j \lambda_j P_j$ and $\varrho_2 = \sum_k \mu_k Q_k$. The condition $\langle \varrho_1 | \varrho_2 \rangle_{\mathrm{HS}} = 0$ is equivalent to $P_j Q_k = 0$ for all j, k (see Example 1.57). We have $\phi \in$ supp(ϱ_1) if and only if $\left(\sum_j P_j \right) \phi = \phi$, and similarly $\varphi \in$ supp(ϱ_2) if and only if $\left(\sum_k Q_k \right) \varphi = \varphi$ (see Example 2.6). Therefore, for two vectors $\phi \in$ supp(ϱ_1) and $\varphi \in$ supp(ϱ_2) we obtain

$$\langle \phi | \varphi \rangle = \left\langle \left(\sum_j P_j \right) \phi \bigg| \left(\sum_k Q_k \right) \varphi \right\rangle = \sum_{j,k} \langle \phi | P_j Q_k \varphi \rangle = 0.$$

This means that supp$(\varrho_1) \perp$ supp(ϱ_2).   △

In order to express density operators in a convenient way we choose a selfadjoint orthogonal basis for $\mathcal{T}(\mathcal{H})$ such that the identity operator $E_0 = I$ is a basis element. The orthogonality requirement then implies that the remaining basis elements $E_1, \ldots, E_{d^2-1}$ are traceless. Moreover, we choose the normalization to be $\langle E_j | E_k \rangle_{\mathrm{HS}} = d \delta_{jk}$. Consequently, a general quantum state can be expressed as

$$\varrho = \frac{1}{d}(I + \vec{r} \cdot \vec{E}), \tag{2.13}$$

where $\vec{E} = (E_1, \ldots, E_{d^2-1})$ and the vector $\vec{r} = (\mathrm{tr}[\varrho E_1], \ldots, \mathrm{tr}[\varrho E_{d^2-1}])$ is a $(d^2 - 1)$-dimensional real vector called the *Bloch (state) vector*. Using this parametrization, the set of quantum states can be viewed as a convex subset of

$\mathbb{R}^{d^2-1}$. The whole space $\mathbb{R}^{d^2-1}$ is identified with the set of all selfadjoint operators of unit trace.

Let us note that, in general, a linear combination of Bloch vectors is not reflected in a corresponding linear combination of density operators. Suppose that ϱ_1, ϱ_2 are states and $\vec{r}_1$, $\vec{r}_2$ are the corresponding Bloch vectors. A Bloch vector $\vec{r} = a\vec{r}_1 + b\vec{r}_2$ is associated with the operator

$$R = \frac{1}{d}(I + \vec{r} \cdot \vec{E}) = \frac{1}{d}(I + a\vec{r}_1 \cdot \vec{E} + b\vec{r}_2 \cdot \vec{E}).$$

However, the linear combination $a\varrho_1 + b\varrho_2$ is not, in general, associated with any Bloch vector, since $\text{tr}[a\varrho_1 + b\varrho_2] = a+b$. Only if the coefficients satisfy $a+b = 1$, is the correspondence one-to-one, meaning that

$$a\varrho_1 + b\varrho_2 = \frac{1}{d}[I + (a\vec{r}_1 + b\vec{r}_2) \cdot \vec{E}].$$

Example 2.21 (*Bloch sphere*)
Let us consider the two-dimensional Hilbert space $\mathcal{H} = \mathbb{C}^2$ with an orthonormal basis $\{\varphi, \varphi_\perp\}$ (see Figure 2.4). The standard operator basis consists of the identity operator $\sigma_0 = I$ and the *Pauli operators*

$$\sigma_x = |\varphi\rangle\langle\varphi_\perp| + |\varphi_\perp\rangle\langle\varphi|,$$
$$\sigma_y = -i|\varphi\rangle\langle\varphi_\perp| + i|\varphi_\perp\rangle\langle\varphi|,$$
$$\sigma_z = |\varphi\rangle\langle\varphi| - |\varphi_\perp\rangle\langle\varphi_\perp|.$$

They satisfy the orthogonality relations $\text{tr}[\sigma_0\sigma_j] = 0$ and $\text{tr}[\sigma_j\sigma_k] = 2\delta_{jk}$ for $j, k = x, y, z$. In this basis a general qubit state can be written as

Figure 2.4 A Bloch sphere represents the state space of a two-dimensional quantum system. The orthogonal vectors φ and $\varphi_\perp$ are associated with antipodal Bloch vectors.

$$\varrho = \tfrac{1}{2}(I + \vec{r} \cdot \vec{\sigma}),$$

(2.14)

where $\vec{r} \in \mathbb{R}^3$. The eigenvalues of the operator ϱ in (2.14) are $\lambda_{\pm} = \tfrac{1}{2}(1 \pm \|\vec{r}\|)$. It follows that the positivity of the operator ϱ is equivalent to the condition $\|\vec{r}\| \leq 1$. Hence the states form a unit sphere in $\mathbb{R}^3$, the so-called *Bloch sphere*. We have $\lambda_{+} = 1$ and $\lambda_{-} = 0$ exactly when $\|\vec{r}\| = 1$. Hence, the pure states correspond to vectors $\vec{r}$ of unit length. Clearly the vectors of unit length also form the boundary of the unit sphere; hence for two-dimensional Hilbert space the boundary of the state space consists of pure states only.   △

Exercise 2.22 This exercise is related to Example 2.21 and provides an insight into the dimensionality of the state space. Consider a set $\{\varrho_x, \varrho_y, \varrho_z\}$ with $\varrho_k = \tfrac{1}{2}(I + \sigma_k)$. Find states $\varrho_1, \ldots, \varrho_n$ (and determine n) such that the set $\{\varrho_x, \varrho_y, \varrho_z, \varrho_1, \ldots, \varrho_n\}$ forms a (nonorthogonal) basis of $T(\mathbb{C}^2)$. [Hint: Express I as a linear combination of $\varrho_x, \varrho_y, \varrho_z$. Then consider $\tfrac{1}{2}(I - \sigma_x)$.]

The purity of a state

$$\varrho = \frac{1}{d}(I + \vec{r} \cdot \vec{E})$$

is given by

$$\mathrm{tr}[\varrho^2] = \frac{1}{d^2}\mathrm{tr}\left[(I + \vec{r} \cdot \vec{E})(I + \vec{r} \cdot \vec{E})\right] = \frac{1}{d}(1 + \|\vec{r}\|^2),$$

(2.15)

and hence $\|\vec{r}\| \leq \sqrt{d-1}$. This means that the state space is embedded in the $\sqrt{d-1}$-radius sphere of $\mathbb{R}^{d^2-1}$. However, not every operator with $\|\vec{r}\| \leq \sqrt{d-1}$ is positive, and the particular form of the state space is more complicated. To see that not every vector with $\|\vec{r}\| \leq \sqrt{d-1}$ corresponds to a state, let ψ, φ be two unit vectors, P_ψ, P_φ the corresponding pure states and $\vec{r}_\psi, \vec{r}_\varphi$ the associated Bloch vectors. Then

$$|\langle \psi | \varphi \rangle|^2 = \mathrm{tr}[P_\psi P_\varphi] = \frac{1}{d}(1 + \vec{r}_\psi \cdot \vec{r}_\varphi).$$

(2.16)

Consequently, the vectors ψ and φ are orthogonal if and only if $\vec{r}_\psi \cdot \vec{r}_\varphi = -1$. It follows that the angle between the Bloch vectors corresponding to orthogonal pure states is

$$\theta = \arccos\left(\frac{1}{1-d}\right).$$

(2.17)

Interestingly, if $\vec{r}_\psi$ (with $\|\vec{r}_\psi\| = \sqrt{d-1}$) defines a pure quantum state then its antipodal vector $\vec{t} = -\vec{r}_\psi$ does not correspond to any quantum state if $d > 2$. In fact, since $\|\vec{t}\| = \sqrt{d-1}$ the candidate state should be pure. However,

$$\frac{1}{d}(1 + \vec{r}_\psi \cdot \vec{t}) = \frac{1}{d}(1 - d + 1) = \frac{2 - d}{d} < 0$$

for $d > 2$, which is in contradiction with the fact that $|\langle \psi | \varphi \rangle|^2 \geq 0$ for all vectors $\varphi \in \mathcal{H}$.

Exercise 2.23 Fix a unit vector φ. Consider the traceless selfadjoint operator

$$E_1 = \frac{1}{\sqrt{d - 1}} (d \, |\varphi\rangle\langle\varphi| - I),$$

which is an element of some orthogonal operator basis consisting of I and traceless selfadjoint operators E_j such that $\text{tr}[E_j^2] = d$. The operator

$$\varrho = \frac{1}{d}(I + \sqrt{d - 1} \, E_1) = |\varphi\rangle\langle\varphi|$$

describes a pure state with Bloch vector $\vec{r} = (\sqrt{d - 1}, 0, \dots, 0)$. According to the previous paragraph the operator

$$A = \frac{1}{d}(I - \sqrt{d - 1} \, E_1) = \frac{2}{d}I - |\varphi\rangle\langle\varphi|$$

associated with the antipodal Bloch vector $\vec{t} = -\vec{r}$ is not positive. In fact,

$$A\varphi = \left(\frac{2}{d} - 1\right)\varphi$$

and the eigenvalue $(2 - d)/d$ is clearly negative for $d > 2$. Find the maximal value x for which the operator

$$A = \frac{1}{d}(I - x\sqrt{d - 1} \, E_1) = \frac{1 + x}{d}I - x|\varphi\rangle\langle\varphi|$$

is positive and hence a density operator. In order to see how mixed the 'antipode' state A is, evaluate its purity.

We have seen that each state can be represented as a Bloch vector but not every such vector is associated with some state. Let us also recall that, in a sphere, each point can be expressed as a convex combination of at most two extremal elements (see Example 2.15). This is not true for quantum states when $d \geq 3$. The simple picture of the Bloch sphere for the two-dimensional case (Example 2.21) is in many respects exceptional and its properties are not valid for larger dimensions.

Example 2.24 (*Uniformly random state*)
Consider a preparation procedure for a random quantum state. Before we proceed we must make clear what this means. Usually we say that an event or a process is random if all the alternatives have the same chance of happening, i.e. they are uniformly sampled. For an uncountable set this rough idea requires careful formulation.

The group $\mathcal{U}(\mathcal{H})$ of all unitary operators is a compact group if $\mathcal{H}$ is finite dimensional. It follows that $\mathcal{U}(\mathcal{H})$ is endowed with an *invariant Haar measure*, which is unique if we normalize it to be a probability measure (see e.g. [60] for an explanation). We can use the Haar measure on $\mathcal{U}(\mathcal{H})$ to define a uniformly random state. Let us fix a state $\varrho \in \mathcal{S}(\mathcal{H})$. We denote by S_ϱ the set of all states ϱ' such that $\varrho' = U\varrho U^*$ for some $U \in \mathcal{U}(\mathcal{H})$. In other words, the set S_ϱ is the *orbit* of ϱ under the action of the unitary group. We will average over all states in this orbit; hence the random state of the orbit S_ϱ is defined as

$$\overline{\varrho} = \int_{\mathcal{U}(\mathcal{H})} U\varrho U^* dU.$$

It follows from the invariance of the Haar measure that $\overline{\varrho}$ commutes with all unitary operators. By Schur's lemma we then conclude that $\overline{\varrho} = cI$ for some $c \in \mathbb{R}$. The trace condition fixes the constant $c = \frac{1}{d}$, where $d = \dim \mathcal{H}$. Thus $\overline{\varrho} = \frac{1}{d}I$ and we observe that the averaging procedure does not depend on in which state ϱ we start. We will see in Example 2.28 that this state $\frac{1}{d}I$ is indeed the unique maximally mixed state, as one would expect for the uniformly random state. Let us note that the corresponding Bloch vector is $\vec{r} = \vec{0}$.   △

We end this subsection with some remarks on entropy. In subsection 2.1.2 we introduced the purity function $\mathcal{P}$ to quantify the degree of mixedness of quantum states. In classical physics a mixture of two (or several) states can be interpreted as a lack of knowledge in the preparation process. This interpretation is problematic for mixtures of quantum states, since the convex decomposition of a mixed state into pure states is highly nonunique. In classical theory the mixedness of the probability distribution can be clearly quantified. There are several measures, but the so-called *Shannon entropy* is the most suitable for the quantification of the information lost in the preparation process. We will define the *von Neumann entropy*, which is considered as the quantum counterpart of the Shannon entropy although there are significant differences in their roles in classical and quantum information theory, respectively. Detailed expositions of the von Neumann entropy can be found in the books by Ohya and Petz [105] and Petz [116].

Definition 2.25 The *von Neumann entropy* $S(\varrho)$ of a state ϱ is defined as follows:

$$S(\varrho) := -\text{tr}\big[\varrho \log \varrho\big] = -\sum_j \lambda_j \log \lambda_j ,$$

where the λ_j are the nonzero eigenvalues of ϱ (including their multiplicities).

Example 2.26 Consider a state $\varrho = tP_\varphi + (1-t)P_\psi$, where P_φ, P_ψ are one-dimensional projections defined by the vectors φ and $\psi = a\varphi + b\varphi_\perp$, $\langle \varphi | \varphi_\perp \rangle = 0$. The operator ϱ has two nonzero eigenvalues,

$$\lambda_\pm = \tfrac{1}{2}\left(1 \pm \sqrt{1 - 4t(1-t)|b|^2}\right),$$

and the von Neumann entropy is therefore $S(\varrho) = -\lambda_+ \log \lambda_+ - \lambda_- \log \lambda_-$. We see that $S(\varrho) = 0$ if and only if $P_\varphi = P_\psi = \varrho$. If we consider $S(\varrho)$ as a function of t, the maximum is achieved for $t = \tfrac{1}{2}$. $\triangle$

Since the eigenvalues of a state are between 0 and 1, we see that $S(\varrho) \geq 0$ for every state ϱ. In the following we list some other important properties of the von Neumann entropy that are needed for our purposes.

Proposition 2.27 The von Neumann entropy S has the following properties:

(a) S is concave, i.e. $S(t\varrho_1 + (1-t)\varrho_2) \geq tS(\varrho_1) + (1-t)S(\varrho_2)$;
(b) S is invariant under unitary conjugation, i.e. $S(U\varrho U^*) = S(\varrho)$;
(c) $S(\varrho) = 0$ if and only if ϱ is a pure state.

Proof (a): The function $f(x) = -x \log x$ defined for $x \geq 0$ is concave, i.e. $f(tx_1 + (1-t)x_2) \geq tf(x_1) + (1-t)f(x_2)$ for $t \in [0, 1]$ and $x_1, x_2 \geq 0$. Every selfadjoint operator X has a spectral decomposition $X = \sum_k x_k |\psi_k\rangle\langle\psi_k|$, where $\{\psi_k\}$ is an orthonormal basis of $\mathcal{H}$ (see subsection 1.3.3). We write

$$f(X) := \sum_k f(x_k)|\psi_k\rangle\langle\psi_k|,$$

and then for every $\psi \in \mathcal{H}$ we obtain

$$f(\langle\psi|X\psi\rangle) = f\left(\sum_k |\langle\psi_k|\psi\rangle|^2 x_k\right)$$
$$\geq \sum_k |\langle\psi_k|\psi\rangle|^2 f(x_k) = \langle\psi|f(X)\psi\rangle. \tag{2.18}$$

If we apply (2.18) in the case $X = t\varrho_1 + (1-t)\varrho_2$, this gives

$$S(t\varrho_1 + (1-t)\varrho_2) = \sum_j f\left(\langle\psi_j|(t\varrho_1 + (1-t)\varrho_2)\psi_j\rangle\right)$$
$$\geq \sum_j tf\left(\langle\psi_j|\varrho_1\psi_j\rangle\right) + (1-t)f(\langle\psi_j|\varrho_2\psi_j\rangle)$$
$$\geq \sum_j t\langle\psi_j|f(\varrho_1)\psi_j\rangle + (1-t)\langle\psi_j|f(\varrho_2)\psi_j\rangle$$
$$= tS(\varrho_1) + (1-t)S(\varrho_2).$$

Here we have used (2.18) and the concavity of f.

(b): Two operators ϱ and $U\varrho U^*$ have the same eigenvalues (including multiplicities), hence $S(\varrho) = S(U\varrho U^*)$.

(c): Since a pure state has a single nonzero eigenvalue 1, it follows that its von Neumann entropy is 0. However, if ϱ is mixed and hence has an eigenvalue $0 < \lambda_k < 1$ then $S(\varrho) = -\sum_j \lambda_j \log \lambda_j \geq -\lambda_k \log \lambda_k > 0$. $\square$

The purity and the von Neumann entropy each induce an order in the set of states. As shown by Wei *et al.* [140], these orders are different. Namely, there are states ϱ_1 and ϱ_2 such that $\mathcal{P}(\varrho_1) > \mathcal{P}(\varrho_2)$ but not $S(\varrho_1) < S(\varrho_2)$. However, pure states are minimally mixed with respect to both purity and von Neumann entropy. In the following example we show that the maximally mixed element also is the same for both orders.

Example 2.28 (*Maximally mixed state*)
Which state is maximally mixed? It turns out that there exists a quantum state, called a *complete* or *total mixture*, which is the unique maximally mixed state with respect to the both purity and von Neumann entropy. This state is $\frac{1}{d}I$, where d is the dimension of the Hilbert space.

To show that $\frac{1}{d}I$ is a maximally mixed state, we first recall that two states ϱ and $U\varrho U^*$ have the same value of purity and of von Neumann entropy for all unitary operators U. If U_k is a collection of unitary operators and we consider a mixture of the states $\varrho_k = U_k \varrho U_k^*$, we obtain a state $\varrho' = \sum_k p_k \varrho_k$ with $\mathcal{P}(\varrho') \leq \sum_k p_k \mathcal{P}(\varrho_k) = \mathcal{P}(\varrho)$ and $S(\varrho') \geq \sum_k p_k S(\varrho_k) = S(\varrho)$. This means that with respect to both quantities the state ϱ' is at least as mixed as ϱ.

Let ϱ be a state and $\varrho = \sum_j \lambda_j |\varphi_j\rangle\langle\varphi_j|$ be its canonical convex decomposition. We complete the set $\{\varphi_j\}$, so that it becomes an orthonormal basis; then $\varrho = \sum_{j=1}^d \lambda_j |\varphi_j\rangle\langle\varphi_j|$. We define the cyclic shift operator

$$U_{\text{shift}} = \sum_{j=1}^d |\varphi_j\rangle\langle\varphi_{j\oplus1}|,$$

with summation modulo d, i.e. $d \oplus 1 = 1$. Applying the shift operator to ϱ we get the state $\varrho_1 = U_{\text{shift}}\varrho U_{\text{shift}}^* = \sum_j \lambda_{j\oplus1}|\varphi_j\rangle\langle\varphi_j|$. Recursively, $\varrho_k = U_{\text{shift}}\varrho_{k-1}U_{\text{shift}}^* = \sum_j \lambda_{j\oplus k}|\varphi_j\rangle\langle\varphi_j|$. An equal mixture of the states $\varrho_1, \ldots, \varrho_d$ gives

$$\varrho' = \frac{1}{d}\sum_{k=1}^d \varrho_k = \frac{1}{d}\sum_{j=1}^d \left(\sum_{k=1}^d \lambda_{j\oplus k}\right)|\varphi_j\rangle\langle\varphi_j| = \frac{1}{d}I,$$

because $\sum_{k=1}^d \lambda_{j\oplus k} = \lambda_1 + \cdots + \lambda_d = \text{tr}[\varrho] = 1$. We conclude that $\frac{1}{d}I$ is at least as mixed as any other state, hence it is maximally mixed. The proof of uniqueness constitutes the next exercise. △

Exercise 2.29 Show that the state $\frac{1}{d}I$ is the *unique* maximally mixed state with respect to the purity and the von Neumann entropy. [Hint: The uniqueness of the maximally mixed state follows from the fact that the functions $\sum_{j=1}^d x_j^2$ and $\sum_{j=1}^d x_j \log x_j$ have unique minima under the constraint $\sum_{j=1}^d x_j = 1$.]

2.1.4 From states to effects

In subsection 2.1.1 we argued that physical experiments can be described by the concepts of states and effects. In the basic statistical framework the state space is a convex set, and in subsection 2.1.2 we identified the states of a quantum system with the positive trace class operators of trace 1 on a Hilbert space $\mathcal{H}$. Following the basic statistical framework the effects are affine mappings from $\mathcal{S}(\mathcal{H})$ to $[0, 1]$. More conveniently, effects can be represented as suitable Hilbert space operators and in the following we find out how this is done.

Suppose that T is a bounded selfadjoint operator on $\mathcal{H}$. The mapping

$$\varrho \mapsto \text{tr}[\varrho T] , \tag{2.19}$$

defined on $\mathcal{S}(\mathcal{H})$, is an affine functional. For each unit vector $\psi \in \mathcal{H}$ we have

$$\text{tr}\big[|\psi\rangle\langle\psi|T\big] = \langle\psi|T\psi\rangle .$$

It follows that T is a positive operator if and only if $\text{tr}[PT] \geq 0$ for every pure state P. But all other states are mixtures of pure states, hence $T \geq O$ if and only if $\text{tr}[\varrho T] \geq 0$ for every state ϱ. Since $\text{tr}[\varrho(I - T)] = 1 - \text{tr}[\varrho T]$, we may also conclude that $T \leq I$ if and only if $\text{tr}[\varrho T] \leq 1$ for every state ϱ. As a consequence, any selfadjoint operator T satisfying the operator inequalities $O \leq T \leq I$ determines an effect by formula (2.19). We will prove next that all effects arise in this way.

Proposition 2.30 Let E be an effect, i.e. an affine mapping from $\mathcal{S}(\mathcal{H})$ to $[0, 1]$. There exists a bounded selfadjoint operator $\hat{E}$ such that

$$E(\varrho) = \text{tr}[\varrho \hat{E}] \qquad \forall \varrho \in \mathcal{S}(\mathcal{H}).$$

The operator $\hat{E}$ satisfies $O \leq \hat{E} \leq I$.

Proof We will prove that every effect E has an extension $\tilde{E}$ to a continuous linear functional on $\mathcal{T}(\mathcal{H})$. The rest then mostly follows from the duality $\mathcal{T}(\mathcal{H})^* = \mathcal{L}(\mathcal{H})$ explained in subsection 1.3.4.

We start with an effect E and extend it in steps to $\mathcal{T}(\mathcal{H})$. First we set $\tilde{E}(O) := 0$ and, for each positive trace class operator $T \neq O$, we define

$$\tilde{E}(T) := \text{tr}[T] \, E(\text{tr}[T]^{-1}T) . \tag{2.20}$$

In this way, the extension $\tilde{E}$ of E is defined for all positive trace class operators. If T is a positive trace class operator and $s \geq 0$ then

$$\tilde{E}(sT) = \text{tr}[sT] \, E(\text{tr}[sT]^{-1}sT) = s\tilde{E}(T) .$$

Moreover, for two positive (nonzero) trace class operators S and T we obtain

$$\tilde{E}(S+T) = \text{tr}[S+T]\, E(\text{tr}[S+T]^{-1}(S+T))$$

$$= \text{tr}[S+T]\, E\left(\frac{\text{tr}[S]}{\text{tr}[S+T]}\frac{S}{\text{tr}[S]} + \frac{\text{tr}[T]}{\text{tr}[S+T]}\frac{T}{\text{tr}[T]}\right)$$

$$= \tilde{E}(S) + \tilde{E}(T). \tag{2.21}$$

For the second step we recall that every selfadjoint operator T can be written as $T = T^+ - T^-$, where $T^+ = \frac{1}{2}(|T|+T)$ and $T^- = \frac{1}{2}(|T|-T)$ are the positive and negative parts of T, respectively. If T is a trace class operator then T^+ and T^- are also trace class operators. Both T^+ and T^- are positive operators, and we can thus define

$$\tilde{E}(T) := \tilde{E}(T^+) - \tilde{E}(T^-) \tag{2.22}$$

for every $T \in \mathcal{T}_s(\mathcal{H})$. To see that

$$\tilde{E}(S+T) = \tilde{E}(S) + \tilde{E}(T) \tag{2.23}$$

for $S, T \in \mathcal{T}_s(\mathcal{H})$, notice that

$$S+T = (S+T)^+ - (S+T)^- \quad \text{but also} \quad S+T = S^+ - S^- + T^+ - T^-.$$

Therefore

$$(S+T)^+ + S^- + T^- = (S+T)^- + S^+ + T^+.$$

Applying $\tilde{E}$ on both sides and using (2.21) and (2.22), we obtain (2.23).

In the third step we use the fact that any operator $T \in \mathcal{T}(\mathcal{H})$ can be written as a linear combination of its real and imaginary parts, $T = T_R + iT_I$, which are selfadjoint operators (see subsection 1.2.2). Both T_R and T_I are linear combinations of T and T^*, hence they are trace class operators. We can now extend $\tilde{E}$ to all trace class operators by defining

$$\tilde{E}(T) := \tilde{E}(T_R) + i\tilde{E}(T_I)$$

for every $T \in \mathcal{T}(\mathcal{H})$. Since $(S+T)_R = S_R + T_R$ and $(S+T)_I = S_I + T_I$ for two operators S and T, we see that $\tilde{E}$ is a linear mapping on $\mathcal{T}(\mathcal{H})$.

We have thus constructed a linear functional $\tilde{E}$ on $\mathcal{T}(\mathcal{H})$. It is bounded (hence continuous), since

$$|\tilde{E}(T)| = |\tilde{E}(T_R^+ - T_R^- + iT_I^+ - iT_I^-)|$$

$$\leq |\tilde{E}(T_R^+)| + |\tilde{E}(T_R^-)| + |\tilde{E}(T_I^+)| + |\tilde{E}(T_I^-)|$$

$$\leq \text{tr}[T_R^+] + \text{tr}[T_R^-] + \text{tr}[T_I^+] + \text{tr}[T_I^-]$$

$$= \tfrac{1}{2}\text{tr}[|T+T^*|] + \tfrac{1}{2}\text{tr}[|T-T^*|] \leq 2\|T\|_{\text{tr}}$$

for every $T \in \mathcal{T}(\mathcal{H})$. From the duality relation between trace class operators and bounded operators (see Theorem 1.68) follows that there exists a unique operator $\hat{E} \in \mathcal{L}(\mathcal{H})$ such that

$$\tilde{E}(T) = \mathrm{tr}[\hat{E}T] \qquad \forall T \in \mathcal{T}(\mathcal{H}) . \tag{2.24}$$

From the construction of $\tilde{E}$ it follows that $\tilde{E}(T) \in \mathbb{R}$ for every $T \in \mathcal{T}_s(\mathcal{H})$, therefore $\hat{E}$ is selfadjoint (see Proposition 1.69). The operator inequalities $O \le \hat{E} \le I$ follow from the discussion before the proposition. $\qquad\square$

To complete the identification of affine mappings on states with the operators E satisfying $O \le E \le I$, we need to check that two different operators E_1 and E_2 have different actions on states. (We will no longer use carets to distinguish between these two different representations of effects.) Suppose that $\mathrm{tr}[\varrho E_1] = \mathrm{tr}[\varrho E_2]$ for all $\varrho \in \mathcal{S}(\mathcal{H})$. Choosing ϱ to be a pure state $\varrho = |\psi\rangle\langle\psi|$, we conclude that $\langle\psi|E_1\psi\rangle = \langle\psi|E_2\psi\rangle$ for all vectors $\psi \in \mathcal{H}$. It then follows from Proposition 1.21 that $E_1 = E_2$. Therefore, $E_1 \ne E_2$ implies that $\mathrm{tr}[\varrho E_1] \ne \mathrm{tr}[\varrho E_2]$ at least for some state ϱ.

From now on, an effect will mean an operator E satisfying $O \le E \le I$. We denote by $\mathcal{E}(\mathcal{H})$ the set of all effects, i.e.,

$$\mathcal{E}(\mathcal{H}) = \{E \in \mathcal{L}_s(\mathcal{H}) \mid O \le E \le I\} .$$

The set $\mathcal{E}(\mathcal{H})$ is a convex subset of the real vector space $\mathcal{L}_s(\mathcal{H})$. On the one hand, all projections $P \in \mathcal{P}(\mathcal{H})$ are effects since both $P \ge O$ and $P^\perp \ge O$. On the other hand, an operator tI with $0 \le t \le 1$ is an effect but not a projection, unless $t = 0$ or $t = 1$. We thus have the strict inclusions

$$\mathcal{P}(\mathcal{H}) \subset \mathcal{E}(\mathcal{H}) \subset \mathcal{L}_s(\mathcal{H}) .$$

In the following exercises we will establish some basic properties of $\mathcal{E}(\mathcal{H})$. The hint for both the following exercises is the fact that $O \le E \le I$ is equivalent to the requirement that $0 \le \langle\psi|E\psi\rangle \le 1$ for all unit vectors $\psi \in \mathcal{H}$.

Exercise 2.31 Show that $E \in \mathcal{E}(\mathcal{H})$ if and only if $I - E \in \mathcal{E}(\mathcal{H})$.

Exercise 2.32 Show that $E \in \mathcal{E}(\mathcal{H})$ if and only if $tE \in \mathcal{E}(\mathcal{H})$ for all $t \in [0, 1]$.

Example 2.33 (*Qubit effects*)
In Example 2.21 we discussed the Bloch sphere and gave a parametrization for the qubit states in terms of the Pauli operators. The operators I, σ_x, σ_y, σ_z are selfadjoint, and any selfadjoint operator $A \in \mathcal{L}_s(\mathcal{H})$ can be written as a real linear combination of them:

$$A = \tfrac{1}{2}(\alpha I + \vec{a} \cdot \vec{\sigma}) , \tag{2.25}$$

where $\alpha \in \mathbb{R}$ and $\vec{a} \in \mathbb{R}^3$. (The factor $\frac{1}{2}$ is included just for convenience.) The eigenvalues of A are $\lambda_{\pm} = \frac{1}{2}(\alpha \pm \|\vec{a}\|)$. Therefore, the operator A is an effect if and only if $0 \leq \lambda_-$ and $\lambda_+ \leq 1$. These requirements can be written in the form

$$\|\vec{a}\| \leq \alpha \leq 2 - \|\vec{a}\| . \tag{2.26}$$

We also notice that (2.26) implies that $\|\vec{a}\| \leq 1$.

In $\mathbb{C}^2$ a projection is either one dimensional or trivial (O or I). The operator A in (2.25) is a one-dimensional projection if and only if its eigenvalues are $\lambda_- = 0$ and $\lambda_+ = 1$, or equivalently $\alpha = \|\vec{a}\| = 1$. (The condition $\alpha = \|\vec{a}\| = 1$ can be obtained also by requiring that $A^2 = A$ and tr$[A] = 1$.) $\qquad\triangle$

For each pair consisting of a state ϱ and an effect E, the number $\text{tr}[\varrho E]$ is the probability that the measurement event represented by E occurs when the system is prepared in the state ϱ. This is the fundamental link between physical experiments and the mathematical formalism of quantum theory. Sometimes this trace formula for calculating probabilities is referred to as the *Born rule*.

It is obvious that for most choices of a state ϱ and an effect E, the number $\text{tr}[\varrho E]$ is strictly less than 1. The important special case of certain prediction can be characterized mathematically as follows.

Proposition 2.34 Let E be an effect and P a pure state. The following conditions are equivalent:

(i) $\text{tr}[PE] = 1$;

(ii) $EP = P$;

(iii) $P \leq E$.

Proof We will prove this proposition by showing that (i)$\Rightarrow$(ii)$\Rightarrow$(iii)$\Rightarrow$(i). In the following, we fix a unit vector $\psi \in \mathcal{H}$ such that $P = |\psi\rangle\langle\psi|$.

Suppose that (i) holds. Then

$$1 = \text{tr}[PE] = \langle\psi|E\psi\rangle = |\langle\psi|E\psi\rangle| \leq \|\psi\|\,\|E\psi\| \leq \|E\| \leq 1 ,$$

where the first inequality follows from the Cauchy–Schwarz inequality and the second from (1.19). This shows that the equality contained in the first inequality can be true only if the vectors $E\psi$ and ψ are collinear, i.e. $E\psi = c\psi$ for some $c \in \mathbb{C}$. Moreover, $1 = \langle\psi|E\psi\rangle = c$ and hence $E\psi = \psi$. Thus, $EP = P$ and (ii) holds.

Suppose that (ii) holds. Since

$$PE = P^*E^* = (EP)^* = P^* = P = EP ,$$

we conclude that the operators E and P commute. By Theorem 1.33 the square-root $E^{\frac{1}{2}}$ also commutes with P. Hence (ii) implies that $P = E^{\frac{1}{2}} P E^{\frac{1}{2}}$. For every vector $\phi \in \mathcal{H}$ we then obtain

$$\langle \phi | P \phi \rangle = \langle E^{\frac{1}{2}} \phi | P E^{\frac{1}{2}} \phi \rangle \leq \| P \| \, \| E^{\frac{1}{2}} \phi \|^2 \leq \langle \phi | E \phi \rangle .$$

The first inequality is an application of the Cauchy–Schwarz inequality; in the second we used (1.20) and the third follows from the fact that $\| P \| = 1$. This shows that $P \leq E$.

Finally, suppose that (iii) holds. Then

$$\mathrm{tr}[PE] = \langle \psi | E \psi \rangle \geq \langle \psi | P \psi \rangle = 1 .$$

Hence, (i) holds. $\square$

Exercise 2.35 Let E be an effect and ϱ a mixed state. Prove that $\mathrm{tr}[\varrho E] = 1$ is equivalent to $E\varrho = \varrho$ but not equivalent to $\varrho \leq E$. [Hint: Take a canonical convex decomposition for ϱ and use Proposition 2.34. Further, choose, for instance, $E = \varrho = \frac{1}{d} I$. Then $\varrho \leq E$ but $\mathrm{tr}[\varrho E] = \frac{1}{d}$.]

2.1.5 From effects to states

In subsection 2.1.4 we saw that if the set of states is chosen to be $\mathcal{S}(\mathcal{H})$ then the mathematical form for effects follows from the basic statistical framework. It is also possible to fix the mathematical form for effects first and take this as a starting point. In this subsection we explain this line of thought.

Let us forget subsection 2.1.2 for a moment and start again from the general framework of subsection 2.1.1. In particular, suppose that we have not fixed the specific mathematical form of states and effects. Starting from the definition of an effect as an affine mapping on the set of states, we can define a partial binary operation $\boxplus$ on the set of effects. If E_1, E_2, E_3 are effects and, for all states ϱ, the equality

$$E_1(\varrho) + E_2(\varrho) = E_3(\varrho)$$

holds then we write $E_1 \boxplus E_2 = E_3$. Notice that $E_1 \boxplus E_2$ exists only if $E_1(\varrho) + E_2(\varrho) \leq 1$ for all states ϱ. Therefore, the operation $E_1 \boxplus E_2$ is not defined for all pairs of effects and for this reason $\boxplus$ is called a partial operation. (For instance, $I \boxplus I$ is not defined.)

Exercise 2.36 Verify the following four properties of the partial binary operation $\boxplus$.

(a) If $E_1 \boxplus E_2$ exists then $E_2 \boxplus E_1$ exists and $E_1 \boxplus E_2 = E_2 \boxplus E_1$.
(b) If $E_1 \boxplus E_2$ and $(E_1 \boxplus E_2) \boxplus E_3$ exist then $E_2 \boxplus E_3$ and $E_1 \boxplus (E_2 \boxplus E_3)$ exist and $(E_1 \boxplus E_2) \boxplus E_3 = E_1 \boxplus (E_2 \boxplus E_3)$.

(c) For every E, there is a unique E' such that $E \boxplus E' = I$.

(d) If $E \boxplus I$ exists then $E = O$.

Generally, a set $\mathcal{E}$ with two distinct elements O, I and equipped with a partial binary operation $\boxplus$ is called an *effect algebra* if the conditions (a)–(d) hold.

Since effects are affine mappings from the set of states to the interval $[0, 1]$, each state ϱ defines a function f_ϱ from the set of effects to the interval $[0, 1]$ by the formula

$$f_\varrho(E) := E(\varrho).$$

It is clear from this definition that f_ϱ satisfies

$$f_\varrho(O) = O(\varrho) = 0, \qquad f_\varrho(I) = I(\varrho) = 1, \tag{2.27}$$

where O and I are the zero and identity effects introduced in Example 2.3. Moreover, assume that E_1, E_2 are effects and $E_1 \boxplus E_2$ exists. Then we get

$$f_\varrho(E_1 \boxplus E_2) = (E_1 \boxplus E_2)(\varrho) = E_1(\varrho) + E_2(\varrho) = f_\varrho(E_1) + f_\varrho(E_2). \tag{2.28}$$

In conclusion, each function f_ϱ is normalized and additive (with respect to $\boxplus$).

The properties (2.27) and (2.28) are immediate consequences of our definitions. To proceed, we need to require something more. First let us observe that the partial binary operation $\boxplus$ defines a partial ordering: for two effects E_1 and E_2, we write $E_1 \le E_2$ if there exists an effect E_1' such that $E_1 \boxplus E_1' = E_2$. It follows from (2.28) that

$$E_1 \le E_2 \Rightarrow f_\varrho(E_1) \le f_\varrho(E_2) \tag{2.29}$$

for all effects E_1, E_2.

With partial ordering defined, we can treat infinite sequences. Namely, suppose that $E_1, E_2, \dots$ is a sequence of effects such that the sum $E_1 \boxplus \cdots \boxplus E_n$ exists for each finite n. By $E_1 \boxplus E_2 \boxplus \cdots$ we mean the least upper bound of the increasing sequence $E_1, E_1 \boxplus E_2, E_1 \boxplus E_2 \boxplus E_3, \dots$, if it exists.

Definition 2.37 A mapping f from the set of effects to the interval $[0, 1]$ is a *generalized probability measure* if it satisfies $f(O) = 0$, $f(I) = 1$ and

$$f(E_1 \boxplus E_2 \boxplus \cdots) = f(E_1) + f(E_2) + \cdots \tag{2.30}$$

whenever $E_1 \boxplus E_2 \boxplus \cdots$ exists.

We now require that states correspond to the generalized probability measures on effects. This can be seen as an additional assumption in the basic statistical framework. Note, however, that (2.27) and (2.28) follow from the basic framework and that (2.30) is just a slight generalization of (2.28).

Let us then see the consequences that ensue if the set $\mathcal{E}(\mathcal{H})$ is taken to represent the set of effects. We will identify the binary operation $\boxplus$ with the usual addition of operators, in which case the partial ordering is the usual partial ordering of selfadjoint operators. Our proof of the following result is based on that in [23].

Proposition 2.38 Let f be a generalized probability measure on $\mathcal{E}(\mathcal{H})$. There exists an operator $\varrho_f \in \mathcal{S}(\mathcal{H})$ such that

$$f(E) = \text{tr}[\varrho_f E] \qquad \forall E \in \mathcal{E}(\mathcal{H}).$$

Proof The main idea of the proof is to show that f extends to a positive normal linear functional on $\mathcal{L}_s(\mathcal{H})$, the real vector space of bounded selfadjoint operators. The rest then follows from Theorem 1.71.

We will first show that f satisfies $f(rE) = rf(E)$ for every $E \in \mathcal{E}(\mathcal{H})$ and $r \in [0, 1]$. Let $E \in \mathcal{E}(\mathcal{H})$. If $mE \in \mathcal{E}(\mathcal{H})$ for some $m \in \mathbb{N}$ then (2.28) gives

$$f(mE) = f(E + \cdots + E) = f(E) + \cdots + f(E) = mf(E).$$

Moreover, for each $n \in \mathbb{N}$ we split E into n effects $\dfrac{1}{n}E$, and hence

$$\frac{1}{n}f(E) = \frac{1}{n}f\left(n\frac{1}{n}E\right) = \frac{1}{n}nf\left(\frac{1}{n}E\right) = f\left(\frac{1}{n}E\right).$$

We conclude that $f(qE) = qf(E)$ for every rational number $q \in [0, 1]$. Suppose then that $r \in [0, 1]$ is an irrational number. We choose an increasing sequence $\{q_i\}$ and a decreasing sequence $\{p_j\}$ of rational numbers such that $\lim_i q_i = \lim_j p_j = r$. From (2.29) follows that

$$q_i f(E) = f(q_i E) \le f(rE) \le f(p_j E) = p_j f(E).$$

In the limit we have $rf(E) \le f(rE) \le rf(E)$, hence $f(rE) = rf(E)$.

If $T \ne O$ is a positive operator then $T/\|T\|$ is an effect (see Exercise 1.32). Therefore we can extend f to all positive operators by setting

$$f(T) := \|T\| f(T/\|T\|).$$

For every positive operator T and a number $r \ge 0$ we then have

$$f(rT) = \|rT\| f(rT/\|rT\|) = r\|T\| f(T/\|T\|) = rf(T).$$

For two (nonzero) positive operators S, T, we obtain

$$\begin{aligned}
f(S+T) &= \|S+T\| f((S+T)/\|S+T\|) \\
&= \|S+T\| f(S/\|S+T\|) + \|S+T\| f(T/\|S+T\|) \\
&= f(S) + f(T).
\end{aligned}$$

In the last step we extend f to all bounded selfadjoint operators. Every $T \in \mathcal{L}_s(\mathcal{H})$ can be written as the difference of its positive and negative parts, $T = T^+ - T^-$. We define

$$f(T) := f(T^+) - f(T^-).$$

The fact that $f(S + T) = f(S) + f(T)$ for all $S, T \in \mathcal{L}_s(\mathcal{H})$ can be shown in a similar way to that in which we proved (2.23). We have thus extended f to a positive linear functional on $\mathcal{L}_s(\mathcal{H})$.

We need to show that f is normal, i.e. that for any norm-bounded increasing sequence $\{T_j\}_{j=1}^{\infty} \subset \mathcal{L}_s(\mathcal{H})$ with limit $T \in \mathcal{L}_s(\mathcal{H})$ it follows that $\{f(T_j)\}_{j=1}^{\infty}$ converges to $f(T)$. To show normality we obviously have to use (2.30), and for this reason we scale the operators T_j to be effects. We have

$$T_j + \|T_1\| I \geq T_1 + \|T_1\| I \geq O,$$

and therefore the operators

$$F_j := \frac{1}{\|T + \|T_1\| I\|}(T_j + \|T_1\| I), \qquad F := \frac{1}{\|T + \|T_1\| I\|}(T + \|T_1\| I)$$

are effects. Since $F_j \leq F_{j+1}$, the operators $E_1 := F_1$ and $E_j := F_{j+1} - F_j$ for $j = 2, \ldots$ are also effects. They satisfy

$$E_1 + E_2 + \cdots + E_j = F_j.$$

The least upper bound for the sequence $E_1, E_1 + E_2, \ldots$ is F. Hence, from (2.30) follows that

$$f(F) = f(E_1 + E_2 + \cdots) = \sum_{j=1}^{\infty} f(E_j) = \lim_{j \to \infty} f(F_j).$$

Since f is linear we obtain $\lim_{j \to \infty} f(T_j) = f(T)$.

We need to extend f to a positive normal linear functional on $\mathcal{L}_s(\mathcal{H})$. By Theorem 1.71 there exists a positive trace class operator ϱ_f such that $f(T) = \mathrm{tr}[\varrho_f T]$ for all $T \in \mathcal{L}_s(\mathcal{H})$. From the normalization $1 = f(I) = \mathrm{tr}[\varrho_f]$ it then follows that $\varrho_f \in \mathcal{S}(\mathcal{H})$. $\qquad\square$

Exercise 2.39 Let $\varrho, \varrho' \in \mathcal{S}(\mathcal{H})$ be two different states. Show that they determine different generalized probability measures, i.e. that there exists an effect E such that $\mathrm{tr}[\varrho E] \neq \mathrm{tr}[\varrho' E]$. Conclude that the correspondence $f \leftrightarrow \varrho_f$ in Proposition 2.38 is one-to-one. [Hint: Use a similar approach to that in the uniqueness part of the proof of Proposition 2.30.]

We conclude that the choices $\mathcal{S}(\mathcal{H})$ and $\mathcal{E}(\mathcal{H})$ for the mathematical description of states and effects, respectively, are compatible in the sense that one implies the other under the assumption in the basic statistical framework. It should be emphasized, perhaps, that the discussion here and in the earlier subsections is not a derivation of the Hilbert space structure of quantum theory. We have simply shown how the Hilbert space structure, where states and effects are represented by certain types of operator, fits the basic statistical framework.

2.1.6 Dispersion-free states and Gleason's theorem

Let us continue the discussion of subsection 2.1.5 from a slightly different perspective. As a starting point, we recall that effects themselves describe binary measurements, i.e. measurements with only two outcomes. Each state assigns a probability for all effects. This raises the following question: can we make a consistent assignment in such a way that the results in all binary measurements are predictable with probability 1? In other words, we are searching for a generalized probability measure which takes only the values 0 and 1. A generalized probability measure having this feature is called *dispersion-free*.

It is easy to see that a generalized probability measure f_ϱ on $\mathcal{E}(\mathcal{H})$, related to the state ϱ via $f_\varrho(E) = \text{tr}[\varrho E]$, is not dispersion-free. For instance, setting $E = \frac{1}{2}\varrho$ we obtain $f_\varrho(\frac{1}{2}\varrho) = \frac{1}{2}\text{tr}[\varrho^2]$ and hence $0 < f_\varrho(\frac{1}{2}\varrho) < 1$. Since by Proposition 2.38 all generalized probability measures on $\mathcal{E}(\mathcal{H})$ are of the form f_ϱ for some state $\varrho \in \mathcal{S}(\mathcal{H})$, we conclude that there are no dispersion-free generalized probability measures on $\mathcal{E}(\mathcal{H})$.

This fact is not unexpected and we can also understand it without Proposition 2.38. Let f be a generalized probability measure on $\mathcal{E}(\mathcal{H})$ and suppose that E_0 and E_1 are two effects such that $f(E_0) = 0$ and $f(E_1) = 1$ (for instance $E_0 = O$ and $E_1 = I$). Then $f(\frac{1}{2}E_0 + \frac{1}{2}E_1) = \frac{1}{2}$, implying that f is not dispersion-free. This argument clearly relates to the fact that $\mathcal{E}(\mathcal{H})$ is a convex set.

Even if there are no dispersion-free generalized probability measures on the set of all effects, we can wonder whether they could be defined on some subset. In particular, can we make certain predictions for effects that are not convex mixtures of other effects?

Example 2.40 (*States and effects in classical mechanics*)
This is a good place to think about the mathematical representation of states and effects in classical mechanics. Briefly, we can say that classical states are probability distributions on a suitable phase space Ω and classical effects are associated with functions on Ω taking values between 0 and 1. In the usual physical situations the phase space Ω is infinite. For example, the phase space for a moving particle is the six-dimensional manifold $\mathbb{R}^3 \times \mathbb{R}^3$ consisting of position and momentum vectors.

To keep things simple, we will focus only on finite phase spaces; these are sufficient for illustration. In the case of a finite phase space Ω (say, containing d elements), the states can be represented as probability vectors $\vec{p} = (p_1, \ldots, p_d)$ with $0 \leq p_j \leq 1$ and $\sum_j p_j = 1$. The effects can be represented by vectors $\vec{e} = (e_1, \ldots, e_d)$ satisfying $0 \leq e_j \leq 1$. The probability of measuring an effect $\vec{e}$ if the system is in a state $\vec{p}$ is $\vec{p} \cdot \vec{e} = \sum_j p_j e_j$. As we saw in Example 2.9, the extremal elements of classical state space are probability vectors $\vec{\delta}_k$ with all entries vanishing except the kth. Similarly, the extremal effects are vectors $\vec{e}$ such that each entry e_j is either 0 or 1.

It is easy to see that there are no dispersion-free states. However, there are states that are dispersion-free when restricted to extremal effects. Namely, if we have an extremal state $\vec{\delta}_k$ and an extremal effect $\vec{e}$ then the number $\vec{\delta}_k \cdot \vec{e}$ is either 0 or 1. $\triangle$

In order to tackle the quantum case, we recall the following result of Davies [51] on extremal elements in $\mathcal{E}(\mathcal{H})$.

Proposition 2.41 The extremal elements of the convex set $\mathcal{E}(\mathcal{H})$ are the projections.

Proof We first prove that every projection is an extremal element. Let P be a projection and assume that E_1, E_2 are effects such that

$$P = \lambda E_1 + (1 - \lambda)E_2 \tag{2.31}$$

for some $0 < \lambda < 1$. Suppose that $\psi \in \mathcal{H}$ satisfies $P\psi = 0$. Since E_1 and E_2 are positive operators, we get

$$0 = \langle \psi | P\psi \rangle = \lambda \langle \psi | E_1 \psi \rangle + (1 - \lambda)\langle \psi | E_2 \psi \rangle \geq \lambda \langle \psi | E_1 \psi \rangle \geq 0.$$

Hence $\langle \psi | E_1 \psi \rangle = 0$. By Proposition 1.35 this implies that $E_1 \psi = 0$. Furthermore, equation (2.31) gives

$$I - P = \lambda(I - E_1) + (1 - \lambda)(I - E_2),$$

and similar reasoning shows that $P\psi = \psi$ implies that $E_1 \psi = \psi$. By Proposition 1.40 every vector in $\mathcal{H}$ can be written as a sum of eigenvectors of P. Hence, E_1 and P act identically on all vectors $\psi \in \mathcal{H}$ and we conclude that $E_1 = P$. Thus, P does not have a nontrivial convex decomposition and it is extremal.

To prove that there are no other extremal effects than projections, suppose that $A \in \mathcal{E}(\mathcal{H})$ is not a projection, so that $A \neq A^2$. We define $E_1 = A^2 \neq A$ and $E_2 = 2A - A^2 \neq A$. The operator E_1 is an effect by Proposition 1.34. Similarly, since $I - E_2 = (I - A)^2$, $I - E_2$ is an effect also. This implies that E_2 itself is an effect (see Exercise 2.31). The equal-weight convex combination of these effects gives

$$\tfrac{1}{2}(E_1 + E_2) = A,$$

and we conclude that A is not an extreme element of $\mathcal{E}(\mathcal{H})$.   □

Exercise 2.42 (*Extremal qubit effects*)
We discussed qubit effects in Example 2.33. Write the qubit effect

$$A = \tfrac{1}{2}(I + \vec{a} \cdot \vec{\sigma}), \qquad \|\vec{a}\| \leq 1,$$

as a convex combination of two projections. [Hint: Choose two orthogonal projections.]

In the light of Proposition 2.41 we will focus on the potential existence of dispersion-free generalized probability measures on the set of projections $\mathcal{P}(\mathcal{H})$. Before proceeding with the main results let us make some elementary observations. We recall from subsection 1.2.3 that the sum $P + Q$ of two projections P and Q is a projection if and only if P and Q are orthogonal. The identity operator I can be written as a sum of orthogonal one-dimensional projections,

$$I = \sum_{k=1}^{d} |\varphi_k\rangle\langle\varphi_k|,$$

where $\{\varphi_k\}_{k=1}^{d}$ is an orthonormal basis of $\mathcal{H}$ (see subsection 1.3.2). Therefore, a generalized probability measure f on $\mathcal{P}(\mathcal{H})$ satisfies

$$\sum_{k=1}^{d} f(|\varphi_k\rangle\langle\varphi_k|) = 1. \tag{2.32}$$

Suppose now that η and ϕ are two orthogonal unit vectors. First, if $f(P_\eta) = 1$ then the orthogonality of η and ϕ combined with the condition (2.32) implies that $f(P_\phi) = 0$. Second, if $f(P_\eta) = f(P_\phi) = 0$ then every linear combination $a\eta + b\phi$, with $|a|^2 + |b|^2 = 1$, also satisfies $f(P_{a\eta+b\phi}) = 0$. One can see this by noticing that

$$|a\eta + b\phi\rangle\langle a\eta + b\phi| + |\bar{b}\eta - \bar{a}\phi\rangle\langle\bar{b}\eta - \bar{a}\phi| = |\eta\rangle\langle\eta| + |\phi\rangle\langle\phi|.$$

Since $a\eta + b\phi$ and $\bar{b}\eta - \bar{a}\phi$ are orthogonal unit vectors, the additivity property of f implies that

$$f(|a\eta + b\phi\rangle\langle a\eta + b\phi|) + f(|\bar{b}\eta - \bar{a}\phi\rangle\langle\bar{b}\eta - \bar{a}\phi|) = f(|\eta\rangle\langle\eta|) + f(|\phi\rangle\langle\phi|) = 0,$$

and hence $f(P_{a\eta+b\phi}) = f(P_{\bar{b}\eta-\bar{a}\phi}) = 0$.

With these preliminary observations we are ready to prove that a generalized probability measure on $\mathcal{P}(\mathcal{H})$ cannot be dispersion-free if $\dim \mathcal{H} \geq 3$. We follow the proof of John Bell [12].

Proposition 2.43 Suppose that dim $\mathcal{H} \geq 3$. There are no dispersion-free generalized probability measures on $\mathcal{P}(\mathcal{H})$.

Proof We make the counter-assumption that there exists a dispersion-free generalized probability measure f on $\mathcal{P}(\mathcal{H})$. Let $\varphi \in \mathcal{H}$ be a unit vector such that $f(P_\varphi) = 1$. We will show that there are no unit vectors $\phi \in \mathcal{H}$ satisfying $f(P_\phi) = 0$ and $\|\varphi - \phi\| < \frac{1}{3}$. But since f has only two values, 0 and 1, and every unit vector is assigned one of these values, there must be arbitrarily close unit vectors with different values of f. This means that the existence of dispersion-free generalized probability measures is impossible.

Let $\phi \in \mathcal{H}$ be a unit vector such that $f(P_\phi) = 0$ and $\|\varphi - \phi\| < \frac{1}{3}$. As we will see, this leads to a contradiction. Without loss of generality we may write ϕ in the form $\phi = (\varphi + \epsilon\varphi')/\sqrt{1 + \epsilon^2}$ for some nonnegative $\epsilon \in \mathbb{R}$ and a unit vector φ' orthogonal to φ. From $\|\varphi - \phi\| < \frac{1}{3}$ it follows that $\epsilon < \frac{1}{2}$. We see also that $\epsilon \neq 0$ since $f(P_\phi) \neq f(P_\varphi)$. We will fix a unit vector φ'' that is orthogonal to both φ and φ' (here we need the assumption dim $\mathcal{H} \geq 3$). Since $f(P_\varphi) = 1$ it follows that $f(P_{\varphi'}) = f(P_{\varphi''}) = 0$.

For any nonzero real number γ, we define

$$\psi_\gamma := \frac{1}{N}(-\varphi' + \gamma\varphi''),$$

$$\psi_\gamma' := \frac{1}{N'}\left(\phi + \frac{\epsilon}{\gamma\sqrt{1 + \epsilon^2}}\varphi''\right),$$

where N and N' are normalization constants that make ψ_γ and ψ_γ' unit vectors, respectively. It is straightforward to verify that ψ_γ and ψ_γ' are orthogonal. Since $\psi_\gamma, \psi_\gamma'$ are linear combinations of φ', φ'' and ϕ, it follows that $f(P_{\psi_\gamma}) = f(P_{\psi_\gamma'}) = 0$.

Since $|\gamma + \gamma^{-1}| \geq 2$ and $\epsilon < \frac{1}{2}$ it follows that there exist real numbers $\gamma_\pm$ such that $\epsilon(\gamma_\pm + \gamma_\pm^{-1}) = \pm 1$. Let us define another pair of unit vectors,

$$\eta_\pm := \frac{\epsilon\sqrt{1 + \gamma_\pm^2}\psi_{\gamma_\pm} + \sqrt{1 + \epsilon^2(1 + \gamma_\pm^{-2})}\psi_{\gamma_\pm}'}{\sqrt{1 + \epsilon^2(\gamma_\pm + \gamma_\pm^{-1})^2}}$$

$$= \frac{\varphi + \epsilon(\gamma_\pm + \gamma_\pm^{-1})\varphi''}{\sqrt{1 + \epsilon^2(\gamma_\pm + \gamma_\pm^{-1})^2}} = \frac{1}{\sqrt{2}}(\varphi \pm \varphi'').$$

Since $\eta_\pm$ are linear combinations of $\psi_{\gamma_\pm}$ and $\psi_{\gamma_\pm}'$, we conclude that $f(\eta_\pm) = 0$. Moreover, we observe that $\varphi = \frac{1}{\sqrt{2}}(\eta_+ + \eta_-)$ and therefore $f(P_\varphi) = 0$. This is in contradiction with our original assumption that $f(P_\varphi) = 1$ and so completes the proof. $\square$

The deepest answer to our speculative inquiry about generalized probability measures on $\mathcal{P}(\mathcal{H})$ is provided by the celebrated theorem of Andrew Gleason declaring that all generalized probability measures on $\mathcal{P}(\mathcal{H})$ are associated with density operators. The proof given by Gleason [66] is long and complicated. We recommend to the interested reader the article [47] by Cooke *et al.*, where an elementary (but still long) proof of Gleason's theorem is presented. An extensive discussion of Gleason's theorem can be found in the book of Dvurečenskij [57].

Theorem 2.44 (*Gleason's theorem*)
Suppose that $\dim \mathcal{H} \geq 3$. Let f be a generalized probability measure on $\mathcal{P}(\mathcal{H})$. There exists a unique operator $\varrho_f \in \mathcal{S}(\mathcal{H})$ such that

$$f(P) = \mathrm{tr}\big[\varrho_f P\big] \tag{2.33}$$

for all $P \in \mathcal{P}(\mathcal{H})$.

From Gleason's theorem we conclude that when $\dim \mathcal{H} \geq 3$ there seems to be no way to escape the probabilistic nature of quantum theory (at least in its usual Hilbert space formulation). All generalized probability measures on $\mathcal{P}(\mathcal{H})$ are obtained from states with the usual trace formula and they are never dispersion-free.

Gleason's theorem does not make any claim on $\dim \mathcal{H} = 2$. This opens up the possibility of constructing generalized probability measures that are dispersion-free, and an example is given below. So far these dispersion-free generalized probability measures have not found any physical interpretation, however.

Example 2.45 (*Generalized probability measures in two dimensions*)
The set of projections on a two-dimensional Hilbert space consists of the operators O, I and $P_{\vec{n}} \equiv \frac{1}{2}(I + \vec{n} \cdot \vec{\sigma})$ with $\vec{n} \in \mathbb{R}^3$, $\|\vec{n}\| = 1$. Therefore, a generalized probability measure f can be understood as a function on the unit vectors in $\mathbb{R}^3$. Let us note that a sum $P_{\vec{n}} + P_{\vec{m}}$ is a projection if and only if $\vec{m} = -\vec{n}$. The additivity constraint on f gives only a single relation,

$$f(P_{\vec{n}}) + f(P_{-\vec{n}}) = f(I) = 1 \,,$$

that is required to hold for all unit vectors $\vec{n}$.

To construct a dispersion-free generalized probability measure we fix a unit vector $\vec{x} \in \mathbb{R}^3$ (see Figure 2.5). We then define $f(P_{\vec{n}}) = 0$ if $\vec{n} \cdot \vec{x} < 0$ and $f(P_{\vec{n}}) = 1$ if $\vec{n} \cdot \vec{x} > 0$. The property $f(P_{\vec{n}}) + f(P_{-\vec{n}}) = 1$ obviously holds for all $\vec{n} \cdot \vec{x} \neq 0$.

To define f also on unit vectors $\vec{m}$ satisfying $\vec{m} \cdot \vec{x} = 0$, we fix one such vector and denote it by $\vec{y}$. We then define $f(P_{\vec{m}}) = 0$ if $\vec{m} \cdot \vec{y} < 0$ and $f(P_{\vec{m}}) = 1$ if $\vec{m} \cdot \vec{y} > 0$. There are exactly two unit vectors, denoted by $\pm\vec{z}$, which are orthogonal to both $\vec{x}$ and $\vec{y}$. We define $f(\vec{z}) = 1$ and $f(-\vec{z}) = 0$, and in this way f becomes a generalized probability measure. $\triangle$

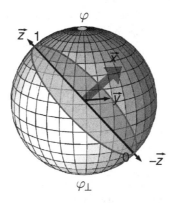

Figure 2.5 Illustration of a probability measure f taking the value $+1$ on Bloch vectors $\vec{r}$ of length 1 and satisfying either $\vec{x}\cdot\vec{r} > 0$ or $\vec{x}\cdot\vec{r} = 0$ and, simultaneously, $\vec{y}\cdot\vec{r} > 0$ or $\vec{r} = \vec{z}$, where $\vec{x}, \vec{y}, \vec{z}$ determines some orthonormal coordinate system (see Example 2.45).

Exercise 2.46 Let f be the generalized probability measure constructed in Example 2.45. We already know that f cannot be of the form $f = f_\varrho$ for some state ϱ, since f is dispersion-free, but we can easily check this directly in this particular instance. As explained in subsection 2.1.3, a qubit state ϱ has the form $\frac{1}{2}(I + \vec{r}\cdot\vec{\sigma})$ for some vector $\vec{r}$ with $\|\vec{r}\| \leq 1$. Calculate $f_\varrho(P_{\vec{n}})$ and conclude that it is impossible to produce the correct values $f(P_{\vec{n}})$ for the constructed function f.

2.2 Superposition structure of pure states

Pure states form an exceptional subset of states, and not only because they are extremal elements of the state space. They have their own special structure called *superposition*, which is often regarded as the core quantum feature. There is no analogy for superposition in classical state space and this is one reason why we meet with problems when we want to explain this phenomenon in words. Often it is stated that matter exhibits wave-like properties. This statement may give an intuitive description of quantum superposition, but it can also cause confusion if taken too literally. Even if the superposition structure of pure states may be conceptually intriguing, the mathematics behind it is very simple and the phenomenon has been verified in many experiments and with different types of quantum objects.

2.2.1 Superposition of two pure states

To explain the mathematical formalism of superposition, we will first adopt a slightly different but equivalent description of pure states. We saw earlier that pure states are described by one-dimensional projections and that each unit vector η defines a one-dimensional projection P_η. However, this correspondence is not injective. Namely, two unit vectors η and η' define the same projection if there is a complex number z of unit modulus such that $\eta' = z\eta$. The converse holds also: if two unit vectors η and η' differ in some other way than by a scalar multiple then they define different projections. We conclude that pure states can be alternatively described as equivalence classes $[\eta]$ of unit vectors $\eta \in \mathcal{H}$, where the equivalence relation is defined in the following way:

$$\eta \sim \eta' \quad \Leftrightarrow \quad \eta' = z\eta \text{ for some } z \in \mathbb{C}, |z| = 1.$$

An equivalence class $[\eta]$ is commonly called a *ray* by physicists. If $\mathcal{H}$ has finite dimension $d+1$ then a mathematician would recognize the set of pure states as the complex projective d-space.

Forming a superposition of two pure states is, essentially, the same thing as forming a linear combination of two vectors. Let $\psi, \varphi \in \mathcal{H}$ be two linearly independent unit vectors. Choose two nonzero complex numbers a, b, and write

$$\omega = \frac{1}{\|a\psi + b\varphi\|}(a\psi + b\varphi). \tag{2.34}$$

Then ω is a unit vector, and the pure state $[\omega]$ is called a *superposition* of the pure states $[\psi]$ and $[\varphi]$. In terms of the corresponding one-dimensional projections P_ψ, P_φ and P_ω, the superposition condition (2.34) reads

$$P_\omega = \frac{1}{\|a\psi + b\varphi\|^2}\left(|a|^2 P_\psi + |b|^2 P_\varphi + \bar{a}b|\varphi\rangle\langle\psi| + a\bar{b}|\psi\rangle\langle\varphi|\right), \tag{2.35}$$

which does not look as simple as (2.34). In fact, the cross terms give the difference between a superposition and a mixture; this will be discussed further in subsection 2.2.2.

Let us note that although pure states are identified with rays, a given superposition depends on the chosen representatives of the equivalence class. In particular, the superpositions of vectors ψ, φ and of vectors $\psi, -\varphi$ (with the same coefficients a, b) give vectors ω, ω' that belong to different equivalence classes.

Proposition 2.47 Let P_1 and P_2 be two different pure states (described as one-dimensional projections). If P_1 and P_2 are not orthogonal, i.e.

$$\text{tr}[P_1 P_2] \neq 0, \tag{2.36}$$

then P_1 can be written as a superposition of P_2 and some other pure state (say P_3), and P_3 can be chosen such that $\text{tr}[P_2 P_3] = 0$.

Proof Suppose that (2.36) holds. We fix unit vectors ω and ψ such that $P_1 = P_\omega$ and $P_2 = P_\psi$. Then $\text{tr}[P_1 P_2] = |\langle\psi|\omega\rangle|^2 \leq 1$. If $|\langle\psi|\omega\rangle| = 1$ then

$P_1 = P_2$. Hence we assume that $|\langle \psi | \omega \rangle| < 1$. We define a unit vector $\psi_\perp = (\omega - y\psi)/\sqrt{1 - |y|^2}$ with $y = \langle \psi | \omega \rangle$; $y \neq 0$ by (2.36). This gives $\langle \psi | \psi_\perp \rangle = 0$. Therefore, $\omega = y\psi + \sqrt{1 - |y|^2}\psi_\perp$, and this proves that P_1 is a superposition of P_2 and $P_3 = P_{\psi_\perp}$. $\qquad\square$

As convexity gives us a way of forming a new state out of two other states, superposition gives us a way of forming a new pure state out of two pure states. However, the superposition structure of pure states is of a character different from the convex structure of states in general. For instance, it can be seen from Proposition 2.47 that any pure state can be written as a superposition of two other pure states. There are no 'extremal' pure states with respect to the superposition structure.

How can one realize a superposition of two states? While there is a clear recipe for realizing a mixture of pure states, there is no clear general recipe for making a superposition. In the former case it is sufficient to mix preparators, whereas a superposition of preparators does not have any meaning. Superposition as developed earlier in this subsection is an abstract mathematical construction and its physical implementation depends on the system in question. There are devices that are considered to be superposition machines. Typically they transform a given input state into a superposition of this state and some other (orthogonal) state.

Example 2.48 (*Hadamard gate*)
Let us consider a two-dimensional Hilbert space with an orthonormal basis φ_0, φ_1. The *Hadamard gate* is a device accepting a vector state ω and transforming it into a vector state ω' such that

$$\omega = a\varphi_0 + b\varphi_1 \qquad \mapsto \qquad \omega' = a\varphi_+ + b\varphi_- ,$$

where $\varphi_\pm = \frac{1}{\sqrt{2}}(\varphi_0 \pm \varphi_1)$ and $|a|^2 + |b|^2 = 1$. If the input vector state is either φ_0 or φ_1 then the output vector state is an equal-amplitude superposition of φ_0 and φ_1. In this sense the Hadamard gate 'creates' the superposition.

If the two-dimensional Hilbert space is used to represent the polarization of a photon then a polarizing beamsplitter (see Example 2.2) constitutes the Hadamard gate for φ_0 and φ_1, which represent the horizontal and the vertical polarization, respectively. $\qquad\triangle$

Exercise 2.49 Calculate the action of the Hadamard gate on the vector states $\varphi_\pm = \frac{1}{\sqrt{2}}(\varphi_0 \pm \varphi_1)$.

2.2.2 Interference

Let $\psi, \varphi \in \mathcal{H}$ be orthogonal unit vectors and let ω be a superposed vector, as in (2.34). The orthogonality of ψ and φ implies that $\|a\psi + b\varphi\|^2 = |a|^2 + |b|^2$.

The probability related to the measurement of an effect E when the system is in the pure state P_ω is then

$$\text{tr}[E\,P_\omega] = \frac{1}{\|\omega\|^2}\Big(|a|^2\text{tr}\big[E\,P_\psi\big] + |b|^2\text{tr}\big[E\,P_\varphi\big] + 2\,\text{Re}\{a^*b\,\langle\psi|E\varphi\rangle\}\Big),$$

where $\|\omega\|^2 = |a|^2 + |b|^2$. The first two terms indicate a convex combination. Indeed, write $t = |a|^2/(|a|^2 + |b|^2)$, so that $1 - t = |b|^2/(|a|^2 + |b|^2)$. Let us then define a mixture of P_ψ and P_φ as $\varrho = t\,P_\psi + (1 - t)\,P_\varphi$. The probability of measuring effect E in ϱ is given by

$$\text{tr}[E\varrho] = \frac{1}{\|\omega\|^2}\left(|a|^2\text{tr}\big[E\,P_\psi\big] + |b|^2\text{tr}\big[E\,P_\varphi\big]\right). \tag{2.37}$$

Comparing the above two probabilities we find that the difference between the superposition P_ω and the mixture ϱ of the orthogonal pure states is given by the *interference term*

$$I_\omega(E) := \frac{2}{\|\omega\|^2}\,\text{Re}\{a^*b\,\langle\psi|E\varphi\rangle\}. \tag{2.38}$$

The number $I_\omega(E)$ is real and satisfies $-1 \leq I_\omega(E) \leq 1$. Loosely speaking we can say that the interference term reflects the difference between a superposition (a purely quantum structure) and a mixture (a general statistical structure) of pure states.

The interference term $I_\omega(E)$ can be detected in experiments but the observed interference obviously depends on the particular choice of the effect E. Each pure state can be expressed as a superposition of other pure states. It is only a question of the proper choice of experiment in which the interference can be seen. For example, if the superposed vectors ψ, φ are orthogonal then $I_\omega(P_\psi) = I_\omega(P_\varphi) = 0$ and $|I_\omega(P_\omega)| = 2|ab|^2/(|a|^2 + |b|^2)$.

Example 2.50 (*Double-slit experiment*)
The double-slit experiment is an elegant and simple demonstration of quantum interference and superposition. Let us briefly recall its basic features.

The situation is as follows: a source produces quantum particles impinging perpendicularly on a screen with two slits that defines the x-axis. It is assumed that the particles have equal probability of approaching any point in some finite area containing the two slits. Thus there is a nonzero probability that from time to time a particle will pass beyond the screen; in a sense, the screen with slits is filtering the incoming particles. Particles that pass the slits evolve freely and are registered on a second screen. Passage through the screen with slits followed by free evolution between the screens is considered to amount to the preparation process of some state. Registration on the second screen is understood as the measurement part of the experiment. For our purposes it is sufficient to consider the one-dimensional version of the problem (i.e. the case where the slits are effectively infinite in the direction perpendicular to the x-axis), in which case a probability density distribution p that depends only on x is measured. We consider three different experimental settings.

- Both slits are open and the probability q is measured.
- The lower slit is closed and the upper is open. We measure the probability p_+.
- The upper slit is closed and the lower is open. We measure the probability p_-.

As indicated in Figure 2.6, after the experiments are complete we end up with probability densities $p_\pm$ and q. Performing the same experiment with particles that are classical, one will find exactly the same probability densities for $p_\pm$; however, the distribution $q = q_{\text{class}}$ will be completely different from the quantum case and will satisfy the identity

$$q_{\text{class}} = \tfrac{1}{2}p_+ + \tfrac{1}{2}p_- \,.$$

The interpretation is that classically the double-slit experiment can be seen as an equal mixture of single-slit experiments in which, with equal probabilities one slit is closed. This reasoning is based on the fact that in each run of the double-slit experiment the particle must go through exactly one slit; the second slit is closed. However, for quantum particles this type of relation does not hold. In the quantum case we have

$$q = \tfrac{1}{2}(p_+ + p_-) + I \,,$$

where I is the interference term.

In probability theory, the probability of a joint event $P(a \cup b)$ is given by a formula $P(a \cup b) = P(a) + P(b) - P(a \cap b)$, where the last term reflects the dependence of the events a and b. Suppose that both slits are open and that event a

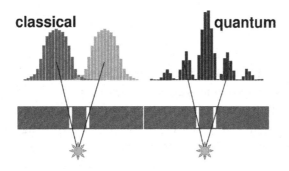

Figure 2.6 Double slit experiment with classical and quantum systems.

consists of the registration at position x of a particle that has passed through the first slit. Similarly, event b is identified with the registration at position x of a particle $P(a \cup b)$ that has passed through the second slit. Could the above formula for the joint probability be used to explain the interference term? If yes, it would mean that the two events are not independent. That is, the particles although separated in time are not passing the slits independently and there is a nonvanishing probability that particles pass through both slits simultaneously. Unfortunately, the situation is even more weird and the probability formula does not really help. Whereas the probability $P(a \cap b)$ is always positive, the interference term can be both negative and positive. Thus, probability theory fails to explain the interference term. This means that the probabilities $P(a \cup b)$, $P(a)$ and $P(b)$ are unrelated within the language of probability theory; hence, in the quantum case the double-slit experiment cannot be understood in terms of superimposed single-slit experiments. We conclude that quantum particles are passing through the slits in some curious way. Registration of a particle on the second screen does not tell us anything about the slit the particle passed through.

The situation is, of course, properly described within the quantum theory. As we said earlier, the first screen is used to prepare the particular pure state of the particle associated with vectors $\phi, \psi_+, \psi_- \in \mathcal{H} = L^2(\mathbb{R})$, respectively. The probability densities of the position measurement are given as $p_\pm(x) = |\psi_\pm(x)|^2$ and $q(x) = |\phi(x)|^2$. The state of the particles prepared by the double-slit screen is a superposition of the states prepared by single-slit screens, i.e.

$$\phi(x) = \tfrac{1}{\sqrt{2}}[\psi_-(x) + \psi_+(x)].$$

The calculation

$$|\phi(x)|^2 = \tfrac{1}{2}|\psi_-(x) + \psi_+(x)|^2$$
$$= \tfrac{1}{2}[p_-(x) + p_+(x)] + \tfrac{1}{2}[\psi_-(x)^*\psi_+(x) + \psi_+(x)^*\psi_-(x)]$$

shows that $I(x) = \mathrm{Re}(\psi_-(x)^*\psi_+(x))$. $\triangle$

2.3 Symmetry

Two physicists, Alice and Bob, are in a laboratory. Suppose that Alice prepares a quantum system in a specific way and describes this preparation procedure as state ϱ. The probability of the occurrence of some experimental event, described by an effect E, is $\mathrm{tr}[\varrho E]$. Bob is watching what Alice is doing and he also wants to predict the probability of the same event. Bob knows all the necessary details of the experiment and uses ϱ' and E' to describe the state and effect, respectively. If Alice and Bob are able to give the same predictions for the measurement outcome probabilities, does it mean that they are actually using exactly the same mathematical description? In other words, does the equality $\text{tr}[\varrho E] = \text{tr}[\varrho' E']$ imply that $\varrho = \varrho'$ and $E = E'$?

The potential freedom in the choice of a particular mathematical representation of physical objects is loosely called *symmetry*. In mathematical terms, physical symmetries are intimately related to groups in the sense that symmetry transformations form a group.

The mathematical description of quantum systems is based on Hilbert spaces, and unitary operators represent isomorphisms of a Hilbert space. Therefore, one might expect that the group of unitary operators (perhaps with antiunitary operators added) has a direct connection with the representation of symmetry in the quantum formalism. We will see with no difficulty that unitary and antiunitary operators indeed determine symmetry transformations (subsection 2.3.1). A much deeper result is that there are no other symmetry transformations. We will first show this by concentrating on the state space (subsection 2.3.2). We then discuss a stronger variant known as Wigner's theorem (subsection 2.3.3)

Our discussion is based on the books of Davies [51] and Cassinelli *et al.* [38]. We also refer to the book [101] by Molnár for a detailed study of this subject.

2.3.1 Unitary and antiunitary transformations

Let U be a unitary operator and suppose that the connection between Alice's and Bob's descriptions is

$$\varrho' = U \varrho U^*, \qquad E' = U E U^*.$$

Then

$$\text{tr}[\varrho' E'] = \text{tr}[U \varrho U^* U E U^*] = \text{tr}[\varrho E],$$

which means that Alice and Bob make the same predictions about measurement outcome probabilities. Since only these probabilities are related to physical predictions and comparable with experiments, we conclude that there is a *unitary freedom* in the choice of the mathematical description of states and effects.

Example 2.51 (*Rotations of the Bloch sphere*)
We saw in Example 2.21 that qubit states are in one-to-one correspondence with points in the Bloch sphere. Since the eigenvalues of any unitary operator are complex numbers with modulus 1, a unitary operator U on $\mathbb{C}^2$ can be written in the spectral decomposition form

$$U = e^{i\alpha} |\varphi\rangle \langle \varphi| + e^{i\beta} |\varphi_\perp\rangle \langle \varphi_\perp|,$$

where $\varphi, \varphi_\perp$ are eigenvectors of U forming an orthonormal basis of $\mathbb{C}^2$ and α, β are real parameters (subsection 1.3.3). We know that

$$|\varphi\rangle\langle\varphi| = \tfrac{1}{2}(I + \vec{n} \cdot \vec{\sigma}), \qquad |\varphi_\perp\rangle\langle\varphi_\perp| = \tfrac{1}{2}(I - \vec{n} \cdot \vec{\sigma}),$$

for some unit vector $\vec{n}$. Moreover, we can set $\alpha = \kappa + \eta$ and $\beta = \kappa - \eta$ and then obtain

$$U = e^{i\kappa}\left(\frac{e^{i\eta} + e^{-i\eta}}{2}I + \frac{e^{i\eta} - e^{-i\eta}}{2}\vec{n} \cdot \vec{\sigma}\right)$$

$$= e^{i\kappa}(\cos\eta I + i\sin\eta(\vec{n} \cdot \vec{\sigma})). \tag{2.39}$$

We conclude that unitary operators on $\mathbb{C}^2$ are determined by four real parameters.

In the action $\varrho \mapsto U\varrho U^*$ the parameter κ vanishes. In the Bloch vector formalism this action induces a transformation $\vec{r} \mapsto \vec{r}'$ with

$$r'_j = \mathrm{tr}[\sigma_j U\varrho U^*] = \sum_k r_k \tfrac{1}{2}\mathrm{tr}[\sigma_j U\sigma_k U^*] \equiv \sum_k R_{jk}r_k. \tag{2.40}$$

A direct calculation gives

$$R_{jk} = \tfrac{1}{2}\mathrm{tr}[\sigma_j U\sigma_k U^*] = \delta_{jk}\cos 2\eta + (1 - \cos 2\eta)n_j n_k + \epsilon_{jkl}n_l \sin 2\eta,$$

and therefore

$$\vec{r}' = \vec{r}\cos 2\eta + (1 - \cos 2\eta)(\vec{n} \cdot \vec{r})\vec{n} + \sin 2\eta(\vec{r} \times \vec{n}).$$

The reader may notice that the transformation relating to the matrix R corresponds to an orthogonal rotation (R is a 3×3 real matrix satisfying $\det R = 1$ and $R^{-1} = R^T$) around the axis $\vec{n}$ by an angle 2η. In conclusion, unitary operators induce rotations of the Bloch sphere. $\triangle$

Let us now take a more detailed look at unitary freedom. For each unitary operator U on $\mathcal{H}$, we define a linear mapping σ_U on $\mathcal{L}(\mathcal{H})$ by

$$\sigma_U(T) := UTU^*. \tag{2.41}$$

Since $UU^* = U^*U = I$, it follows that σ_U is a bijective mapping.

Proposition 2.52 For the mapping σ_U defined in (2.41), the following statements hold:

(a) $\sigma_U(E) \in \mathcal{E}(\mathcal{H})$ if and only if $E \in \mathcal{E}(\mathcal{H})$;
(b) $\sigma_U(\varrho) \in \mathcal{S}(\mathcal{H})$ if and only if $\varrho \in \mathcal{S}(\mathcal{H})$;
(c) $\sigma_U(P) \in \mathcal{S}^{\mathrm{ext}}(\mathcal{H})$ if and only if $P \in \mathcal{S}^{\mathrm{ext}}(\mathcal{H})$.

(See the list of symbols, which can be found before the list of references, near the end of the book.) Thus the restrictions of σ_U to $\mathcal{E}(\mathcal{H})$, $\mathcal{S}(\mathcal{H})$ and $\mathcal{S}^{\mathrm{ext}}(\mathcal{H})$ are bijective mappings.

Proof (a): Let us first show that the mapping σ_U preserves positivity, which means that $\sigma_U(T) \geq O$ if and only if $T \geq O$. For each $T \in \mathcal{L}(\mathcal{H})$, we have $\langle \psi | UTU^* \psi \rangle = \langle U^* \psi | T(U^* \psi) \rangle$ for all $\psi \in \mathcal{H}$. Since unitary operators are bijections on $\mathcal{H}$, the range of U^* is the whole Hilbert space $\mathcal{H}$. Thus $UTU^* \geq O$ if and only if $T \geq O$. Consequently, $\sigma_U(E) \geq O$ and $\sigma_U(I - E) \geq O$ if and only if $E \geq O, I - E \geq O$; thus the statement holds.

(b): Since $\mathcal{S}(\mathcal{H}) \subset \mathcal{E}(\mathcal{H})$, it is sufficient to show that σ_U is trace preserving. This is a direct consequence of Exercise 1.54.

(c): Since $\mathcal{S}^{\text{ext}}(\mathcal{H}) \subset \mathcal{S}(\mathcal{H})$, it is sufficient to show that σ_U preserves purity (recall Proposition 2.11). For a state ϱ, we obtain

$$\text{tr}\big[\sigma_U(\varrho)^2\big] = \text{tr}\big[U\varrho U^* U \varrho U^*\big] = \text{tr}\big[\varrho^2\big] ,$$

thus the purity of $\sigma_U(\varrho)$ is equal to the purity of ϱ. $\qquad\square$

It is an important fact that two mappings σ_U and σ_V may be the same even if the unitary operators U and V are different. This is specified in Proposition 2.53 below.

Proposition 2.53 Let U and V be unitary operators. The following conditions are equivalent:

(i) $\sigma_U = \sigma_V$;
(ii) $\sigma_U(E) = \sigma_V(E)$ for every $E \in \mathcal{E}(\mathcal{H})$;
(iii) $\sigma_U(\varrho) = \sigma_V(\varrho)$ for every $\varrho \in \mathcal{S}(\mathcal{H})$;
(iv) $\sigma_U(P) = \sigma_V(P)$ for every $P \in \mathcal{S}^{\text{ext}}(\mathcal{H})$;
(v) $U = zV$ for some $z \in \mathbb{T} = \{z \in \mathbb{C} : |z| = 1\}$.

Proof Trivially, (i)$\Rightarrow$(ii). Moreover, $\mathcal{S}(\mathcal{H})$ is a subset of $\mathcal{E}(\mathcal{H})$ and $\mathcal{S}^{\text{ext}}(\mathcal{H})$ is a subset of $\mathcal{S}(\mathcal{H})$, hence (ii)$\Rightarrow$(iii)$\Rightarrow$(iv). Since

$$\sigma_{zV}(T) = zVT(zV)^* = z\bar{z}VTV^* = \sigma_V(T),$$

we conclude that (v)$\Rightarrow$(i). Therefore, to complete the proof we need to show that (iv)$\Rightarrow$(v).

Let us first note that, on the one hand, for a one-dimensional projection P_ψ determined by a unit vector $\psi \in \mathcal{H}$ we have $\sigma_U(P_\psi) = P_{U\psi}$. On the other hand, two projections P_ψ and P_φ are the same if and only if there is a complex number $z \in \mathbb{T}$ such that $\psi = z\varphi$. Therefore, (iv) implies that there exists a mapping c from the unit vectors of $\mathcal{H}$ to $\mathbb{T}$ such that $U\psi = c(\psi)V\psi$ for all unit vectors $\psi \in \mathcal{H}$. To prove (v), we need to show that c is a constant function.

Let $\psi, \varphi \in \mathcal{H}$ be two unit vectors. We then have

$$\langle \psi | \varphi \rangle = \langle U\psi | U\varphi \rangle = \overline{c(\psi)}c(\varphi) \langle V\psi | V\varphi \rangle = \overline{c(\psi)}c(\varphi) \langle \psi | \varphi \rangle .$$

Therefore, in the case $\langle \psi | \varphi \rangle \neq 0$ we get $c(\psi) = c(\varphi)$.

Suppose then that $\langle\psi|\varphi\rangle = 0$. We define a unit vector

$$\phi = \|\psi + \varphi\|^{-1} (\psi + \varphi).$$

Since $\langle\psi|\phi\rangle \neq 0$, our earlier observation implies that $c(\phi) = c(\psi)$. Thus,

$$U\psi + U\varphi = U(\psi + \varphi) = \|\psi + \varphi\| \, U\phi = \|\psi + \varphi\| \, c(\psi)V\phi$$
$$= c(\psi)V(\psi + \varphi) = c(\psi)V\psi + c(\psi)V\varphi$$
$$= U\psi + c(\psi)V\varphi.$$

Comparing the first and the last expressions, we conclude that $U\varphi = c(\psi)V\varphi$, implying that $c(\psi) = c(\varphi)$. ☐

We will use the notation σ_U also for the restrictions of σ_U to $\mathcal{E}(\mathcal{H})$, $\mathcal{S}(\mathcal{H})$ or $\mathcal{S}^{\text{ext}}(\mathcal{H})$. We have just seen that any of these restrictions determines σ_U completely, hence this use of notation will not cause any confusion.

In the following we are going to need the antiunitary operators introduced at the end of subsection 1.2.4. For each antiunitary operator A, we define a mapping σ_A with the same formula as for unitary operators:

$$\sigma_A(T) := ATA^*. \tag{2.42}$$

Let us recall that for antiunitary operators $AA^* = A^*A = I$. It is easy to see that if T is a linear operator then ATA^* is also a linear operator. In fact, for every $\varphi, \psi \in \mathcal{H}$ and $c \in \mathbb{C}$ we obtain

$$ATA^*(\varphi + c\psi) = AT(A^*\varphi + \bar{c}A^*\psi) = A((TA^*)\varphi + \bar{c}(TA^*)\psi)$$
$$= (ATA^*)\varphi + c(ATA^*)\psi.$$

Moreover, we have

$$\|ATA^*\psi\| = \sqrt{\langle ATA^*\psi|ATA^*\psi\rangle} = \|TA^*\psi\| \leq \|T\| \, \|A^*\psi\| = \|T\| \, \|\psi\|.$$

This shows that $\sigma_A(T) \in \mathcal{L}(\mathcal{H})$ for all $T \in \mathcal{L}(\mathcal{H})$.

Exercise 2.54 Verify that the results of Propositions 2.52 and 2.53 are true when unitary operators are replaced by antiunitary operators.

Antiunitary operators describe different symmetries than unitary operators. Namely, we have the following observation.

Proposition 2.55 Let U be a unitary operator and A an antiunitary operator. Then $\sigma_U \neq \sigma_A$ on $\mathcal{S}^{\text{ext}}(\mathcal{H})$.

Proof We make the counter-assumption that $\sigma_U(P) = \sigma_A(P)$ for all $P \in \mathcal{S}^{\text{ext}}(\mathcal{H})$. This can be true only if for every unit vector $\psi \in \mathcal{H}$ there is a complex number $c(\psi)$ such that $U\psi = c(\psi)A\psi$ and $|c(\psi)| = 1$. Let $\psi, \varphi \in \mathcal{H}$ be linearly independent unit vectors, and write $\phi = \|\psi + i\varphi\|^{-1} (\psi + i\varphi)$. Then on the one hand

$$U(\psi + i\varphi) = U\psi + iU\varphi = c(\psi)A\psi + ic(\varphi)A\varphi \qquad (2.43)$$

and on the other hand

$$U(\psi + i\varphi) = \|\psi + i\varphi\| U\phi = \|\psi + i\varphi\|c(\phi)A\psi - i \|\psi + i\varphi\|c(\phi)A\varphi . \quad (2.44)$$

Since A is antiunitary, the vectors $A\psi$ and $A\varphi$ are linearly independent. Comparison of (2.43) and (2.44) gives

$$c(\psi) = \|\psi + i\varphi\|c(\phi) \qquad \text{and} \qquad c(\varphi) = -\|\psi + i\varphi\|c(\phi).$$

Thus, $c(\psi) = -c(\varphi)$. Repeating the same calculation with the vector $\phi = \|\psi + \varphi\|^{-1} (\psi + \varphi)$ leads to $c(\psi) = c(\varphi)$. This means that $c(\psi) = 0$, which conflicts with the fact that $|c(\psi)| = 1$. We conclude that the counter-assumption is false and therefore $\sigma_U \neq \sigma_A$ on $\mathcal{S}^{\text{ext}}(\mathcal{H})$. $\qquad\square$

Even if antiunitary operators define symmetry transformations in much the same way as unitary operators, these two classes have crucial differences. Perhaps the most important is related to complete positivity, which will be discussed in subsection 4.1.1. It will then become clear that antiunitary operators can describe only abstract symmetries (e.g. time inversion), not physically realizable symmetries such as rotations or translations.

Example 2.56 *(Quantum NOT gate)*
In classical information theory a NOT gate flips the value of a bit. Hence, its action is $0 \mapsto 1$ and $1 \mapsto 0$ or, more briefly, $b \mapsto b + 1 \pmod 2$. A quantum analogue of the NOT gate is defined on $\mathbb{C}^2$ as a mapping transforming every pure state P into its complement $P^\perp = I - P$. The quantum NOT gate is thus defined as

$$\mathcal{E}_{\text{NOT}}(T) = \text{tr}[T]\, I - T$$

for every $T \in \mathcal{L}(\mathbb{C}^2)$. The quantum NOT gate satisfies $\mathcal{E}_{\text{NOT}}(I) = I$. Moreover, it is invertible and

$$\begin{aligned}
\text{tr}\big[\mathcal{E}_{\text{NOT}}(E)\mathcal{E}_{\text{NOT}}(\varrho)\big] &= \text{tr}\big[(\text{tr}[E]\, I - E)(I - \varrho)\big] \\
&= \text{tr}[E]\,\text{tr}[I] + \text{tr}\big[E\varrho\big] - \text{tr}[E]\,\text{tr}\big[\varrho\big] - \text{tr}[E] \\
&= \text{tr}\big[E\varrho\big] , \qquad\qquad\qquad\qquad\qquad\qquad (2.45)
\end{aligned}$$

for any effect $E \in \mathcal{E}(\mathcal{H})$ and state $\varrho \in \mathcal{S}(\mathcal{H})$.

If $T = \frac{1}{2}(I + \vec{r} \cdot \vec{\sigma})$ then $\mathcal{E}_{\text{NOT}}(T) = \frac{1}{2}(I - \vec{r} \cdot \vec{\sigma})$. Therefore, in Bloch-sphere language $\mathcal{E}_{\text{NOT}}$ induces a transformation $\vec{r} \mapsto -\vec{r}$, which is space inversion. On the basis of Example 2.51 we conclude that there is no unitary operator U such that

$\mathcal{E}_{\text{NOT}} = \sigma_U$, because unitary operators induce only orthogonal rotations and space inversion is not an orthogonal rotation. △

Exercise 2.57 Find an antiunitary operator A such that $\mathcal{E}_{\text{NOT}} = \sigma_A$. [Hint: Recall that each antiunitary operator can be written as a product of a unitary operator and its complex conjugate.]

For each unitary or antiunitary operator U, we denote by $[U]$ the equivalence class consisting of all operators of the form zU, $z \in \mathbb{T}$. As we learned earlier, the equivalence class $[U]$ consists of those operators that define the same mapping σ_U. We denote by Σ the set of all equivalence classes $[U]$.

For each $z \in \mathbb{T}$, the operator zI is unitary and the set $\mathbb{T}I = \{zI : z \in \mathbb{T}\}$ is a subgroup of the group $\mathcal{U}(\mathcal{H}) \cup \overline{\mathcal{U}}(\mathcal{H})$. An operator zI clearly commutes with all unitary operators. However, if A is an antiunitary operator then $AzI = \bar{z}A$ but $zIA = zA$. Even if A and zI do not commute, we see that $A(zI)A^{-1} \in \mathbb{T}I$. In mathematical language, this means that $\mathbb{T}I$ is a *normal* subgroup of $\mathcal{U}(\mathcal{H}) \cup \overline{\mathcal{U}}(\mathcal{H})$. As a consequence, we can define multiplication in Σ as follows:

$$[U] \cdot [V] = [UV],$$

and in this way Σ becomes a group. In algebraic terminology Σ is the quotient group of $\mathcal{U}(\mathcal{H}) \cup \overline{\mathcal{U}}(\mathcal{H})$ by $\mathbb{T}I$; this is written as

$$\Sigma = \mathcal{U}(\mathcal{H}) \cup \overline{\mathcal{U}}(\mathcal{H})/\mathbb{T}I .$$

As we will see in the following two subsections, Σ describes the innate symmetry of quantum theory.

2.3.2 State automorphisms

Symmetry transformations, by definition, preserve all the essential features and properties of the object under consideration. The characteristic feature for a set of states is the possibility of forming mixtures. This is reflected as the convex structure of $\mathcal{S}(\mathcal{H})$. The symmetry transformations on $\mathcal{S}(\mathcal{H})$, also called the automorphisms of $\mathcal{S}(\mathcal{H})$, are therefore taken to be the convex structure preserving bijections.

Definition 2.58 A function $s : \mathcal{S}(\mathcal{H}) \to \mathcal{S}(\mathcal{H})$ is a *state automorphism* if:

- s is a bijection;
- $\forall \varrho_1, \varrho_2 \in \mathcal{S}(\mathcal{H}), \lambda \in [0, 1] : s(\lambda\varrho_1 + (1 - \lambda)\varrho_2) = \lambda s(\varrho_1) + (1 - \lambda)s(\varrho_2)$.

We denote by $\text{Aut}(\mathcal{S}(\mathcal{H}))$ the set of all state automorphisms.

It is clear that a composition of two state automorphisms is also a state automorphism. This fact taken together with Proposition 2.59 below means that $\text{Aut}(\mathcal{S}(\mathcal{H}))$ is a group.

Proposition 2.59 Let s be a state automorphism. Then the inverse mapping s^{-1} is also a state automorphism.

Proof The state automorphism s is a bijection on $\mathcal{S}(\mathcal{H})$. Hence, the inverse mapping s^{-1} exists and is a bijection. For all $\varrho_1, \varrho_2 \in \mathcal{S}(\mathcal{H})$, $\lambda \in [0, 1]$, we obtain

$$s(\lambda s^{-1}(\varrho_1) + (1 - \lambda)s^{-1}(\varrho_2)) = \lambda\varrho_1 + (1 - \lambda)\varrho_2 = s(s^{-1}(\lambda\varrho_1 + (1 - \lambda)\varrho_2)).$$

Since s is a bijection, it is injective. Therefore,

$$\lambda s^{-1}(\varrho_1) + (1 - \lambda)s^{-1}(\varrho_2) = s^{-1}(\lambda\varrho_1 + (1 - \lambda)\varrho_2).$$

This shows that s^{-1} is a state automorphism. $\qquad\qquad\qquad\qquad\qquad\square$

It follows from our discussion in subsection 2.3.1 that mappings σ_U, where U is either a unitary or an antiunitary operator, are state automorphisms. We are now going to prove that each state automorphism s is of the form σ_U for some unitary or antiunitary operator U. This will be a somewhat lengthy and technical effort, but the basic ingredient of our proof is the following simple property of state automorphisms.

Proposition 2.60 Let s be a state automorphism. Then a state ϱ is pure if and only if $s(\varrho)$ is pure.

Proof We prove the claim by showing that a state ϱ is mixed if and only if $s(\varrho)$ is mixed. Let ϱ be a mixed state, so that $\varrho = \lambda\varrho_1 + (1 - \lambda)\varrho_2$ for some $\varrho_1 \neq \varrho_2$ and $0 < \lambda < 1$. Then $s(\varrho) = \lambda s(\varrho_1) + (1 - \lambda)s(\varrho_2)$ and $s(\varrho_1) \neq s(\varrho_2)$. Thus, $s(\varrho)$ is mixed. Employing the inverse automorphism s^{-1} we can prove in the same way that if $s(\varrho)$ is mixed then the state $s^{-1}s(\varrho) = \varrho$ is mixed. $\qquad\square$

Further, we will need the following technical results.

Lemma 2.61 Let $s \in \mathrm{Aut}(\mathcal{S}(\mathcal{H}))$. There exists a unique linear mapping $\tilde{s} : \mathcal{T}(\mathcal{H}) \to \mathcal{T}(\mathcal{H})$ such that $\tilde{s}(\varrho) = s(\varrho)$ for every $\varrho \in \mathcal{S}(\mathcal{H})$. Moreover $\tilde{s}$ is positive and invertible and preserves the trace of any selfadjoint trace class operator.

Proof We can extend s to a linear mapping $\tilde{s}$ on $\mathcal{T}(\mathcal{H})$ in a way similar way to that in which E was extended to $\tilde{E}$ in the proof of Proposition 2.30. For each nonzero positive operator $T \in \mathcal{T}(\mathcal{H})$, we define

$$\tilde{s}(T) := \mathrm{tr}[T]\, s(T/\mathrm{tr}[T])$$

and the steps to selfadjoint and general operators follow the steps of the proof of Proposition 2.30. It follows from the construction that $\tilde{s}$ is positive.

To see that the extension is unique, we recall that every $T \in \mathcal{T}(\mathcal{H})$ can be written as a sum $T = T_R + iT_I$ of two selfadjoint linear operators, and every selfadjoint trace class operator is a sum of one-dimensional projections (see subsection 1.3.3). Therefore, a linear function on $\mathcal{T}(\mathcal{H})$ is completely determined by its action on states.

Since $s^{-1} \in \text{Aut}(\mathcal{S}(\mathcal{H}))$, we can extend s^{-1} to a unique linear mapping on $\mathcal{T}(\mathcal{H})$. By our earlier observation this type of extension is unique, and it follows that the extension of s^{-1} is the inverse of $\tilde{s}$. In particular, $\tilde{s}$ is invertible.

Finally, we need to show that $\tilde{s}$ preserves the trace of any selfadjoint trace class operator. If $T \in \mathcal{T}_s(\mathcal{H})$ then

$$\begin{aligned}
\|\tilde{s}(T)\|_{\text{tr}} &= \|\tilde{s}(T^+ - T^-)\|_{\text{tr}} = \|\tilde{s}(T^+) - \tilde{s}(T^-)\|_{\text{tr}} \\
&\leq \|\tilde{s}(T^+)\|_{\text{tr}} + \|\tilde{s}(T^-)\|_{\text{tr}} = \text{tr}[\tilde{s}(T^+)] + \text{tr}[\tilde{s}(T^-)] \\
&= \text{tr}[T^+] + \text{tr}[T^-] = \text{tr}[T^+ + T^-] = \text{tr}[|T|] = \|T\|_{\text{tr}}.
\end{aligned}$$

A similar derivation gives $\|\tilde{s}^{-1}(T)\|_{\text{tr}} \leq \|T\|_{\text{tr}}$. These inequalities taken together give

$$\|T\|_{\text{tr}} = \|\tilde{s}^{-1}(\tilde{s}(T))\|_{\text{tr}} \leq \|\tilde{s}(T)\|_{\text{tr}} \leq \|T\|_{\text{tr}},$$

which means that $\|\tilde{s}(T)\|_{\text{tr}} = \|T\|_{\text{tr}}$ and hence proves the claim.   □

Lemma 2.62 Let $P_1, P_2 \in \mathcal{S}^{\text{ext}}(\mathcal{H})$. Then $0 \leq \text{tr}[P_1 P_2] \leq 1$ and

$$\|P_1 - P_2\|_{\text{tr}} = 2\sqrt{1 - \text{tr}[P_1 P_2]}. \tag{2.46}$$

The pure states P_1 and P_2 are orthogonal if and only if $\|P_1 - P_2\|_{\text{tr}} = 2$.

Proof If $P_1 = P_2$ the claim is trivially true. Hence, we will assume that $P_1 \neq P_2$. Fix unit vectors $\psi, \varphi \in \mathcal{H}$ such that $P_1 = P_\psi$ and $P_2 = P_\varphi$. Since $\text{tr}[P_1 P_2] = |\langle \psi | \varphi \rangle|^2$, it follows that $\text{tr}[P_1 P_2] \geq 0$ and the Cauchy–Schwarz inequality implies that $\text{tr}[P_1 P_2] \leq 1$.

For a vector $\phi \in \mathcal{H}$, we obtain

$$(P_1 - P_2)\phi = (P_\psi - P_\varphi)\phi = \langle \psi | \phi \rangle \psi - \langle \varphi | \phi \rangle \varphi.$$

This shows that if ϕ is an eigenvector of $P_1 - P_2$, it must be of the form $\phi = c_1 \psi + c_2 \varphi$ for some $c_1, c_2 \in \mathbb{C}$. Inserting this expression into the eigenvalue equation

$$(P_1 - P_2)\phi = \lambda \phi$$

gives

$$\begin{aligned}
\lambda c_1 &= c_1 + c_2 \langle \psi | \varphi \rangle, \\
\lambda c_2 &= -c_2 - c_1 \langle \varphi | \psi \rangle.
\end{aligned}$$

This leads to the two solutions

$$\lambda_\pm = \pm\sqrt{1 - |\langle \psi | \varphi \rangle|^2}.$$

Therefore

$$\| P_1 - P_2 \|_{\text{tr}} = \text{tr}[|P_1 - P_2|] = |\lambda_+| + |\lambda_-|$$
$$= 2\sqrt{1 - |\langle \psi | \varphi \rangle|^2} = 2\sqrt{1 - \text{tr}[P_1 P_2]}.$$

We saw in subsection 1.2.5 that two projections P_1 and P_2 are orthogonal if and only if $\text{tr}[P_1 P_2] = 0$ and this fact proves the last claim. $\qquad \square$

With this preparation we are ready to prove the main theorem of this section. We follow the proof of Davies [51].

Theorem 2.63 Each state automorphism s is of the form $s = \sigma_U$ for a unitary or antiunitary operator U. The operator U is unique up to the equivalence class $[U]$.

Proof Let $f : \mathcal{T}(\mathcal{H}) \to \mathcal{T}(\mathcal{H})$ be a linear mapping that is positive, trace-norm-preserving on $\mathcal{T}_s(\mathcal{H})$ and invertible and for which $f(\mathcal{S}^{\text{ext}}(\mathcal{H})) \subseteq \mathcal{S}^{\text{ext}}(\mathcal{H})$. We will show that $f = \sigma_U$ for some unitary or antiunitary operator U and that U is unique up to the equivalence class $[U]$. The first claim then follows from Proposition 2.60 and Lemma 2.61, while the second claim follows from Proposition 2.53.

Let $\varphi, \varphi' \in \mathcal{H}$ be two orthogonal unit vectors. Since f maps pure states into pure states, there are unit vectors $\psi, \psi' \in \mathcal{H}$ such that

$$f(|\varphi\rangle\langle\varphi|) = |\psi\rangle\langle\psi|, \qquad f(|\varphi'\rangle\langle\varphi'|) = |\psi'\rangle\langle\psi'|. \qquad (2.47)$$

Moreover, as f preserves the trace of any selfadjoint trace class operator, it follows from Lemma 2.62 that ψ and ψ' are orthogonal. We can identify the two dimensional subspaces $\text{span}\{\varphi, \varphi'\}$ and $\text{span}\{\psi, \psi'\}$ with $\mathbb{C}^2$. The function f restricted to the trace class operators on $\text{span}\{\varphi, \varphi'\}$ is hence a linear mapping on $\mathcal{T}(\mathbb{C}^2)$. It induces a transformation $\vec{r} \mapsto \vec{r}'$ on the Bloch sphere, and from the properties of f it follows that this transformation is invertible and affine and maps the surface of the Bloch sphere into itself. But this kind of transformation is either orthogonal rotation, space inversion or their combination. We may then conclude from Examples 2.51 and 2.56 that the restriction of f is of the form σ_V for some unitary or antiunitary operator V on $\mathbb{C}^2$.

Suppose that V is a unitary operator. Inserting $f = \sigma_V$ into (2.47) gives $V\varphi = \langle \psi | V\varphi \rangle \psi$ and $V\varphi' = \langle \psi' | V\varphi' \rangle \psi'$. Since V preserves the norm, we have $|\langle \psi | V\varphi \rangle| = |\langle \psi' | V\varphi' \rangle| = 1$. We obtain

$$f(|\varphi\rangle\langle\varphi'|) = V|\varphi\rangle\langle\varphi'|V^* = z|\psi\rangle\langle\psi'| \qquad (2.48)$$

for some $z \in \mathbb{C}$ with $|z| = 1$. However, if V is antiunitary then a similar derivation leads to

$$f(|\varphi\rangle\langle\varphi'|) = z|\psi'\rangle\langle\psi| \tag{2.49}$$

for some $z \in \mathbb{C}$ with $|z| = 1$. Now let $\{\varphi_i\}$ be an orthonormal basis for $\mathcal{H}$. For each i, there is a unit vector $\psi_i \in \mathcal{H}$ such that

$$f(|\varphi_i\rangle\langle\varphi_i|) = |\psi_i\rangle\langle\psi_i|,$$

and two vectors $\psi_i, \psi_j \in \mathcal{H}$ are orthogonal whenever $i \neq j$. Our previous discussion shows that for $\varphi_i \neq \varphi_j$ we have

$$f(|\varphi_i\rangle\langle\varphi_j|) = z_{ij}|\psi_i\rangle\langle\psi_j| \quad \text{or} \quad f(|\varphi_i\rangle\langle\varphi_j|) = z_{ij}|\psi_j\rangle\langle\psi_i| \tag{2.50}$$

for complex numbers z_{ij} with $|z_{ij}| = 1$. As the linear span of the operators $|\varphi_i\rangle\langle\varphi_j|$ is dense in $\mathcal{T}(\mathcal{H})$, the linear span of the operators $|\psi_i\rangle\langle\psi_j|$ is also dense in $\mathcal{T}(\mathcal{H})$ (since f is continuous and surjective). It follows that the set $\{\psi_i\}$ must be an orthonormal basis for $\mathcal{H}$.

We claim that the option in (2.50) (either switching i and j or not) holds for all pairs. To see this, let us assume that there are three different vectors $\varphi_i, \varphi_j, \varphi_k$ such that

$$f(|\varphi_i\rangle\langle\varphi_j|) = z_{ij}|\psi_i\rangle\langle\psi_j| \quad \text{and} \quad f(|\varphi_i\rangle\langle\varphi_k|) = z_{ik}|\psi_k\rangle\langle\psi_i|.$$

The unit vector $\varphi_i' \equiv \frac{1}{\sqrt{2}}(\varphi_j + \varphi_k)$ is orthogonal to φ_i and, by the linearity of f, we get

$$f(|\varphi_i\rangle\langle\varphi_i'|) = \frac{1}{\sqrt{2}}z_{ij}|\psi_i\rangle\langle\psi_j| + \frac{1}{\sqrt{2}}z_{ik}|\psi_k\rangle\langle\psi_i|.$$

The resulting operator is not rank-1, which contradicts (2.48), (2.49). We conclude that if the first index in $|\varphi_i\rangle\langle\varphi_j|$ is kept fixed then the order of the indices in the image $f(|\varphi_i\rangle\langle\varphi_j|)$ is the same for all j. But analogous reasoning shows that if the second index in $|\varphi_i\rangle\langle\varphi_j|$ is kept fixed then the order of the indices in the image $f(|\varphi_i\rangle\langle\varphi_j|)$ is the same for all i. Therefore, either f switches the indices in $f(|\varphi_i\rangle\langle\varphi_j|)$ for all i, j, or for none.

We first consider the nonswitching case, i.e. $f(|\varphi_i\rangle\langle\varphi_j|) = z_{ij}|\psi_i\rangle\langle\psi_j|$ for all i, j. If $\dim\mathcal{H} = d < \infty$, we define a unit vector $\phi = \frac{1}{\sqrt{d}}\sum_{i=1}^{d}\varphi_i$. Then on the one hand

$$f(|\phi\rangle\langle\phi|) = \frac{1}{d}\sum_{i,j} z_{ij}|\psi_i\rangle\langle\psi_j|.$$

On the other hand f satisfies $f(\mathcal{S}^{\text{ext}}(\mathcal{H})) \subseteq \mathcal{S}^{\text{ext}}(\mathcal{H})$, and hence $f(|\phi\rangle\langle\phi|)$ is a one-dimensional projection. Thus there is a unit vector $\eta = \frac{1}{\sqrt{d}}\sum_i c_i\psi_i$ such that

$$f(|\phi\rangle\langle\phi|) = |\eta\rangle\langle\eta| = \frac{1}{d}\sum_{i,j} \bar{c}_i c_j|\psi_i\rangle\langle\psi_j|.$$

A comparison of these two expressions for $f(|\phi\rangle\langle\phi|)$ shows that $z_{ij} = \bar{c}_i c_j$. As $|z_{ij}| = 1$ for all i, j, we also have $|c_i| = 1$ for all i. For each i, we define $U\varphi_i := c_i\psi_i$ and extend U to a linear mapping on $\mathcal{H}$. Since both $\{\varphi_i\}$ and $\{c_i\psi_i\}$ are orthonormal bases, U is a unitary operator on $\mathcal{H}$ (see Proposition 1.49). If $\dim\mathcal{H} = \infty$, we modify the definition of ϕ to $\phi = \sum_{i=1}^{\infty} \frac{1}{2^{i/2}}\varphi_i$. Otherwise the same reasoning applies.

Let us then consider the switching case, i.e. $f(|\varphi_i\rangle\langle\varphi_j|) = z_{ij}|\psi_j\rangle\langle\psi_i|$ for all i, j. Let J be the complex conjugate operator relating to the orthonormal basis $\{\psi_j\}$ (see the discussion at the end of subsection 1.2.4). The composite mapping $T \mapsto Jf(T)J$ is now of the nonswitching type, hence the previous analysis shows that there is a unitary operator U such that $Jf(T)J = UTU^*$. But then $f(T) = (JU)T(JU)^*$, and so JU must be an antiunitary operator. $\square$

Theorem 2.63 means that state automorphisms are in one-to-one correspondence with the elements in Σ. This relation is also a group homomorphism; if s_1 and s_2 are related to equivalence classes $[U_1]$ and $[U_2]$ then the composition $s_1 s_2$ is related to $[U_1 U_2]$. We conclude that the groups $\mathrm{Aut}(\mathcal{S}(\mathcal{H}))$ and Σ are isomorphic.

2.3.3 Pure state automorphisms and Wigner's theorem

On the basis of Section 2.2, it seems reasonable to require that automorphisms of pure states preserve the superposition feature. A numerical quantity related to superpositions is $\mathrm{tr}[P_1 P_2]$ (see Proposition 2.47). One possibility is thus to define automorphisms in the following way.

Definition 2.64 A function $p : \mathcal{S}^{\mathrm{ext}}(\mathcal{H}) \rightarrow \mathcal{S}^{\mathrm{ext}}(\mathcal{H})$ is a *pure-state automorphism* if:

- p is a bijection;
- $\mathrm{tr}\big[p(P_1)p(P_2)\big] = \mathrm{tr}[P_1 P_2]$.

We denote by $\mathrm{Aut}(\mathcal{S}^{\mathrm{ext}}(\mathcal{H}))$ the set of all pure-state automorphisms.

The fundamental result in the theory of automorphism groups is the following statement, known as Wigner's theorem.

Theorem 2.65 (*Wigner's theorem*)
Let $p \in \mathrm{Aut}(\mathcal{S}^{\mathrm{ext}}(\mathcal{H}))$. There is a unitary or antiunitary operator U such that

$$p(P) = \sigma_U(P) \qquad \forall P \in \mathcal{S}^{\mathrm{ext}}(\mathcal{H}).$$

The operator U is unique up to the equivalence class $[U]$.

This result was first stated by Wigner in his book [144]. The first rigorous proof was perhaps given by Bargmann [10], and since then many different proofs have been presented. A reader-friendly proof of Wigner's theorem can be found in e.g. [38].

To underline the fundamental role of Wigner's theorem, we note that it is stronger than Theorem 2.63 in the sense that the latter is quite easy to recover from the former. We need only the following two simple observations.

Proposition 2.66 Let $s_1, s_2 \in \mathrm{Aut}(\mathcal{S}(\mathcal{H}))$ and assume that $s_1(P) = s_2(P)$ for every $P \in \mathcal{S}^{\mathrm{ext}}(\mathcal{H})$. Then $s_1 = s_2$.

Proof Write $s = s_1 s_2^{-1}$. The assumption means that $s(P) = P$ for every $P \in \mathcal{S}^{\mathrm{ext}}(\mathcal{H})$ and we have to prove that s is the identity mapping on $\mathcal{S}(\mathcal{H})$. For each $\varrho \in \mathcal{S}(\mathcal{H})$, we have a canonical convex decomposition $\varrho = \sum_i \lambda_i P_i$ of ϱ into pure states P_i. Since s is continuous, we obtain

$$s(\varrho) = s\left(\sum_i \lambda_i P_i\right) = \sum_i \lambda_i s(P_i) = \sum_i \lambda_i P_i = \varrho.$$

This shows that s is the identity mapping.   □

Proposition 2.67 Let s be a state automorphism. The restriction of s to $\mathcal{S}^{\mathrm{ext}}(\mathcal{H})$ is a pure-state automorphism.

Proof By Proposition 2.60, the restriction of s is a mapping from $\mathcal{S}^{\mathrm{ext}}(\mathcal{H})$ to $\mathcal{S}^{\mathrm{ext}}(\mathcal{H})$. The same is true for the inverse mapping s^{-1}, and therefore the restriction of s is a bijection on $\mathcal{S}^{\mathrm{ext}}(\mathcal{H})$.

Let $\tilde{s}$ be the extension of s to $\mathcal{T}(\mathcal{H})$ as in Lemma 2.61, and let $P_1, P_2 \in \mathcal{S}^{\mathrm{ext}}(\mathcal{H})$. Using Lemma 2.62 we obtain

$$2\sqrt{1 - \mathrm{tr}[P_1 P_1]} = \|P_1 - P_2\|_{\mathrm{tr}} = \|\tilde{s}(P_1 - P_2)\|_{\mathrm{tr}} = \|\tilde{s}(P_1) - \tilde{s}(P_2)\|_{\mathrm{tr}}$$
$$= \|s(P_1) - s(P_2)\|_{\mathrm{tr}} = 2\sqrt{1 - \mathrm{tr}[s(P_1)s(P_1)]}.$$

Hence $\mathrm{tr}[s(P_1)s(P_2)] = \mathrm{tr}[P_1 P_2]$.   □

We conclude that each state automorphism determines a pure-state automorphism, and that two different state automorphisms determine different pure-state automorphisms. Therefore Theorem 2.63 can be recovered from Wigner's theorem.

2.4 Composite systems

Suppose that we have two quantum systems A and B and let $\mathcal{H}_A$ and $\mathcal{H}_B$ be the Hilbert spaces used in the mathematical description of these systems. The question is, what Hilbert space should we associate with the composite system $A + B$?

The answer depends on whether A and B are distinguishable. To make things simpler we will assume that systems A and B are of different kinds and hence distinguishable. For the purposes of this section it is helpful to recall the tensor product introduced in subsection 1.3.5.

2.4.1 System versus subsystems

We assume that A and B are identifiable parts of the compound system $A + B$ and call them *subsystems* of $A + B$. The assumption means, in particular, that we can manipulate systems A and B separately and that we are able to perform experiments addressing the properties of A and B individually. Suppose that we have an effect E_A on $\mathcal{H}_A$ and an effect E_B on $\mathcal{H}_B$. These effects correspond to measurements on the systems A and B, respectively. Hence, by the assumption we should have an effect $\gamma(E_A, E_B)$ on $\mathcal{H}_{AB}$ that describes both these separate measurements on A and B. In other words, we require that there is a mapping γ from $\mathcal{E}(\mathcal{H}_A) \times \mathcal{E}(\mathcal{H}_B)$ to $\mathcal{E}(\mathcal{H}_{AB})$.

In a similar way, for separate preparations of subsystems we have a mapping $\bar{\gamma}$ from $\mathcal{S}(\mathcal{H}_A) \times \mathcal{S}(\mathcal{H}_B)$ to $\mathcal{S}(\mathcal{H}_{AB})$. If the measurements and preparations are made separately, the systems are statistically independent and we should have

$$\text{tr}\big[\bar{\gamma}(\varrho_A, \varrho_B)\gamma(E_A, E_B)\big] = \text{tr}\big[\varrho_A E_A\big] \, \text{tr}\big[\varrho_B E_B\big] . \tag{2.51}$$

The question of the description of a compound system can thus be approached by searching for a suitable Hilbert space $\mathcal{H}_{AB}$ and mappings γ and $\bar{\gamma}$.

Let us make a trial by choosing $\mathcal{H}_{AB} = \mathcal{H}_A \otimes \mathcal{H}_B$ and setting $\gamma(E_A, E_B) = E_A \otimes E_B$ and $\bar{\gamma}(\varrho_A, \varrho_B) = \varrho_A \otimes \varrho_B$. The properties of the tensor product guarantee that condition (2.51) is satisfied. This motivates the choice of the tensor product Hilbert space $\mathcal{H}_A \otimes \mathcal{H}_B$ as a mathematical description of the compound system $A + B$. We accept this choice without further investigation.

A curious thing arises from the properties of the tensor product. Namely, an operator T on $\mathcal{H}_A \otimes \mathcal{H}_B$ need not be of the product form $T = T_A \otimes T_B$, so how should we understand those states and effects of the compound system that are not simple products? Following the picture of the basic statistical framework it is clear that convex combinations (mixtures) of separate preparations and effects are no longer of product form. This is no surprise. However, it is a remarkable consequence of the tensor product algebra on Hilbert spaces that the states and effects of composite Hilbert spaces contain also elements that cannot be understood as mixtures of factorized states and effects, respectively. In turns out that states of the two subsystems can be intertwined in a strange way, in which case the compound state is called *entangled*. Entanglement will be our topic in Chapter 6. In the present section we will focus on elementary properties of the tensor product as a description of compound systems.

The physical meaning of the following definition will become clear shortly.

Definition 2.68 The *partial trace* over the system A is the linear mapping

$$\mathrm{tr}_A : \mathcal{T}(\mathcal{H}_A \otimes \mathcal{H}_B) \to \mathcal{T}(\mathcal{H}_B)$$

satisfying

$$\mathrm{tr}[\mathrm{tr}_A[T]\, E] = \mathrm{tr}[T(I \otimes E)] \tag{2.52}$$

for all $T \in \mathcal{T}(\mathcal{H}_A \otimes \mathcal{H}_B)$ and $E \in \mathcal{L}(\mathcal{H}_B)$. We define the partial trace tr_B over the subsystem B in a similar way.

It may be worthwhile to notice that we use the same notation $\mathrm{tr}[\cdot]$ for the trace in all Hilbert spaces. In particular, the trace on the left-hand side of (2.52) is calculated in $\mathcal{H}_B$ while the trace on the right-hand side is calculated in $\mathcal{H}_A \otimes \mathcal{H}_B$.

Example 2.69 Suppose $T \in \mathcal{T}(\mathcal{H}_A \otimes \mathcal{H}_B)$ is of the product form $T = T_A \otimes T_B$. Then the defining condition (2.52) gives

$$\mathrm{tr}[\mathrm{tr}_A[T_A \otimes T_B]\, E] = \mathrm{tr}[T_A]\, \mathrm{tr}[T_B E] \ .$$

Since this holds for every $E \in \mathcal{L}(\mathcal{H}_B)$, we conclude that

$$\mathrm{tr}_A[T_A \otimes T_B] = \mathrm{tr}[T_A]\, T_B \ , \tag{2.53}$$

and analogously,

$$\mathrm{tr}_B[T_A \otimes T_B] = \mathrm{tr}[T_B]\, T_A \ . \tag{2.54}$$

Note that if T_A and T_B are states then $\mathrm{tr}_A[T_A \otimes T_B] = T_B$ and $\mathrm{tr}_B[T_A \otimes T_B] = T_A$. $\triangle$

Actually, we have not yet shown that the partial trace mapping even exists. To do this, we give a formula for calculating the partial trace. Fix orthonormal bases $\{\psi_j\}$ and $\{\varphi_k\}$ for $\mathcal{H}_A$ and $\mathcal{H}_B$, respectively. Then $\{\psi_j \otimes \varphi_k\}$ is an orthonormal basis for $\mathcal{H}_A \otimes \mathcal{H}_B$. If $T \in \mathcal{T}(\mathcal{H}_A \otimes \mathcal{H}_B)$, we can write

$$T = ITI = \sum_{j,k} \sum_{m,n} |\psi_j \otimes \varphi_k\rangle \langle \psi_j \otimes \varphi_k | T \psi_m \otimes \varphi_n\rangle \langle \psi_m \otimes \varphi_n |$$

$$= \sum_{j,k} \sum_{m,n} \langle \psi_j \otimes \varphi_k | T \psi_m \otimes \varphi_n\rangle\, |\psi_j\rangle\langle\psi_m| \otimes |\varphi_k\rangle\langle\varphi_n| \ . \tag{2.55}$$

We have seen that (2.53) holds for tr_A, and this leads to

$$\mathrm{tr}_A\big[|\psi_j\rangle\langle\psi_m| \otimes |\varphi_k\rangle\langle\varphi_n|\big] = \mathrm{tr}\big[|\psi_j\rangle\langle\psi_m|\big]\, |\varphi_k\rangle\langle\varphi_n| = \delta_{jm}|\varphi_k\rangle\langle\varphi_n| \ .$$

Hence, we obtain

$$\text{tr}_A[T] = \sum_{j,k,n} \langle \psi_j \otimes \varphi_k | T \psi_j \otimes \varphi_n \rangle |\varphi_k\rangle\langle\varphi_n|. \tag{2.56}$$

One can now verify that $\text{tr}_A[T]$, as given in (2.56), satisfies the defining condition (2.52) for every $E \in \mathcal{L}(\mathcal{H}_B)$. Similarly we obtain

$$\text{tr}_B[T] = \sum_{j,k,m} \langle \psi_j \otimes \varphi_k | T \psi_m \otimes \varphi_k \rangle |\psi_j\rangle\langle\psi_m|. \tag{2.57}$$

In many cases one can use the defining condition (2.52) directly to calculate the partial trace; nevertheless, sometimes it is easier to apply formulas (2.56) and (2.57).

Exercise 2.70 Let $T \in \mathcal{T}(\mathcal{H}_A \otimes \mathcal{H}_B)$ and $\psi \in \mathcal{H}_A$. Prove that

$$\langle \psi | \text{tr}_B[T] \psi \rangle = \sum_k \langle \psi \otimes \varphi_k | T \psi \otimes \varphi_k \rangle \tag{2.58}$$

for every orthonormal basis $\{\varphi_k\}$ of $\mathcal{H}_B$. [Hint: Use (2.57).]

The following proposition shows that the partial trace preserves the physically relevant properties of operators. In particular, the partial trace of a state is again a state.

Proposition 2.71 Let $T \in \mathcal{T}(\mathcal{H}_A \otimes \mathcal{H}_B)$. Then:

(a) $\text{tr}[T] = \text{tr}[\text{tr}_A[T]] = \text{tr}[\text{tr}_B[T]]$;
(b) $T \geq O$ implies that $\text{tr}_A[T] \geq O$ and $\text{tr}_B[T] \geq O$.

Proof (a): Choose $E = I$ in (2.52).
(b): Assume that $T \geq O$. Fix a unit vector $\eta \in \mathcal{H}_B$ and choose $E = P_\eta$ in (2.52). We then obtain

$$\langle \eta | \text{tr}_A[T] \eta \rangle = \text{tr}\big[\text{tr}_A[T] P_\eta\big] = \text{tr}\big[T (I \otimes P_\eta)\big] = \text{tr}\big[(I \otimes P_\eta) T (I \otimes P_\eta)\big]$$
$$= \text{tr}\Big[(T^{\frac{1}{2}}(I \otimes P_\eta))^*(T^{\frac{1}{2}}(I \otimes P_\eta))\Big] \geq 0.$$

This shows that $\text{tr}_A[T] \geq O$. $\qquad\square$

Effects of the form $E \otimes I$ are interpreted as corresponding to experiments measuring the properties of subsystem A only. Since the identity $\text{tr}\big[(E \otimes I)\varrho\big] = \text{tr}\big[\text{tr}_B[\varrho] E\big]$ holds for a given state of the composite system ϱ and all effects $E \in \mathcal{E}(\mathcal{H}_A)$ defined on the subsystem A, it is natural to identify the state $\text{tr}_B[\varrho]$ with the state of the subsystem A. Hence, a state ϱ of the composite system determines the states of its subsystems via the partial trace.

Definition 2.72 Let $\varrho \in \mathcal{S}(\mathcal{H}_A \otimes \mathcal{H}_B)$ be a state of the composite system $A + B$. Then the operators $\mathrm{tr}_B[\varrho]$ and $\mathrm{tr}_A[\varrho]$ describe the states of the subsystems A and B, respectively. The states $\mathrm{tr}_B[\varrho]$ and $\mathrm{tr}_A[\varrho]$ are called *reduced states*, and $\varrho_{AB} \equiv \varrho$ is a *joint state*.

Assuming that the subsystems A and B are described by states ϱ_A and ϱ_B, respectively, we can ask the following question: what are the possible joint states ϱ_{AB} of the composite system? Clearly, one possibility is that $\varrho_{AB} = \varrho_A \otimes \varrho_B$ with $\varrho_A = \mathrm{tr}_B[\varrho]$ and $\varrho_B = \mathrm{tr}_A[\varrho]$. Generally, however, there are also other possible choices. This means that the knowledge of the states of the subsystems A and B does not enable one to specify the state of the composite system. The following proposition describes one important exception, i.e. a case when reduced states specify the joint state completely.

Proposition 2.73 Let ϱ_{AB} be a state of the composite system $A + B$. If the reduced states $\varrho_A = \mathrm{tr}_B[\varrho_{AB}]$ and $\varrho_B = \mathrm{tr}_A[\varrho_{AB}]$ are pure states then the joint state ϱ_{AB} is of the product form $\varrho_{AB} = \varrho_A \otimes \varrho_B$.

Proof Suppose that $\varrho_A = P_\psi$ for some unit vector $\psi \in \mathcal{H}_A$ and that $\varrho_B = P_\varphi$ for some unit vector $\varphi \in \mathcal{H}_B$. We want to show that $\varrho_{AB} = P_\psi \otimes P_\varphi$. We choose orthonormal bases $\{\psi_j\}$ and $\{\varphi_k\}$ for $\mathcal{H}_A$ and $\mathcal{H}_B$, respectively, such that $\psi_1 = \psi$ and $\varphi_1 = \varphi$. Using (2.58) and a similar formula for tr_A we obtain

$$1 = \sum_k \langle \psi_1 \otimes \varphi_k | \varrho_{AB} \psi_1 \otimes \varphi_k \rangle = \sum_j \langle \psi_j \otimes \varphi_1 | \varrho_{AB} \psi_j \otimes \varphi_1 \rangle .$$

Since $\langle \psi_j \otimes \varphi_k | \varrho_{AB} \psi_j \otimes \varphi_k \rangle \geq 0$ and $\sum_{j,k} \langle \psi_j \otimes \varphi_k | \varrho_{AB} \psi_j \otimes \varphi_k \rangle = \mathrm{tr}[\varrho_{AB}] = 1$, it follows that

$$\langle \psi_j \otimes \varphi_k | \varrho_{AB} \psi_j \otimes \varphi_k \rangle = \delta_{1j}\delta_{1k} .$$

By Proposition 1.35 we conclude that $\varrho_{AB}\psi_j \otimes \varphi_k = 0$ if $j \neq 1$, $k \neq 1$. Applying this result to (2.55) we get $\varrho_{AB} = \langle \psi_1 \otimes \varphi_1 | \varrho_{AB} \psi_1 \otimes \varphi_1 \rangle P_\psi \otimes P_\varphi$. From $\mathrm{tr}[\varrho_{AB}] = 1$ it follows that $\varrho_{AB} = P_\psi \otimes P_\varphi$.   □

The physical content of Proposition 2.73 is the following. Suppose that two physicists, Alice and Bob, share a composite system in an unknown state ϱ_{AB}. The situation could be, for instance, that they each receive photons from the same unknown source. We assume that both are skilled experimentalists; hence Alice can determine the reduced state $\varrho_A = \mathrm{tr}_B[\varrho_{AB}]$ and Bob can determine the reduced state $\varrho_B = \mathrm{tr}_A[\varrho_{AB}]$. If they both notice that their reduced states are pure then, by communicating this information, they know that the composite system is of product form. This implies that the photon source can be thought of as two uncorrelated and independent sources, one sending photons to Alice and one to Bob.

2.4.2 State purification

Mixed states are convex mixtures of pure states. The concept of composite systems provides an alternative interpretation of mixed states in terms of reduced states. In the following we will see that any mixed state can be seen as a reduced state of some pure state of a composite system.

Reduced states are defined via a partial trace mapping. In opposition to the reduction in one of the subsystems, we can add an extra system, to our description at least. For this purpose, we introduce a so-called *ancillary system*, represented by some additional Hilbert space $\mathcal{H}_{anc}$.

Definition 2.74 Let ϱ be a state on $\mathcal{H}$. A pure state P on a composite system $\mathcal{H} \otimes \mathcal{H}_{anc}$ is a *purification* of ϱ if $\mathrm{tr}_{anc}[P] = \varrho$.

In a sense, purification is an inverse procedure to the partial trace; however, it is highly nonunique. Let us note that the size of the ancillary system is not limited. For instance, if P is a purification of ϱ then any compound state $P \otimes P'$, where P' is a pure state of some additional system $\mathcal{H}'$, is also a purification of ϱ.

To show that every mixed state has purifications, let $\varrho \in \mathcal{S}(\mathcal{H})$ be a mixed state and $\varrho = \sum_{j=1}^{n} p_j |\eta_j\rangle\langle\eta_j|$ a canonical convex decomposition of ϱ into orthogonal pure states. We fix a Hilbert space $\mathcal{H}_{anc}$ with $\dim \mathcal{H}_{anc} = n$ and an orthonormal basis $\{\phi_j\}_{j=1}^{n}$ for $\mathcal{H}_{anc}$. Then $\psi = \sum_{j=1}^{n} \sqrt{p_j}\, \eta_j \otimes \phi_j$ is a unit vector in $\mathcal{H} \otimes \mathcal{H}_{anc}$. The corresponding pure state is

$$P_\psi = |\psi\rangle\langle\psi| = \sum_{j=1}^{n} p_j |\eta_j\rangle\langle\eta_j| \otimes |\phi_j\rangle\langle\phi_j|,$$

and using (2.57) we see that $\mathrm{tr}_{anc}[P] = \varrho$. Therefore, P_ψ is a purification of ϱ.

This construction shows that we can purify a state $\varrho \in \mathcal{S}(\mathcal{H})$ by adding an ancillary system $\mathcal{H}_{anc}$ with dimension equal to the rank of ϱ. In particular, we can purify any state ϱ on $\mathcal{H}$ if we add an ancillary system $\mathcal{H}_{anc}$ with the same dimension as $\mathcal{H}$.

Exercise 2.75 (*Purification of the total mixture*)
Consider a finite d-dimensional Hilbert space $\mathcal{H}_d$. We recall from subsection 2.1.3 that the state $\frac{1}{d}I$ is a total mixture. Suppose that a unit vector $\psi \in \mathcal{H}_d \otimes \mathcal{H}_{anc}$ gives a purification of $\frac{1}{d}I$, i.e. that $\mathrm{tr}_{anc}[P_\psi] = \frac{1}{d}I$. We claim that any vector $\psi' = (U \otimes V)\psi$ also gives a purification of $\frac{1}{d}I$, where U, V are unitary operators acting on $\mathcal{H}$, $\mathcal{H}_{anc}$, respectively. In order to see this, it is sufficient to confirm that the identity

$$\mathrm{tr}_{anc}[(U \otimes V)\varrho(U^* \otimes V^*)] = U(\mathrm{tr}_{anc}[\varrho])U^* \tag{2.59}$$

holds for all joint states ϱ. Since in our case

$$U \operatorname{tr}_{\mathrm{anc}} \left[|\psi\rangle\langle\psi| \right] U^* = U \left(\frac{1}{d} I \right) U^* = \frac{1}{d} I ,$$

it follows that the pure state $|\psi'\rangle\langle\psi'|$ is a purification of the total mixture $\frac{1}{d} I$. Prove the identity (2.59). [Hint: Use (2.57) and recall that unitary operators map orthonormal bases into orthonormal bases.]

3

Observables

The intrinsic randomness of measurement outcomes is a key feature of quantum theory. Any experiment produces a sequence of outcomes, each outcome occurring with a certain probability depending on the particular settings of the measuring and preparation devices. Quantum theory predicts only the probabilities of measurement outcomes, not individual occurrences of particular outcomes. The mathematical concept of an *observable*, which will be introduced and analyzed in this chapter, is used to capture the statistical essence of the measurement process. The monographs of Holevo [76], Busch *et al.* [34] and de Muynck [53] are recommended general references on this subject.

Considering quantum theory as a framework for calculating measurement outcome probabilities, it has aspects which make it both a generalization and a restriction of the usual probability theory. It is a generalization in the sense that observables are described by positive operator-valued measures. These are more general than probability measures, in much the same way as matrices are more general than numbers. However, quantum theory imposes inherent restrictions on the probability distributions that we can find in quantum measurements. In this sense, it can also be seen as a restriction of the usual probability theory.

3.1 Observables as positive operator-valued measures

In Chapter 2 we introduced effects as statistical events. A prototypical example of a relevant event is 'a measurement M gives an outcome x'. This event is associated with an effect E, and the probability for the event to happen is given by the trace formula $\mathrm{tr}[\varrho E]$, where ϱ is the density operator representing the input state of the measured system.

The whole measurement process can be thought of as a collection of events, each associated with an effect. In performing a measurement we want to observe which event is realized. The mathematical description of the possible events is called an

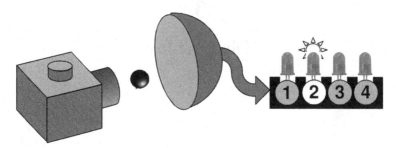

Figure 3.1 A measurement device can be thought of as a box with labelled LEDs representing the individual outcomes of our observations.

observable. In what follows we will argue that observables can be identified with so-called positive operator-valued measures (POVMs).

3.1.1 Definition and basic properties of observables

Let us consider a measurement device with n possible outcomes. We can think of the measurement apparatus as a box with n LEDs or other indicators of measurement outcomes (see Figure 3.1). Each event 'LED j flashes' is represented as an effect E_j, and the probability that this event occurs when the system is in a state ϱ is $\text{tr}[\varrho E_j]$. We will assume that at most one LED can flash at a time, hence

$$\text{tr}[\varrho E_1] + \text{tr}[\varrho E_2] + \cdots + \text{tr}[\varrho E_n] \leq 1$$

is satisfied for all states ϱ. If the inequality is saturated then some LED flashes every time we perform a measurement. In this case the set of effects $E_1, \ldots, E_n$ is complete and no additional outcome is possible. Otherwise, we could add one more effect $E_{n+1} := I - \sum_{j=1}^{n} E_j$ to count missed flashes. Without loss of generality we can therefore assume that the sum of probabilities of all the outcomes of any measurement equals 1.

In summary, we conclude that measurement devices are described by sets of operators $\{E_1, \ldots, E_n\}$ satisfying the relations

$$\forall j = 1, \ldots, n : \quad 0 \leq \text{tr}[\varrho E_j] \leq 1 ; \tag{3.1}$$

$$\sum_{j=1}^{n} \text{tr}[\varrho E_j] = 1, \tag{3.2}$$

for all $\varrho \in \mathcal{S}(\mathcal{H})$. For a fixed state these requirements coincide with conditions on probability distributions defined on the set $\{1, \ldots, n\}$. As they are required to hold for all states, they are equivalent to the following operator relations:

$$\forall j = 1, \ldots, n: \quad O \leq E_j \leq I; \tag{3.3}$$

$$\sum_{j=1}^{n} E_j = I, \tag{3.4}$$

respectively. (The partial ordering $\leq$ is that defined in subsection 1.2.2.)

Exercise 3.1 Confirm that the requirements (3.1), (3.2) hold for all states if and only if (3.3), (3.4) hold, respectively. [Hint: If $\mathrm{tr}[\varrho S] = \mathrm{tr}[\varrho T]$ for all states ϱ then we have $\langle \psi | S \psi \rangle = \langle \psi | T \psi \rangle$ for all vectors $\psi \in \mathcal{H}$, and thus $S = T$ by Proposition 1.21.]

The normalization condition (3.4) is very simple but still it significantly restricts the allowed collections of effects. Suppose, for instance, that P_φ and P_ψ are one-dimensional projections, defined by unit vectors $\varphi, \psi \in \mathcal{H}$. We assume that they correspond to disjoint outcomes of some measurement. If the input state is $\varrho = \frac{1}{2}(|\varphi\rangle\langle\varphi| + |\psi\rangle\langle\psi|)$, the sum of their probabilities is

$$\mathrm{tr}[\varrho P_\varphi] + \mathrm{tr}[\varrho P_\psi] = \langle\varphi|\varrho\varphi\rangle + \langle\psi|\varrho\psi\rangle = 1 + |\langle\varphi|\psi\rangle|^2 \geq 1.$$

As the sum of the outcome probabilities cannot exceed 1, we must have $\langle\varphi|\psi\rangle = 0$. In conclusion, unless P_φ and P_ψ are orthogonal they cannot describe disjoint outcomes of the same measurement.

Exercise 3.2 Confirm that the previous conclusion is true for all projections and not just for one-dimensional projections. [Hint: Recall from Proposition 1.44 that two projections P, Q satisfy $P + Q \leq I$ if and only if they are orthogonal.]

However, if two effects E_1 and E_2 satisfy $E_1 + E_2 \leq I$ then there is no rule prohibiting the existence of a measurement containing E_1 and E_2 in its description. We can thus say in a preliminary way that an observable is a collection of operators $\{E_j\}$ satisfying the formulas (3.3), (3.4).

Example 3.3 (*Ideal Stern–Gerlach apparatus*)
The Stern–Gerlach apparatus is a device designed to measure a spin component of a particle. In this measurement, a beam of particles passes between the opposite poles of a magnet (see Figure 3.2). These poles determine the orientation of the Stern–Gerlach apparatus. The magnetic force depends upon the spin state, and in an idealized picture the beam is split by the magnet into well-separated parts. Therefore, the value of the spin component can be inferred from the observed deflection of the beam. In practice this method is useful only for neutral particles since for charged particles the Lorentz force obscures the deflection.

A beam of spin-$\frac{1}{2}$ particles (e.g. neutrons) splits into two parts. On sending a spin-$\frac{1}{2}$ particle through the Stern–Gerlach apparatus, it is deflected by an angle $\pm\theta$ and is detected on the screen. Each particle is detected on either the upper or the lower half of the screen plane. Let us label these events by $+$ and $-$ and

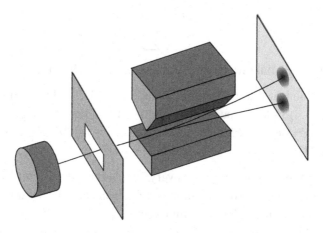

Figure 3.2 The Stern–Gerlach apparatus.

let $E_\pm$ be the corresponding effects. It turns out that it is sufficient to consider two-dimensional Hilbert space. In an ideal case there exist some states ϱ_+ and ϱ_- that are simply deflected either up or down; their beam is not split into two parts. In such case all the particles sent through the Stern–Gerlach apparatus give rise to the same event; thus, $\mathrm{tr}\big[\varrho_+ E_+\big] = \mathrm{tr}\big[\varrho_- E_-\big] = 1$. From Example 2.33 in subsection 2.1.4 it follows that E_+, E_- and ϱ_+, ϱ_- are projections and that $E_\pm = \varrho_\pm = \frac{1}{2}(I \pm \vec{n} \cdot \vec{\sigma})$ for some unit vector $\vec{n} \in \mathbb{R}^3$. The direction of the vector $\vec{n}$ is determined by the direction of the magnetic field in the Stern–Gerlach apparatus. We denote by $\mathsf{S}^{\vec{n}}$ the *(ideal) spin component observable* in the direction $\vec{n}$ and set $\mathsf{S}^{\vec{n}}(\pm) = \frac{1}{2}(I \pm \vec{n} \cdot \vec{\sigma})$. △

Example 3.4 (*Coin-tossing observable*)
Let us define two effects $C_\pm$ by $C_\pm = c_\pm I$, where $c_\pm = \frac{1}{2}(1 \pm b)$ and b is a fixed number from the interval $[0, 1]$. These effects form an observable since $C_\pm \geq O$ and $C_+ + C_- = I$. For any input state ϱ we obtain $\mathrm{tr}\big[\varrho C_\pm\big] = c_\pm$, and so the measurement outcome probabilities are thus independent of the state ϱ. The measurement device described by the effects $C_\pm$ can be simulated by tossing a coin. Namely, we set $+$ to denote heads and $-$ to denote tails. Without manipulating the input state at all we just toss a coin to obtain a measurement outcome. The parameter b is the *bias* of the coin. If $b = 0$ then the coin is fair and both outcomes occur with the same probability. We write $\mathsf{F}(\pm) = \frac{1}{2}I$ and call this the *fair-coin-tossing observable*. △

The previously presented introductory description of an observable as a collection of effects is not general enough to cover all relevant situations. For example, there are measurements with a continuous set of outcomes such as measurements of position and momentum. (One can argue that all actual measurements have always only a finite number of possible measurement outcomes, but in any case measurements with continuous sets of outcomes are very useful idealizations.) We need to generalize our preliminary definition and observables turn out to be a certain kind of operator measure.

The required modification to our description of an observable is similar to the generalization from finite probability distributions to probability measures. Therefore, let us first recall some basic definitions from probability theory.

Let Ω be a nonempty set. A *σ-algebra* on Ω is a collection $\mathcal{F}$ of subsets of Ω that has the following properties:

(i) $\emptyset \in \mathcal{F}$ and $\Omega \in \mathcal{F}$;
(ii) if $X \in \mathcal{F}$ then $\Omega \setminus X \in \mathcal{F}$;
(iii) if $X_1, X_2, \ldots \in \mathcal{F}$ then $\cup_i X_i \in \mathcal{F}$.

The pair $(\Omega, \mathcal{F})$ is called a *measurable space*. A set $X \in \mathcal{F}$ is called an *event* and the σ-algebra $\mathcal{F}$ is thus a collection of all events. Clearly, the collection 2^{Ω} of all subsets of Ω is a σ-algebra. However, it turns out that in some situations it is not possible or practical to choose 2^{Ω}, and therefore we have to deal with a general σ-algebra.

A *probability measure* is a mapping $p : \mathcal{F} \to [0, 1]$ that satisfies the following conditions:

(i) $p(\emptyset) = 0$;
(ii) $p(\Omega) = 1$;
(iii) $p(\cup_i X_i) = \sum_i p(X_i)$ for any sequence $\{X_i\}$ of disjoint sets in $\mathcal{F}$.

The number $p(X)$ is the probability for an event X to occur. The usual properties of the probability can be derived from the defining condition (iii), known as *σ-additivity*. The following concept is essential in the description of observables and represents a direct generalization of the notion of a probability measure.

Definition 3.5 A *positive operator-valued measure* (POVM) is a mapping $\mathsf{A} : \mathcal{F} \to \mathcal{E}(\mathcal{H})$ such that

(i) $\mathsf{A}(\emptyset) = O$;
(ii) $\mathsf{A}(\Omega) = I$;
(iii) $\mathsf{A}(\cup_i X_i) = \sum_i \mathsf{A}(X_i)$ (in the weak sense) for any sequence $\{X_i\}$ of disjoint sets in $\mathcal{F}$.

In other words, a mapping A from $\mathcal{F}$ into $\mathcal{E}(\mathcal{H})$ is a POVM if and only if the 'sandwiching' $X \mapsto \langle \psi | \mathsf{A}(X) \psi \rangle$ is a probability measure for every unit vector $\psi \in \mathcal{H}$. We can develop this statement into a more useful form. First we notice

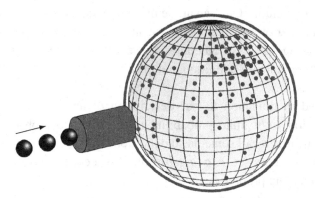

Figure 3.3 Illustration of the spin direction observable. Particles, each in the same spin state, enter a Stern–Gerlach apparatus that is oriented randomly. The orientation of the apparatus is re-randomized each time a particle passes through the apparatus. The dots on the large sphere represents the observed spin directions of the emerging particles.

that $\langle \psi | A(X) \psi \rangle = \mathrm{tr}\big[|\psi\rangle\langle\psi| A(X)\big]$ and that $|\psi\rangle\langle\psi|$ is a pure state. But all states can be decomposed into mixtures of pure states. Hence, a mapping A from $\mathcal{F}$ into $\mathcal{E}(\mathcal{H})$ is a POVM if and only if the mapping $X \mapsto \mathrm{tr}\big[\varrho A(X)\big]$ is a probability measure for every state $\varrho \in \mathcal{S}(\mathcal{H})$.

Definition 3.6 From now on, we will identify observables with POVMs. For an observable A defined on a measurable space $(\Omega, \mathcal{F})$ we say that Ω is the *sample space* of A and $(\Omega, \mathcal{F})$ is the *outcome space* of A.

Example 3.7 (*Measurement of the spin direction*)
In Example 3.3 we discussed an ideal spin-component measurement in a direction $\vec{n}$, described by the projections $S^{\vec{n}}(\pm) = \frac{1}{2}(I \pm \vec{n} \cdot \vec{\sigma})$. Suppose that we choose the direction $\vec{n}$ at random and measure the corresponding spin component of a prepared particle. If the measurement outcome is $+$ we interpret this to mean that the spin direction of the particle is $\vec{n}$, while a measurement outcome $-$ is taken to mean that the spin direction is $-\vec{n}$. The measurement outcome is thus a direction in $\mathbb{R}^3$ (see Figure 3.3). After each measurement we choose another random direction. The eventual output is a probability measure on $\mathbb{S}^2$, the surface of the unit sphere in $\mathbb{R}^3$. If the input state is $\varrho = \frac{1}{2}(I + \vec{r} \cdot \vec{\sigma})$ then

$$\mathrm{tr}\big[\varrho S^{\vec{n}}(+)\big] = \mathrm{tr}\big[\varrho S^{-\vec{n}}(-)\big] = \tfrac{1}{2}(1 + \vec{r} \cdot \vec{n})$$

and the probability measure is thus predicted to be

$$X \mapsto 2 \int_X \tfrac{1}{2}(1 + \vec{r} \cdot \vec{n})\, \frac{1}{4\pi}\, d\vec{n} = \frac{1}{4\pi} \int_X (1 + \vec{r} \cdot \vec{n})\, d\vec{n}.$$

Here X is a (measurable) subset of $\mathbb{S}^2$ and $\frac{1}{4\pi}d\vec{n}$ denotes the usual (invariant) integration on $\mathbb{S}^2$. We conclude that the corresponding POVM is

$$\mathsf{D}(X) = \frac{1}{4\pi} \int_X (I + \vec{n} \cdot \vec{\sigma}) \, d\vec{n}.$$

We say that this is the *spin direction observable*.

From the mathematical point of view, this POVM is a special instance of a *coherent state POVM*. For a detailed survey of this topic, we refer to [2]. △

In the following proposition we list some basic mathematical properties of observables. They may look boring now but will be useful later. It is worth noticing that a probability measure satisfies analogous formulas.

Proposition 3.8 Let A be an observable with an outcome space $(\Omega, \mathcal{F})$ and let $X, Y \in \mathcal{F}$.

(a) If $X \subseteq Y$ then $\mathsf{A}(X) \leq \mathsf{A}(Y)$.
(b) If $X \subseteq Y$ and $\mathsf{A}(Y) = O$ then $\mathsf{A}(X) = O$.
(c) $\mathsf{A}(X \cup Y) + \mathsf{A}(X \cap Y) = \mathsf{A}(X) + \mathsf{A}(Y)$.

Proof (a): Since $X \subseteq Y$ we can write Y as the disjoint union

$$Y = X \cup (Y \smallsetminus X).$$

We then have $\mathsf{A}(Y) = \mathsf{A}(X) + \mathsf{A}(Y \smallsetminus X)$, which implies that $\mathsf{A}(X) \leq \mathsf{A}(Y)$.

(b): This is a direct consequence of (a) since $O \leq \mathsf{A}(X)$ by the definition of an observable.

(c): Since $X \subseteq X \cup Y$, we can write $X \cup Y$ as the disjoint union $X \cup Y = X \cup Z$, where $Z \equiv (X \cup Y) \smallsetminus X$. This gives $\mathsf{A}(X \cup Y) = \mathsf{A}(X) + \mathsf{A}(Z)$. Adding $\mathsf{A}(X \cap Y)$ to both sides of this equation and noting that Y is a disjoint union of $X \cap Y$ and Z, we get (c). □

3.1.2 Observables and statistical maps

In the discussion after Definition 3.5, we noted that a POVM is essentially a collection of probability measures, labelled by states. It is useful to elaborate this viewpoint a bit more.

Let $Prob(\Omega)$ be the set of all probability measures on an outcome space $(\Omega, \mathcal{F})$. An observable A with outcome space $(\Omega, \mathcal{F})$ determines a mapping Φ_{A} from $\mathcal{S}(\mathcal{H})$ to $Prob(\Omega)$ via the formula

$$\Phi_{\mathsf{A}}(\varrho) := \mathrm{tr}\big[\varrho \mathsf{A}(\cdot)\big]. \tag{3.5}$$

The mapping Φ_{A} underlines the statistical nature of the observable A; it can be thought of as an input–output device, which takes states as inputs and gives probability distributions as outputs. We call Φ_A the *statistical map* corresponding to the observable A.

Exercise 3.9 Let A and B be two observables defined on Ω. Show that $\Phi_A \neq \Phi_B$ if and only if A $\neq$ B.

The basic property of Φ_A is that for any $\varrho_1, \varrho_2 \in \mathcal{S}(\mathcal{H})$ and $0 \leq \lambda \leq 1$, we have

$$\Phi_A(\lambda \varrho_1 + (1-\lambda)\varrho_2) = \lambda \Phi_A(\varrho_1) + (1-\lambda)\Phi_A(\varrho_2). \tag{3.6}$$

This is a simple consequence of the linearity of the trace. In other words, Φ_A is an affine mapping.

Proposition 3.10 Let Φ be an affine mapping from $\mathcal{S}(\mathcal{H})$ into $Prob(\Omega)$. Then $\Phi = \Phi_A$ for some observable A.

Proof Fix $X \in \mathcal{F}$. Denote by Φ^X the mapping from $\mathcal{S}(\mathcal{H})$ to $[0, 1]$ defined as $\Phi^X(\varrho) := [\Phi(\varrho)](X)$. This mapping is affine. Therefore, there is an operator $A(X) \in \mathcal{E}(\mathcal{H})$ such that $\Phi^X(\varrho) = \text{tr}[\varrho A(X)]$ for every state ϱ (recall subsection 2.1.4). It is clear that $A(\emptyset) = O$ and $A(\Omega) = I$. If $\{X_i\}$ is a sequence of disjoint sets in $\mathcal{F}$ then

$$\text{tr}[\varrho A(\cup_i X_i)] = [\Phi(\varrho)](\cup_i X_i) = \sum_i [\Phi(\varrho)](X_i) = \sum_i \text{tr}[\varrho A(X_i)].$$

We conclude that A is a POVM and the claim is thus proved.   □

Example 3.11 (*Three-outcome qubit observable*)
A general three-outcome qubit observable F consists of three effects $F(j) = \frac{1}{2}(\alpha_j I + \vec{m}_j \cdot \vec{\sigma})$, where $\alpha_j \in \mathbb{R}$, $\vec{m}_j \in \mathbb{R}^3$ and $\|\vec{m}_j\| \leq \alpha_j \leq 2 - \|\vec{m}_j\|$ (see Example 2.33). Normalization requires that $\sum_j \alpha_j = 2$ and $\sum_j \vec{m}_j = \vec{0}$. This implies that the vectors $\vec{m}_j$ are in a plane.

One particular example, which has an obvious geometrical symmetry, is

$$F(1) = \tfrac{1}{3}(I + \sigma_y), \quad F(2) = \tfrac{1}{3}(I + \tfrac{\sqrt{3}}{2}\sigma_x - \tfrac{1}{2}\sigma_y), \tag{3.7}$$
$$F(3) = \tfrac{1}{3}(I - \tfrac{\sqrt{3}}{2}\sigma_x - \tfrac{1}{2}\sigma_y).$$

In this case the vectors $\vec{m}_j$ satisfy the relation $\vec{m}_j \cdot \vec{m}_k = -\frac{1}{2}$ for $j \neq k$, meaning that the angle between any two equals 120 degrees. The probability of observing outcome j is given by the formula $p_j = \text{tr}[\varrho F(j)] = \frac{1}{3}(1 + \vec{r} \cdot \vec{m}_j)$. The associated statistical map reads

$$\Phi_F : \varrho \quad \mapsto \quad \vec{p} = \tfrac{1}{3}(1 + y, 1 + \tfrac{\sqrt{3}}{2}x - \tfrac{1}{2}y, 1 - \tfrac{\sqrt{3}}{2}x - \tfrac{1}{2}y), \tag{3.8}$$

where x and y are the components of the Bloch vector corresponding to ϱ. Since $\vec{p} \cdot \vec{p} = \frac{1}{3} + \frac{1}{6}(x^2 + y^2)$ and $x^2 + y^2 \leq 1$, it follows that the probability vectors

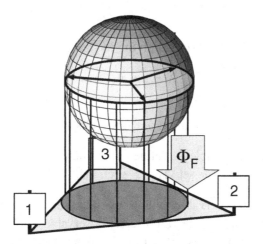

Figure 3.4 Illustration of the statistical map Φ_F corresponding to the three-outcome observable F introduced in Example 3.11. The labels 1, 2, 3 refer to the three outcomes. Each Bloch vector is mapped into a sphere contained inside a triangle representing all possible three-valued probability distributions.

$\vec{p}$ are contained within a sphere of radius $\sqrt{\frac{2}{3}}$. The normalization constraint $p_1 + p_2 + p_3 = 1$ determines a plane in the three-dimensional $\vec{p}$-space, and positivity restricts the probability vectors to a triangle connecting the points $(1, 0, 0)$, $(0, 1, 0)$ and $(0, 0, 1)$. The intersection of this triangle with the sphere $\|\vec{p}\| \leq \sqrt{\frac{2}{3}}$ gives the image $\Phi_\mathsf{F}(\mathcal{S}(\mathcal{H}))$ of the state space (see Figure 3.4). △

3.1.3 Discrete observables

Let us consider the POVM description for measurements with countably many outcomes. Let $\Omega = \{x_1, x_2, \ldots\}$ be a countable set (i.e. either finite or countably infinite) of outcomes. In this case we always choose the σ-algebra $\mathcal{F}$ to be the collection of all subsets of Ω, i.e. $\mathcal{F}$ is the power set $\mathcal{F} = 2^\Omega$. Notice that formally we have to write $\{x\}$ for a set consisting of a single outcome x, since an observable is defined on events, which are sets of outcomes, and not on outcomes as such. However, it is convenient to drop the braces, and the meaning should be clear from the context; hence, we will write $\mathsf{A}(\{x\}) \equiv \mathsf{A}(x)$.

For an arbitrary subset $X \subseteq \Omega$, we have

$$\mathsf{A}(X) = \sum_{x_j \in X} \mathsf{A}(x_j). \tag{3.9}$$

It follows that an observable A with a countable outcome space $(\Omega, 2^\Omega)$ is completely determined by the mapping

$$x_j \mapsto A(x_j), \tag{3.10}$$

from Ω to the set of effects $\mathcal{E}(\mathcal{H})$, satisfying the normalization condition

$$\sum_{x_j \in \Omega} A(x_j) = A(\Omega) = I. \tag{3.11}$$

We adopt the following definition for a discrete observable.

Definition 3.12 An observable A, defined on an outcome space $(\Omega, \mathcal{F})$, is called *discrete* if there is a countable set $\Omega_0 \in \mathcal{F}$ such that $A(\Omega_0) = I$.

On the one hand, all observables with a countable sample space are discrete. On the other hand, the sample space of a discrete observable can be uncountable (e.g. the real line $\mathbb{R}$). However, this is only a trivial difference as the sample space can be redefined to be a countable set Ω_0 without changing the essence of the observable. The terminology in Definition 3.12 reflects the fact that it is convenient sometimes to have freedom in the choice of sample space, and we do not want the property of being discrete to depend on that choice.

Sometimes in the literature one meets with a definition of an observable as a finite collection of effects $E_1, \ldots, E_n$ satisfying the normalization condition $\sum_j E_j = I$. Naturally, one is then restricted to observables with finite outcome sets. This definition should be understood in the sense that we first fix a set of n measurement outcomes (e.g. $\Omega = \{x_1, \ldots, x_n\}$ with $x_j \in \mathbb{R}$) and then adopt the definition corresponding to formula (3.10), which induces the whole POVM.

Example 3.13 (*Discretization of observables*)
Let us consider an observable A that has a sample space Ω and is *not* discrete. Suppose that we cannot distinguish all the original measurement outcomes; for example, it may happen that we can say only that a measurement outcome is in some region $X_j \subseteq \Omega$. We assume that there is a finite number of these disjoint regions and they cover Ω. The modified observable A' describing the overall measurement is then defined on the labels of the regions and given by $A'(j) := A(X_j)$. The observable A' has a finite number of outcomes and is thus discrete. We call it *discretization* of A.

As a concrete demonstration, let us recall Example 3.7. The POVM describing the spin direction measurement is given by

$$D(X) = \frac{1}{4\pi} \int_X (I + \vec{n} \cdot \vec{\sigma}) \, d\vec{n},$$

where X is a (measurable) subset of the surface of the unit sphere in $\mathbb{R}^3$. In the extreme case a discretization D' of D has only two outcomes. For instance, let us fix a direction $\vec{m}$ and split the unit sphere into two hemispheres $X_{\pm\vec{m}}$ with poles $\pm\vec{m}$, respectively. The corresponding discretization D' of D is then

$$\mathsf{D}'(\pm) := \mathsf{D}(X_{\pm\vec{m}}) = \frac{1}{4\pi} \int_{X_{\pm\vec{m}}} (I + \vec{n} \cdot \vec{\sigma})\, d\vec{n} = \tfrac{1}{2} \left(I \pm \tfrac{1}{2}\vec{m} \cdot \vec{\sigma} \right).$$

The effects $\mathsf{D}'(\pm)$ are similar to the projections $\mathsf{S}^{\vec{m}}(\pm)$, but there is an additional factor $\tfrac{1}{2}$ in front of the unit vector $\vec{m}$. $\triangle$

3.1.4 Real observables

If the sample space Ω is $\mathbb{R}$ then it is not convenient to choose as the σ-algebra $\mathcal{F}$ the collection of all subsets of $\mathbb{R}$. Instead, it is common to choose $\mathcal{F}$ as the *Borel σ-algebra* $\mathcal{B}(\mathbb{R})$. This σ-algebra contains all the sets that are needed in calculations. For instance, all open and closed intervals and their complements and countable unions belong to $\mathcal{B}(\mathbb{R})$. For our purposes this is just a technical remark and a reader who is not familiar with $\mathcal{B}(\mathbb{R})$ can simply think of it as a collection that contains all the 'usual' subsets of $\mathbb{R}$ that can be imagined.

Definition 3.14 An observable A is *real*, or *real valued*, if the outcome set of A is either $\mathbb{R}$ or a subset of $\mathbb{R}$. In this case the σ-algebra is chosen to be the corresponding Borel σ-algebra. The Borel σ-algebra is formally defined as the smallest σ-algebra containing all open sets on $\mathbb{R}$.

Let us note that typical observables are usually real or can be redefined to be real. For instance, the Stern–Gerlach observable in Example 3.3 has sample space $\{+, -\}$ but can be made real by using the sample spaces $\{1, -1\}$ or $\{0, 1\}$ instead of the original sample space. The particular labelling of the outcomes is usually not important. Although strictly speaking they determine different POVMs, we do not consider this difference to have any physical relevance.

For real observables we can define some useful statistical quantities. First, we can define the *average value* or *mean value* or *expectation value* of A in a state ϱ as

$$\langle \mathsf{A} \rangle_\varrho := \int_{\mathbb{R}} x\, \mathrm{tr}\!\left[\varrho \mathsf{A}(dx)\right]. \tag{3.12}$$

Here $\mathrm{tr}\!\left[\varrho \mathsf{A}(dx)\right]$ denotes integration with respect to the probability measure $\mathrm{tr}\!\left[\varrho \mathsf{A}(\cdot)\right]$. It may happen that the integral in (3.12) does not converge, and in that case we write $\langle \mathsf{A} \rangle_\varrho = \infty$. The *variance* $\Delta_\varrho(\mathsf{A})$ of A in a state ϱ is defined as the root mean square distance of the measured value from the mean value of A, i.e.

$$\left(\Delta_\varrho(\mathsf{A})\right)^2 = \int_{\mathbb{R}} \left(x - \langle \mathsf{A} \rangle_\varrho\right)^2 \mathrm{tr}\!\left[\varrho \mathsf{A}(dx)\right]. \tag{3.13}$$

Again, it can happen that the integral in (3.13) does not converge for all states.

In a similar way the mean value and the variance can be defined in cases when Ω is a subset of $\mathbb{R}$. If, for instance, Ω is a countable set of real numbers $x_1, x_2, \ldots$ then we have

$$\langle A \rangle_\varrho = \sum_{x_j \in \Omega} x_j \, \mathrm{tr}\big[\varrho A(x_j)\big], \tag{3.14}$$

$$(\Delta_\varrho(A))^2 = \sum_j (x_j - \langle A \rangle_\varrho)^2 \, \mathrm{tr}\big[\varrho A(x_j)\big]. \tag{3.15}$$

Example 3.15 (*Polarization observable*)
In Example 2.1 we introduced a polarization filter. If a photodetector is attached to such a device then their action constitutes a measurement of the polarization of a photon. A photon has a certain probability of passing the polarizer; this depends only on its polarization. Without going into further details let us take it as an experimentally verified fact that the polarization property is associated with the two-dimensional Hilbert space $\mathbb{C}^2$. Therefore, it is a physical implementation of the qubit. In the ideal case the measurement implemented by a polarizer and subsequent photodetector is described by a binary POVM $\mathsf{P} : \{0, 1\} \mapsto \mathcal{E}(\mathbb{C}^2)$, where 0 and 1 correspond to no detection and detection, respectively. Up to the labelling of the outcomes, P coincides with the Stern–Gerlach observable described in Example 3.3. Thus,

$$\mathsf{P}(0) = \tfrac{1}{2}(I - \vec{n} \cdot \vec{\sigma}), \qquad \mathsf{P}(1) = \tfrac{1}{2}(I + \vec{n} \cdot \vec{\sigma}), \tag{3.16}$$

for some unit vector $\vec{n} \in \mathbb{R}^3$.

In the usual convention, a vertically oriented linear polarizer is associated with the vector $\vec{n} = (0, 0, 1)$, i.e. $\mathsf{P}_{vert}(1) = \tfrac{1}{2}(I + \sigma_z)$. Let us stress that the Cartesian coordinates for the polarization do not coincide with the spatial alignment of the polarizer. In fact, circular polarizers are not related to any spatial direction but are nevertheless associated with the unit vectors $\vec{n} = (0, \pm 1, 0)$. For the mean value we have

$$\langle \mathsf{P} \rangle_\varrho = \mathrm{tr}\big[\varrho \mathsf{P}(1)\big] = \tfrac{1}{2}(1 + \vec{n} \cdot \vec{r}),$$

where $\varrho = \tfrac{1}{2}(I + \vec{r} \cdot \vec{\sigma})$. Naturally, the mean value coincides with the probability that a photon will pass the polarizer. △

Exercise 3.16 Calculate the variance $\Delta_\varrho(\mathsf{P})$ of the polarization observable P for photons in the state $\varrho = \tfrac{1}{2}(I + \vec{r} \cdot \vec{\sigma})$.

3.1.5 Mixtures of observables

In Section 2.1 we discussed mixtures of states. The convex structure of the set of states results from the possibility of alternating between preparation procedures.

Similarly, we can think of an experiment where the preparation procedure is kept fixed but different measurement apparatuses are alternated. In this way, we can mix two measurements to get a third.

The difference between states and observables is that a mixture in the latter case can mean several different things, depending on how we alternate between measurement apparatuses. For simplicity, we discuss here mixtures only in the case of observables with a finite number of outcomes.

Mixing while keeping outcomes separated

Let us consider an experiment in which we randomly switch between two observables A and B, so that their relative frequencies are λ and $1 - \lambda$, respectively. Even though we are switching between A and B randomly, we keep track of which observable was measured each time.

To describe this kind of experiment, we assume that the outcome sets are $\Omega_A = \{a_1, \ldots, a_N\}$ and $\Omega_B = \{b_1, \ldots, b_M\}$ and that all these labels are mutually distinguishable. We define a new outcome set $\Omega_C = \Omega_A \cup \Omega_B$ and then extend A and B to this new outcome set by writing

$$A(X) = A(X \cap \Omega_A), \qquad B(X) = B(X \cap \Omega_B) \qquad (3.17)$$

for all $X \subseteq \Omega_C$. In other words, the extension means that we define $A(b_j) = O$ and $B(a_i) = O$. This extension changes the probability distributions associated with A and B only by adding some zeroes.

An observable C with outcome set Ω_C is now defined by

$$C(X) = \lambda A(X) + (1 - \lambda)B(X) \qquad (3.18)$$

for all $X \subseteq \Omega_C$. For singleton sets we thus have $C(a_j) = \lambda A(a_j)$ and $C(b_j) = (1 - \lambda)B(b_j)$. We write $C = \lambda A + (1 - \lambda)B$ and say that the observable C is a *mixture* of the observables A and B.

Example 3.17 (*Mixture of Stern–Gerlach apparatuses*)
Suppose there is a random switch between two Stern–Gerlach apparatuses with different orientations, $\vec{n}$ and $\vec{m}$, as illustrated in Figure 3.5. The switch may or may not be under the control of the experimenter. In any case, on the screen she observes four well-separated areas representing four possible outcomes – *right*, *left*, *up* and *down*. The sample space Ω_C consists of four elements identifying the area into which the measured particle falls, and we will write $\Omega_C = \{R, L, U, D\}$. A typical sequence of outcomes would be U, L, R, D, D, U, R, L. Suppose that the

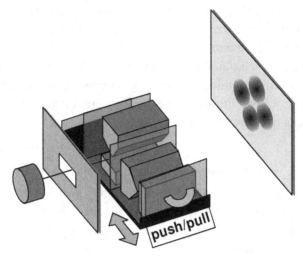

Figure 3.5 For a mixture of two Stern–Gerlach apparatuses, typically their outcomes are separated.

two Stern–Gerlach devices are alternated, so that they are used with probabilities λ and $1 - \lambda$, respectively. The associated effects are given as follows:

$$\begin{aligned} \mathsf{C}(R) &= \lambda \mathsf{S}^{\vec{n}}(+), & \mathsf{C}(L) &= \lambda \mathsf{S}^{\vec{n}}(-), \\ \mathsf{C}(U) &= (1 - \lambda)\mathsf{S}^{\vec{m}}(+), & \mathsf{C}(D) &= (1 - \lambda)\mathsf{S}^{\vec{m}}(-), \end{aligned}$$

and we write $\mathsf{C} = \lambda \mathsf{S}^{\vec{n}} + (1 - \lambda)\mathsf{S}^{\vec{m}}$.   △

Mixing and combining outcome spaces

Let us suppose that the outcomes of two observables A and B are labelled in the same way, i.e. $\Omega_\mathsf{A} = \Omega_\mathsf{B} = \{a_1, \ldots, a_N\}$. Thus we could be in a situation where, after a measurement outcome is registered, we cannot determine whether it pertained to A or to B. In this situation where we cannot say whether the outcome was registered with A or B, the observable C describing this measurement is given as

$$\mathsf{C}(a_j) = \lambda \mathsf{A}(a_j) + (1 - \lambda)\mathsf{B}(a_j) \tag{3.19}$$

for all $a_j \in \Omega_\mathsf{A}$; λ is the probability that the outcome a_j is due to a measurement of observable A.

Example 3.18 (*Mixture of polarization observables*)
Although polarizers and Stern–Gerlach apparatuses can be viewed as different physical implementations of the same observable, the two interpretations of their mixings lead to different mathematical descriptions. The reason is simple. When there is a random switch changing between two polarizers (see Figure 3.6), it does

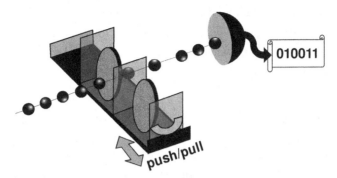

Figure 3.6 For a mixture of two polarizers, typically their outcomes are combined.

not really affect the events we are observing; either the photodetector clicks or it does not. Without any additional knowledge on the actual orientation of the polarizer, the experiment leads only to these two possibilities. Let us denote a pair of polarizers by P_1 and P_2. Their mixture C with probabilities λ, $1 - \lambda$ is described by the convex combination

$$C(j) = \lambda P_1(j) + (1 - \lambda)P_2(j)$$

for $j = 0, 1$. △

Comparing Examples 3.17 and 3.18 we see that the formal convex combination $C = \lambda A + (1 - \lambda)B$ of two observables A and B determines a particular observable C only if the merging of the outcome sets Ω_A and Ω_B is specified. Depending on the system in question, there can also be cases intermediate between the two different mixtures described earlier. For instance, let A and B be two observables with outcome sets $\Omega_A = \{a_1, a_2\}$ and $\Omega_B = \{b_1, a_2\}$, respectively. Suppose that the measurements of A and B are mixed and that, as indicated, the second outcomes in Ω_A and Ω_B are identical. The convex combination $C = \lambda A + (1 - \lambda)B$ is then the three-outcome observable

$$C: \quad a_1 \mapsto \lambda A(a_1), \qquad b_1 \mapsto (1 - \lambda)B(b_1), \qquad (3.20)$$
$$a_2 \mapsto \lambda A(a_2) + (1 - \lambda)B(a_2).$$

We again use the same notation, $C = \lambda A + (1 - \lambda)B$.

Pure observables

For the purpose of the following discussion let us consider only observables defined on a fixed finite outcome set $\Omega = \{x_1, \ldots, x_N\}$. When mixing is understood in the sense of formula (3.19), this subset of observables becomes a convex set. We can define *pure observables* in an analogous way to pure states: they are observables that cannot be written as nontrivial convex combinations of other observables.

Earlier, in subsection 2.1.6, we characterized the extremal elements in the set of all effects, and this leads to a simple sufficient condition for an observable to be pure.

Proposition 3.19 Let A be an observable such that $A(x)$ is a projection for every $x \in \Omega$. Then A is pure.

Proof Let us make the counter-assumption that A is not pure, which would mean that $A = \lambda B + (1 - \lambda)C$ for some number $0 < \lambda < 1$ and some observables B and C different from A. Thus, there must be at least one element $x \in \Omega$ such that $A(x) \neq B(x)$, implying that the convex combination $A(x) = \lambda B(x) + (1 - \lambda)C(x)$ of effects is nontrivial. But by Proposition 2.41 the extremal elements in $\mathcal{E}(\mathcal{H})$ are projections; hence the counter-assumption is false.   □

It is an interesting fact that there are also pure observables other than those consisting of projections. In the following we present another class of pure observables.

Proposition 3.20 Let A be an observable such that $A(x)$ is a rank-1 operator for every $x \in \Omega$ and the operators $A(x_1), \ldots, A(x_N)$ are linearly independent. Then A is pure.

Proof Suppose that $A = \lambda B + (1 - \lambda)C$ for some number $0 < \lambda < 1$ and some observables B and C. This implies that $\lambda B(x_j) \leq A(x_j)$ and $(1 - \lambda)C(x_j) \leq A(x_j)$ for every $x_j \in \Omega$. By Proposition 1.63 in subsection 1.3.2 we conclude that $B(x_j) = b_j A(x_j)$ and $C(x_j) = c_j A(x_j)$ for some nonnegative numbers b_j, c_j. Since B and C satisfy the normalization conditions $\sum_j B(x_j) = \sum_j C(x_j) = I$, we obtain

$$\sum_j b_j A(x_j) = \sum_j c_j A(x_j) = I.$$

Therefore

$$O = I - I = \sum_{j=1}^{N}(1 - b_j)A(x_j) = \sum_{j=1}^{N}(1 - c_j)A(x_j).$$

Since the operators $A(x_1), \ldots, A(x_N)$ are linearly independent we conclude that $b_j = c_j = 1$. But this means that $A = B = C$, hence A does not have a nontrivial convex decomposition and it is pure.   □

It is possible to derive mathematical characterizations for pure observables, but they are not as neat and handy as for pure states. We refer the reader to [50], [111], [114] for further details on the characterization of pure observables.

3.1.6 Coexistence of effects

This chapter started with a discussion on the compatibility of the effects in a given collection. Let us recall that two outcomes occurring in the same measurement setup can be associated with two different effects E_1 and E_2 only if $E_1 + E_2 \leq I$. Otherwise the effects cannot be observed as different outcomes of the same experiment. The concept of *coexistence* extends this notion of the compatibility of effects. The rough idea is that two effects can emerge from the same measurement even if they do not correspond to disjoint events.

Let A be an observable with an outcome space $(\Omega, \mathcal{F})$. We denote by ran(A) the range of A, that is,

$$\text{ran(A)} := \{ \text{A}(X) \mid X \in \mathcal{F} \} \subseteq \mathcal{E}(\mathcal{H}).$$

The range of A is the set of all effects that we could test for and record in the measurement of A.

Definition 3.21 The effects $E, F, \ldots$ are *coexistent* if there exists an observable A such that $\{E, F, \ldots\} \subseteq \text{ran(A)}$.

Obviously, the effects $E_1, E_2, \ldots, E_n$ satisfying $\sum_{i=1}^{n} E_i \leq I$ are coexistent, but the following example illustrates that this inequality is not a necessary condition for coexistence.

Example 3.22 (*Coexistence of all spin directions*)
A measurement of the spin direction observable D (defined in Example 3.7) allows us to make a prediction of the spin component in any direction. In this sense, it contains some information on all spin components. What is then the relation of D to the ideal spin component observables $\text{S}^{\vec{n}}$?

To understand the connection, it is useful to recall the discretization of D into binary observables D′ (see Example 3.13). Since the discretization can be performed in any chosen direction $\vec{n}$, we conclude that all the effects $\frac{1}{2}(I + \frac{1}{2}\vec{n}\cdot\vec{\sigma})$ with $\|\vec{n}\| = 1$ are in the range of D and hence coexistent. The price of having all the spin directions in a single observable is additional noise, which manifests itself as the extra factor $\frac{1}{2}$ in front of each $\vec{n}$. To make this additional noise more transparent, we express D′ as

$$\text{D}'(\pm 1) = \tfrac{1}{2}\text{F}(\pm 1) + \tfrac{1}{2}\text{S}^{\vec{n}}(\pm 1);$$

this shows that D' is a mixture of the fair-coin-tossing observable F and the ideal spin component observable $S^{\vec{n}}$. $\triangle$

From now on we will concentrate mostly on the coexistence of two effects.

Proposition 3.23 Let $E, F \in \mathcal{E}(\mathcal{H})$. The following conditions are equivalent:

(i) E and F are coexistent;
(ii) $E, I - E, F$ and $I - F$ are coexistent;
(iii) there exists an observable A with outcome space $\{1, 2, 3, 4\}$ such that $A(\{1, 2\}) = E$ and $A(\{1, 3\}) = F$;
(iv) there exists an effect G that is the common lower bound of E and F (i.e. $G \le E, G \le F$) and satisfies $E + F - I \le G$.

Proof Trivially, (ii)$\Rightarrow$(i). We will show that (i)$\Rightarrow$(iv)$\Rightarrow$(iii)$\Rightarrow$(ii), and this will complete the proof.

Assume that (i) holds. By definition, there exists an observable A such that $A(X) = E$ and $A(Y) = F$ for some $X, Y \in \Omega_A$. If we set $G = A(X \cap Y)$ then $G \le E, G \le F$ by Proposition 3.8(a) and $E + F - I \le G$ by Proposition 3.8(c). Hence, (iv) follows.

Assume that (iv) holds. We define $A(1) = G, A(2) = E - G, A(3) = F - G$ and $A(4) = I - E - F + G$. It follows from the inequalities $G \le E, G \le F$ and $E + F - I \le G$ that all the effects $A(j)$ are positive. Clearly, $\sum_{j=1}^{4} A(j) = I$. Therefore, A is an observable and (iii) holds.

Assume that (iii) holds. We then have $A(\{3, 4\}) = I - A(\{1, 2\}) = I - E$ and $A(\{2, 4\}) = I - A(\{1, 3\}) = I - F$. Hence (ii) holds. $\square$

Suppose that we want to find out whether two effects E and F are coexistent. Proposition 3.23 suggests the following method: take G to be a lower bound of E and F and then check the operator inequality $E + F - I \le G$. Naturally, the bigger the G is, the easier it is for the latter inequality to be fulfilled. Therefore, to find the *greatest lower bound* of E and F seems to be the best candidate for testing their coexistence. The problem with this method, however, is that the greatest lower bound for two effects need not exist. In other words, the partially ordered set of effects is not a lattice (unlike the subset of projections; see subsection 1.2.3). However, even if the greatest lower bound of two effects does not exist, these effects can still be coexistent. We demonstrate this fact below and refer the reader to [68], [69], [91] for further details.

Example 3.24 (*Nonexistence of the greatest lower bound for two effects*)
We recall from Example 2.33 that a convenient parametrization for qubit effects is

$$A = \tfrac{1}{2}(\alpha I + \vec{a} \cdot \vec{\sigma}), \qquad \|\vec{a}\| \le \alpha \le 2 - \|\vec{a}\|. \qquad (3.21)$$

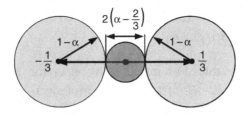

Figure 3.7 The three balls have a common point if and only if $\alpha = \frac{2}{3}$.

In particular, the positivity of A is equivalent to $\|\vec{a}\| \leq \alpha$.

We choose $E = \frac{1}{2}(I + \frac{1}{3}\sigma_z)$ and $F = \frac{1}{2}(I - \frac{1}{3}\sigma_z)$. Since $E + F = I$, they are coexistent effects. A lower bound for E and F is $G = \frac{1}{3}I$. Suppose that an effect A (written in the form (3.21)) is also a lower bound for E and F and that A satisfies $G \leq A$. This means that the three operators $E - A$, $F - A$ and $A - G$ are positive. These operator inequalities translate into the following conditions:

$$1 - \alpha \geq \|\vec{a} - \tfrac{1}{3}\hat{z}\|$$
$$1 - \alpha \geq \|\vec{a} + \tfrac{1}{3}\hat{z}\|$$
$$\alpha - \tfrac{2}{3} \geq \|\vec{a}\|,$$

where $\hat{z}$ is the unit vector $\hat{z} = (0, 0, 1)$. If α is kept fixed and $\vec{a} \in \mathbb{R}^3$ is a free parameter then each inequality defines a ball in $\mathbb{R}^3$. The balls have a common point if and only if $\alpha = \frac{2}{3}$, in which case $\vec{a} = 0$. (See Figure 3.7.) Therefore $A = G$. We conclude that G is a maximal lower bound for E and F. However, G cannot be the greatest lower bound since not all lower bounds for E and F are comparable with it. For instance, the effect $G' = \frac{1}{5}(I + \sigma_x)$ is a lower bound for E and F, but neither $G' \leq G$ nor $G \leq G'$ is true. △

The complicated order structure of effects (in particular, the nonexistence of a greatest lower bound) can be seen as the underlying reason that the question whether two effects are coexistent is not as easy to answer as one might expect. There are, however, some simple necessary or sufficient criteria for coexistence to occur.

Proposition 3.25 Let E, $F \in \mathcal{E}(\mathcal{H})$.

(a) If $[E, F] = 0$ then E and F are coexistent.
(b) If $[E, F] \neq 0$ and either E or F is a projection then E and F are not coexistent.

Proof (a): To prove the claim, we need to find an observable A such that $E, F \in$ ran(A). A possible choice for A is

$$A(1) = EF, \qquad A(2) = E(I - F), \qquad A(3) = (I - E)F,$$
$$A(4) = (I - E)(I - F).$$

This defines an observable. For instance, the operator EF is positive since $EF = E^{1/2}FE^{1/2} = (F^{1/2}E^{1/2})^*(F^{1/2}E^{1/2})$. Moreover, it is easy to verify that $\sum_j A(j) = I$. Both E and F are contained in the range of A since $A(\{1, 2\}) = E$ and $A(\{1, 3\}) = F$.

(b): Suppose that E and F are coexistent and that E is a projection. To prove the claim, we need to show that $[E, F] = 0$. By Proposition 3.23 there exists an observable A with outcome space $\{1, 2, 3, 4\}$ such that $A(\{1, 2\}) = E$ and $A(\{1, 3\}) = F$. We conclude that $A(1) \leq A(\{1, 2\}) = E$ and $A(3) \leq A(\{3, 4\}) = I - E$. Since both E and $I - E$ are projections, Proposition 1.46 implies that $[A(1), E] = 0$ and $[A(3), I - E] = 0$. The latter implies further that $[A(3), E] = 0$. Therefore,

$$[E, F] = [E, A(\{1, 3\})] = [E, A(1) + A(3)] = [E, A(1)] + [E, A(3)] = 0.$$

$\square$

Example 3.26 (*Minimal noise in coexistent spin direction effects*)
Let us, once again, investigate spin direction measurements. Two projections $\mathsf{S}^{\vec{n}}(+)$ and $\mathsf{S}^{\vec{m}}(+)$ with $\vec{n} \neq \pm\vec{m}$ do not commute and by Proposition 3.25 are not coexistent. It was demonstrated in Example 3.22 that the coexistence of two effects is possible even if they do not commute; however, neither can then be a projection. Motivated by the same example, we can ask how much noise has to be added in order to make noisy versions of the noncommuting projections $\mathsf{S}^{\vec{n}}(+)$ and $\mathsf{S}^{\vec{m}}(+)$ coexistent. This question was first investigated by Busch [22].

A mixture of $\mathsf{S}^{\vec{n}}$ and the fair-coin-tossing observable F is given by

$$\lambda\mathsf{S}^{\vec{n}}(\pm) + (1 - \lambda)\mathsf{F}(\pm) = \tfrac{1}{2}\lambda(I \pm \vec{n} \cdot \vec{\sigma}) + \tfrac{1}{2}(1 - \lambda)I = \tfrac{1}{2}(I \pm \lambda\vec{n} \cdot \vec{\sigma}).$$

Hence, we are searching for the greatest number $0 \leq \lambda \leq 1$ such that the effects

$$\tfrac{1}{2}(I + \lambda\vec{n} \cdot \vec{\sigma}) \equiv E \qquad \text{and} \qquad \tfrac{1}{2}(I + \lambda\vec{m} \cdot \vec{\sigma}) \equiv F \qquad (3.22)$$

are coexistent. From Example 3.22 we conclude that $\lambda = \tfrac{1}{2}$ leads to coexistence. However, as we are now interested in only two spin directions rather than all spin directions, we expect that less noise (i.e. a larger value of λ) is enough to make coexistence possible.

Suppose that the effects in (3.22) are coexistent and let G be as in Proposition 3.23 (iv). We parametrize the qubit effect G as usual: $G = \tfrac{1}{2}(\gamma I + \vec{g} \cdot \vec{\sigma})$. The four requirements $G \geq O$, $G \leq E$, $G \leq F$ and $E + F - I \leq G$ then translate into the parameter inequalities

$$\|\vec{g}\| \leq \gamma, \tag{3.23}$$

$$\|\lambda\vec{n} - \vec{g}\| \leq 1 - \gamma, \tag{3.24}$$

$$\|\lambda\vec{m} - \vec{g}\| \leq 1 - \gamma, \tag{3.25}$$

$$\|\lambda\vec{n} + \lambda\vec{m} - \vec{g}\| \leq \gamma. \tag{3.26}$$

If γ is kept fixed and $\vec{g} \in \mathbb{R}^3$ is a free parameter then each inequality defines a ball in $\mathbb{R}^3$. Therefore, the coexistence of E and F is equivalent to the statement that there exists a number $\gamma \geq 0$ such that the four balls (3.23)–(3.26) have a common point. Two balls have a common point exactly when the distance of their centers is not greater than the sum of their radii. Applying this criterion to the pairs (3.23), (3.26) and (3.24), (3.25) we obtain

$$\|\lambda\vec{n} + \lambda\vec{m}\| \leq 2\gamma \quad \text{and} \quad \|\lambda\vec{n} - \lambda\vec{m}\| \leq 2 - 2\gamma.$$

Adding these two inequalities together we see that

$$\lambda\|\vec{n} + \vec{m}\| + \lambda\|\vec{n} - \vec{m}\| \leq 2, \tag{3.27}$$

which is thus a necessary condition for E and F to coexist. The fact that (3.27) is also a sufficient condition can be seen by choosing $\vec{g} = \frac{1}{2}(\lambda\vec{n} + \lambda\vec{m})$ and $\gamma = \|\vec{g}\|$. With these choices the requirements (3.23)–(3.26) reduce to (3.27).

We conclude that the effects E and F are coexistent if and only if

$$\lambda \leq \frac{2}{\|\vec{n} + \vec{m}\| + \|\vec{n} - \vec{m}\|} = \frac{1}{\sqrt{1 + \sin\theta}}, \tag{3.28}$$

where $0 < \theta < \pi$ is the angle between $\vec{n}$ and $\vec{m}$. We see that the greatest amount of additional noise is required when the vectors $\vec{n}$ and $\vec{m}$ are orthogonal, and in this case the largest admissible value of λ is $\lambda = \frac{1}{\sqrt{2}}$. In summary, any pair of spin directions become coexistent if $\lambda \leq \frac{1}{\sqrt{2}}$. △

A complete characterization of coexistent pairs of qubit effects was presented in [29], [132], [149]. In more complicated situations (more effects or higher dimensions) only partial results are known. However, in a given instance the coexistence of two effects in a finite-dimensional Hilbert space can be numerically tested; a semidefinite program for this purpose was given in [146].

3.2 Sharp observables

Sharp observables form a prominent class of observables, which deserves a special treatment. One should notice that in some contexts (especially in the older literature) the concept of an observable refers only to what we call here a sharp observable.

3.2.1 Projection-valued measures

Let us recall that projections are a special kind of effect. As we saw in subsection 2.1.6, they are exactly the extremal elements of the convex set of all effects. This motivates the following terminology.

Definition 3.27 A POVM A is a *projection-valued measure* (PVM) if $A(X)$ is a projection for every $X \in \mathcal{F}$. An observable such as A is called a *sharp observable*.

The following example of a class of sharp observables is frequently discussed. Such observables arise in many applications and we are going to use them later on various occasions.

Example 3.28 (*Sharp observable associated with an orthonormal basis*)
Let $\mathcal{H}$ be a d-dimensional Hilbert space (either $d < \infty$ or $d = \infty$) and $\{\varphi_j\}_{j=1}^{d}$ be an orthonormal basis for $\mathcal{H}$. For each $j = 1, \ldots, d$, we write $A(j) = |\varphi_j\rangle\langle\varphi_j|$. Thus, $A(j)$ is a one-dimensional projection, and the probability of obtaining measurement outcome j for a state ϱ is

$$\text{tr}[\varrho A(j)] = \langle\varphi_j|\varrho\varphi_j\rangle.$$

Since $\{\varphi_j\}_{j=1}^{d}$ is an orthonormal basis, we obtain

$$\sum_{j=1}^{d} A(j) = \sum_{j=1}^{d} |\varphi_j\rangle\langle\varphi_j| = I$$

and the normalization condition holds (see Example 1.61 in subsection 1.3.2). The mapping $j \mapsto A(j)$ defines a discrete observable with sample space $\Omega_A = \{1, \ldots, d\}$. We say that A is the sharp observable associated with the orthonormal basis $\{\varphi_j\}_{j=1}^{d}$.   △

The following proposition gives two useful criteria for an observable to be sharp. For each $X \in \mathcal{F}$, we denote the complement set $\Omega \setminus X$ by $\neg X$.

Proposition 3.29 Let A be an observable with outcome space $(\Omega, \mathcal{F})$. The following conditions are equivalent.

(i) A is sharp.

(ii) $A(X)A(Y) = A(X \cap Y)$ for every $X, Y \in \mathcal{F}$.

(iii) $A(X)A(\neg X) = O$ for every $X \in \mathcal{F}$.

Proof We will prove this proposition by showing that (i)$\Rightarrow$(ii)$\Rightarrow$(iii)$\Rightarrow$(i).

Assume that (i) holds and let $X, Y \in \mathcal{F}$. By Proposition 3.8(a), we have $A(X \cap Y) \leq A(X) \leq A(X \cup Y)$. As all these operators are projections, Proposition 1.42 implies that $A(X)A(X \cap Y) = A(X \cap Y)$ and $A(X)A(X \cup Y) = A(X)$. Hence, multiplying the equation in Proposition 3.8(c) by $A(X)$ gives $A(X) + A(X \cap Y) = A(X) + A(X)A(Y)$, whence (ii) follows.

Assume then that (ii) holds and let $X \in \mathcal{F}$; choosing $Y = \neg X$ in (ii), we obtain

$$A(X \cap \neg X) = A(\emptyset) = O.$$

Finally, assume that (iii) holds and let $X \in \mathcal{F}$. We then find that

$$O = A(X)A(\neg X) = A(X)(I - A(X)) = A(X) - A(X)^2$$

and hence $A(X) = A(X)^2$. This means that A is sharp. $\square$

The following result follows from Proposition 3.25, but it is also an easy consequence of Proposition 3.29.

Proposition 3.30 The range of a sharp observable A consists of mutually commuting projections.

Proof Let $A(X)$ and $A(Y)$ be projections from the range of A. Condition (ii) in Proposition 3.29 implies that

$$A(X)A(Y) = A(X \cap Y) = A(Y \cap X) = A(Y)A(X),$$

which proves the claim. $\square$

A fundamental difference between finite- and infinite-dimensional Hilbert spaces is that all sharp observables in the first case are necessarily discrete, while in the latter case there are also other sharp observables.

Proposition 3.31 Let $\mathcal{H}$ be a finite-dimensional Hilbert space of dimension d. If A is a sharp observable on $\mathcal{H}$ then A is discrete and the sample space Ω_A of A has at most d elements such that $A(j) \neq O$.

Proof By Proposition 3.29 two projections $A(X)$ and $A(Y)$ are orthogonal whenever $X \cap Y = \emptyset$. Hence, to prove the claim it is enough to show that there are at most d pairwise orthogonal projections.

Let us make the counter-assumption that $P_1, \ldots, P_r$ are pairwise orthogonal projections and $r \geq d+1$. Each projection P_j has an eigenvector φ_j with eigenvalue 1. If $i \neq j$ then the vectors φ_i and φ_j are orthogonal:

$$\langle \varphi_i | \varphi_j \rangle = \langle P_i \varphi_i | P_j \varphi_j \rangle = \langle \varphi_i | P_i P_j \varphi_j \rangle = 0.$$

This means that there are r orthogonal vectors in $\mathcal{H}$. This cannot be true as the dimension of $\mathcal{H}$ is d. Hence, we conclude that the counter-assumption is false. $\square$

The archetypal example of a sharp observable with uncountably many outcomes is the canonical position observable on $\mathbb{R}$, which we recall in the following example.

Example 3.32 (*The canonical position observable*)
Let $\mathcal{H} = L^2(\mathbb{R})$, the space of square integrable functions on $\mathbb{R}$ (see subsection 1.1.3). The *canonical position observable* Q has outcome space $(\mathbb{R}, \mathcal{B}(\mathbb{R}))$ and is defined by

$$Q(X)\psi(x) = \chi_X(x)\psi(x),$$

where χ_X is the characteristic function of a set X, i.e. $\chi_X(x) = 1$ if $x \in X$ and $\chi_X(x) = 0$ otherwise.

For a pure state $\varrho = |\psi\rangle\langle\psi|$, we thus obtain

$$\mathrm{tr}[\varrho Q(X)] = \langle \psi | Q(X)\psi \rangle = \int_X |\psi(x)|^2 \, dx.$$

This is the probability that a particle in the state ϱ is localized within X. It is straightforward to see that

$$Q(X)Q(Y) = Q(X \cap Y) = Q(Y)Q(X)$$

for all $X, Y \in \mathcal{B}(\mathbb{R})$; thus, according to Proposition 3.29 the canonical position observable Q is sharp. $\triangle$

Although sharp observables form a distinct class among all observables, it is not easy to point out a meaningful and operationally clear criterion which is valid *only* for sharp observables. One property of any sharp observable A is that, for every $X \in \mathcal{F}$ with $A(X) \neq O$, there exists a state ϱ such that $\mathrm{tr}[\varrho A(X)] = 1$. This means that we can make deterministic predictions if suitably chosen states are measured. However, this condition does not require an observable to be sharp; it is enough that each effect has eigenvalue 1. For instance, let $\varphi_1, \varphi_2, \varphi_3$ form an orthonormal basis of the three dimensional Hilbert space $\mathbb{C}^3$. Consider an observable A defined by

$$1 \mapsto A(1) = |\varphi_1\rangle\langle\varphi_1| + t|\varphi_2\rangle\langle\varphi_2|,$$
$$2 \mapsto A(2) = |\varphi_3\rangle\langle\varphi_3| + (1-t)|\varphi_2\rangle\langle\varphi_2|,$$

where $0 < t < 1$. Obviously, this observable is not sharp, but both $A(1)$ and $A(2)$ have eigenvalue 1.

3.2.2 Sharp observables and selfadjoint operators

In elementary courses on quantum mechanics, observables are defined and understood as selfadjoint operators. In this section we clarify briefly the relation between that kind of formalism and the current definitions.

The link between observables and selfadjoint operators is given by the spectral theorem (see subsection 1.3.3). In the case of a finite-dimensional Hilbert space $\mathcal{H}$, the spectral theorem tells us that every selfadjoint operator is diagonalizable. Let us shortly recall the diagonalization procedure. First, the spectrum of a selfadjoint operator $A \in \mathcal{L}_s(\mathcal{H})$ consists of eigenvalues λ satisfying the equation $\det(A - \lambda I) = 0$. The eigenvectors corresponding to eigenvalue λ are obtained by solving the equation $A\psi = \lambda\psi$. For a given eigenvalue λ, the associated eigenvectors form a linear subspace $\mathcal{H}_\lambda \subset \mathcal{H}$. For different eigenvalues the subspaces $\mathcal{H}_\lambda$ are mutually orthogonal. The dimension d_λ of $\mathcal{H}_\lambda$ is called the degeneracy of the eigenvalue λ. If we denote by P_λ the projection such that $\mathrm{ran}(P_\lambda) = \mathcal{H}_\lambda$ then the selfadjoint operator A can be expressed in the diagonal form $A = \sum_j \lambda_j P_{\lambda_j}$. In order to design a sharp observable associated with the selfadjoint operator A, the eigenvalues of A are used to define the outcome space Ω and the projection-valued measure A is defined as $\mathsf{A}(\{\lambda_j\}) = P_{\lambda_j}$. This construction can also be reversed: starting from a sharp discrete (real) observable A defined on an outcome space $\Omega = \{x_1, \ldots, x_n\}$, we can define a selfadjoint operator $A = \sum_j x_j \mathsf{A}(\{x_j\})$.

Example 3.33 (*Sharp qubit observables*)
By Proposition 3.31, any sharp observable A on $\mathbb{C}^2$ can have only two outcomes, which we can label ± 1. The corresponding operators $\mathsf{A}(1)$ and $\mathsf{A}(-1)$ are either O, I or one-dimensional projections. In the latter case there exists a unit vector $\vec{a} \in \mathbb{R}^3$ such that

$$\mathsf{A}(1) = \tfrac{1}{2}(I + \vec{a} \cdot \vec{\sigma}), \qquad \mathsf{A}(-1) = \tfrac{1}{2}(I - \vec{a} \cdot \vec{\sigma}).$$

The selfadjoint operator A corresponding to the sharp observable A is

$$A = \sum_{j=\pm 1} j\mathsf{A}(j) = \vec{a} \cdot \vec{\sigma}.$$

In this way we recover the usual description of sharp qubit observables in the form $\vec{a} \cdot \vec{\sigma}$.

However, if we label the outcomes $x, y \in \mathbb{R}$ instead of ± 1 then the related selfadjoint operator is

$$x\tfrac{1}{2}(I + \vec{a}\cdot\vec{\sigma}) + y\tfrac{1}{2}(I - \vec{a}\cdot\vec{\sigma}) = \tfrac{1}{2}\big((x+y)I + (x-y)\vec{a}\cdot\vec{\sigma}\big).$$

Therefore, by choosing x, y and $\vec{a}$ appropriately we can obtain any real linear combination of the operators I, σ_x, σ_y, σ_z and hence any selfadjoint operator on $\mathbb{C}^2$.

△

The spectral theorem in an infinite-dimensional Hilbert space has two complications. First, the notion of an eigenvalue is not sufficiently general to formulate a spectral representation of a selfadjoint operator. One needs a more general concept, that of the spectrum of an operator (see subsection 1.2.1), which allows for continuous spectral decompositions. The second point is that not all PVMs correspond to bounded selfadjoint operators; unbounded operators are unavoidable in quantum mechanics. We will not go into the details of this topic but will just discuss some common examples.

The spectral theorem for selfadjoint operators says that for each (bounded or unbounded) selfadjoint operator A, there is a unique projection-valued measure A on the Borel space $(\mathbb{R}, \mathcal{B}(\mathbb{R}))$ such that

$$A = \int_{\mathbb{R}} x\mathsf{A}(dx) . \tag{3.29}$$

This symbolic expression means that, for every vector ψ in the domain of A, we have

$$\langle\psi|A\psi\rangle = \int_{\mathbb{R}} x \,\langle\psi|\mathsf{A}(dx)\psi\rangle . \tag{3.30}$$

In most physical applications of unbounded operators, the domain is a dense subset of $\mathcal{H}$.

Each normalized projection-valued measure A on $\mathbb{R}$ determines a unique selfadjoint operator A through formula (3.29). Thus, in the language introduced in earlier sections, selfadjoint operators give an alternative description for real sharp observables.

Example 3.34 (*Position operator*)
Let us consider the canonical position observable Q introduced in Example 3.32. The *position operator* Q is formally defined as

$$Q = \int_{\mathbb{R}} x\mathsf{Q}(dx).$$

One should note that Q is an unbounded operator and so this expression has to be treated with care. For a vector ψ belonging to the domain of Q, we get

$$\langle\psi|Q\psi\rangle = \int x \,\langle\psi|\mathsf{Q}(dx)\psi\rangle = \int x|\psi(x)|^2\,dx.$$

Thus, the action of the position operator Q is given by

$$Q\psi(x) = x\psi(x). \tag{3.31}$$

This formula makes sense whenever the function $x \mapsto x\psi(x)$ is square integrable, but otherwise $Q\psi \notin L^2(\mathbb{R})$. $\triangle$

Note that the formula (3.30) implies that, for a system in a state ϱ, the mean value of the sharp observable A can be expressed via the associated selfadjoint operator A as follows:

$$\langle A \rangle_\varrho = \mathrm{tr}[\varrho A] =: \langle A \rangle_\varrho. \tag{3.32}$$

Taking into account that

$$\langle \psi | A^2 \psi \rangle = \int_\mathbb{R} x^2 \, \langle \psi | A(dx)\psi \rangle \ ,$$

it follows that the variance of A in ϱ can be written in terms of the operator A in the following way:

$$\Delta_\varrho(A) = \sqrt{\langle A^2 \rangle_\varrho - \langle A \rangle_\varrho^2}.$$

We conclude that the selfadjoint operator A gives a convenient way to calculate mean values and variances of the associated sharp observable. In some cases further statistical information is not even needed. In many situations it is simple and convenient to use selfadjoint operators as representatives of real sharp observables. However, it is untenably restrictive to use only selfadjoint operators and neglect other kinds of observable. As we saw earlier, the probabilistic structure of quantum mechanics leads in a natural way to POVMs as a correct formalization of observables; PVMs describe just one specific class of all observables.

We end this subsection with a brief description of some typical sharp observables and their selfadjoint operators.

Example 3.35 (*Momentum operator*)
The *canonical momentum observable* P is unitarily connected with the canonical position observable Q. It is defined as $P(X) = \mathcal{F}^{-1}Q(X)\mathcal{F}$, where $\mathcal{F} : L^2(\mathbb{R}) \to L^2(\mathbb{R})$ is the unitary extension of the Fourier transform to $L^2(\mathbb{R})$ (see e.g. [121]). Denoting by $\hat\psi$ the Fourier transform of $\psi \in L^2(\mathbb{R})$, we can write

$$\langle \psi | P(X)\psi \rangle = \int_X |\hat\psi(x)|^2 \, dx.$$

The corresponding (unbounded) selfadjoint operator $P = \mathcal{F}^{-1}Q\mathcal{F}$ is called the *momentum operator*. For a function ψ belonging to the Schwartz space (containing smooth functions that decrease rapidly as $x \to \infty$), the momentum operator P acts as follows:

$$P\psi(x) = -i\frac{\partial}{\partial x}\psi(x).$$

One of the most famous properties of Q and P is their *canonical commutation relation*, $QP - PQ = iI$. This relation holds in dense domains of vectors but not in all $L^2(\mathbb{R})$.   △

Exercise 3.36 Some properties of the operators Q and P could not hold if they were not unbounded operators. For instance, the canonical commutation relation $AB - BA = iI$, obeyed by P and Q, does not hold if A and B are bounded operators (a short proof can be found in e.g. [135]). Prove this claim in finite dimensions: two $d \times d$ square matrices M, N cannot satisfy a commutation relation $MN - NM = cI$ for $c \neq 0$. [Hint: Take the trace of both sides.]

Example 3.37 (*Energy operator of the hydrogen atom*)
The analysis of the hydrogen atom was one of the first successful applications of quantum theory. The energy observable of an electron (ignoring its spin degrees of freedom) in the central electric field of a proton is associated with the selfadjoint energy operator H, acting on $L^2(\mathbb{R}^3)$:

$$H\psi(\vec{r}) = -\frac{\hbar^2}{2m}\left(\frac{\partial^2}{\partial x^2} + \frac{\partial^2}{\partial y^2} + \frac{\partial^2}{\partial z^2}\right)\psi(\vec{r}) - \frac{e^2}{4\pi\epsilon_0}\frac{1}{|\vec{r}|}\psi(\vec{r}), \quad (3.33)$$

where m is the mass of electron, e is the elementary electron charge and ϵ_0 is the electrical permittivity of space. The standard analysis that can be found in almost any textbook of quantum mechanics shows that the eigenvalues of H are degenerate; they are given by

$$E_n = -\frac{me^2}{32\pi^2\epsilon_0^2\hbar^2}\frac{1}{n^2} = -13.6\,\text{eV} \times \frac{1}{n^2} \quad (3.34)$$

for $n = 1, 2, \ldots$ For an eigenvalue E_n, we can choose n^2 orthogonal eigenvectors $\psi_{nlm} \in L^2(\mathbb{R}^3)$, where l, m are integers such that $0 \leq l < n - 1$ and $-l \leq m \leq l$; we define $\Pi_n = \sum_{l,m}|\psi_{nlm}\rangle\langle\psi_{nlm}|$. The operator H has not only eigenvalues but also a continuous spectrum of positive energies $E \in [0, \infty)$ representing the situation in which the atom is ionized.

The selfadjoint operator H determines, via the spectral theorem, a selfadjoint operator H, and this operator satisfies

$$\mathsf{H}(\{E_n\}) = \Pi_n, \qquad \mathsf{H}(\mathbb{R}_+) = I - \sum_n \Pi_n \equiv \Pi_+. \quad (3.35)$$

An outcome E_n means that the electron is bounded inside the atom and has energy E_n, while an outcome $x \in \mathbb{R}_+$, where $\mathbb{R}_+$ is the field of the positive reals, indicates that the electron is not bounded inside the atom. △

Example 3.38 (*Emission spectrum of hydrogen atom*)
In this example we will sketch a connection between Example 3.37 and the emission spectrum of a hydrogen atom. In an emission-spectrum experiment a heated ensemble of hydrogen atoms starts to emit photons of some particular frequencies, forming the emission spectrum. The elementary quantum model implies that the hydrogen atom cannot change its energy in an arbitrary way. The observed frequencies are related to energy eigenvalues via the relation

$$\hbar \omega_{n \triangleright n'} = E_1 \left(\frac{1}{n^2} - \frac{1}{n'^2} \right),$$

where $n > n'$ and $E_1 = -13.6$ eV is the ground state energy, i.e. the lowest possible energy of the hydrogen atom. The relation $n \triangleright n'$ is used to indicate a transition from the energy level E_n to energy level $E_{n'}$. The outcome space Ω consists of a countable number of frequencies ω (see Figure 3.8).

Actually, our earlier picture of observables is not directly applicable to the described emission spectrum measurements described above. In fact, the main purpose of emission spectrum measurements is not to analyze the statistics of the sampled atoms but rather to identify the possible frequencies and energy levels. The recorded data are more related to properties of the system's energy operator H than to properties of a quantum state, which is in some sense fixed by the heating

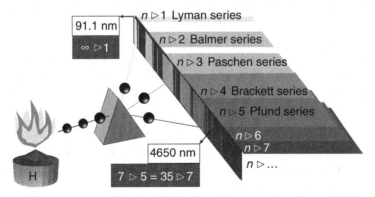

Figure 3.8 Discrete emission spectrum of hydrogen atom starting with the wavelength associated with the transition $\infty \triangleright 1$. Historically, the spectrum was divided into various series of spectral lines that can be nowadays understood as series of frequencies or wavelengths corresponding to transitions $n \triangleright j$, for $n = j + 1, j + 2, \dots$

mechanism. In order to employ the basic statistical framework we will consider a slightly modified experimental setting. Instead of the heated ensemble of hydrogen atoms we will assume that we have access to a source of individual hydrogen atoms (or a heated ensemble of atoms from which we pick a single atom per run of the experiment). When an atom emits a photon, we record only the first photon and ignore all the others emitted by the same atom. After the photon is recorded the measured hydrogen atom is replaced by a fresh one. Collecting the statistics will enable us to draw the emission spectrum and will also reveal properties of the quantum state corresponding to the source of hydrogen atoms.

Let us denote by J_ω the set of all numbers n for which there exists an n' such that the transition frequency $\omega = \omega_{n \triangleright n'}$. Let us stress that in general J_ω contains more than a single value of n and of n', hence, the observed frequency is not associated with a unique energy level E_n. For example, the transitions $7 \triangleright 5$ and $35 \triangleright 7$ both have the value of frequency, given by $\hbar \omega = -E_1 \frac{24}{35^2}$.

Observing a frequency ω, we conclude that the hydrogen atom is in a state ϱ such that at least one eigenvector ψ_{nlm} for some $n \in J_\omega$ belongs to its support. Therefore, the support of the associated effect F_ω consists of all eigenvectors corresponding to the energies E_n for which $n \in J_\omega$, and we have

$$\omega \mapsto F_\omega = \sum_{n \in J_\omega} \sum_{l,m} q_{nlm}(\omega) \Pi_{nlm}.$$

In this expression we assume that l, m are limited by n (as described in Example 3.37), that $\Pi_{nlm} = |\psi_{nlm}\rangle\langle\psi_{nlm}|$ and that each $q_{nlm}(\omega)$ describes the sum of probabilities of transitions from $\psi_{nlm} \triangleright \psi_{n'l'm'}$ such that $\omega = \omega_{n \triangleright n'}$. Let us note that the probabilities $q_{nlm}(\omega)$ depend on the particular experimental setup (temperature, photon detectors) and, moreover, some transitions are forbidden owing to conservation laws. The specification of $q_{nlm}(\omega)$ is not needed for our sketchy consideration. The normalization $\sum_\omega q_{nlm}(\omega) = 1$ guarantees that $\sum_\omega F_\omega = I - \Pi_1 - \Pi_+$, where Π_+ is the projection onto the subspace of vectors with positive energies and Π_1 is the ground state projection. Since there is no emission for a hydrogen atom in the ground state, the effect $\Pi_1 + \Pi_+$ is recorded if no photon is observed. Formally, the effect Π_+ completing the emission-spectrum observable is associated with the presence of unbounded states of the electron. However, it is a reasonable idealization that in emission experiments such states do not occur, hence, the probability of observing Π_+ is vanishingly small. Then the registration of no photons is evidence for the ground state.   △

3.2.3 Complementary observables

The notion of *complementarity* can be traced back to the early days of quantum mechanics. Niels Bohr introduced the concept in his famous Como lecture in

1927. The basic observation behind complementarity is that there exist physical quantities that require mutually exclusive measurements and, in addition to that, give pieces of information completing each other.

Suppose that A and B are sharp observables and that there are at least two effects $A(X)$ and $B(Y)$ that do not commute. Then, as proved in Proposition 3.25, the effects $A(X)$ and $B(Y)$ are not coexistent. This means that there is no observable which would contain both A and B as its parts. Therefore, we need different measurement setups to measure A and B.

Even if A and B require different measurement setups, these can be quite similar in the sense that an A-measurement gives a good estimate of a B-measurement. The key feature of complementary observables, however, is that the precise predictability of one observable implies maximal uncertainty in the other. This feature can be illustrated with spin component observables.

Example 3.39 (*Complementary spin components*)
Suppose that we measure $S^{\vec{n}}$ and we want to predict the measurement outcome distribution of $S^{\vec{m}}$. If on the one hand $\vec{m}$ is close to $\vec{n}$ then the measurement outcome distributions of $S^{\vec{n}}$ and $S^{\vec{m}}$ are very similar for all states ϱ. Namely, if $\varrho = \frac{1}{2}(I + \vec{r} \cdot \vec{\sigma})$ then the difference in the related probabilities $\mathrm{tr}[\varrho S^{\vec{n}}(+)]$ and $\mathrm{tr}[\varrho S^{\vec{m}}(+)]$ is

$$\left| \mathrm{tr}[\varrho S^{\vec{n}}(+)] - \mathrm{tr}[\varrho S^{\vec{m}}(+)] \right| = \tfrac{1}{2} |\vec{r} \cdot (\vec{n} - \vec{m})| \leq \tfrac{1}{2} \|\vec{n} - \vec{m}\| .$$

The difference varies from 0 to $\frac{1}{2} \|\vec{n} - \vec{m}\|$, depending on $\vec{r}$. Thus, we will incur only a small error if we guess the measurement outcome distribution of $S^{\vec{m}}$ from that of $S^{\vec{n}}$.

If on the other hand $\vec{m}$ is orthogonal to $\vec{n}$ then $\frac{1}{2} \|\vec{n} - \vec{m}\| = \frac{1}{\sqrt{2}}$. In this case it is clearly not a good idea to try to guess the measurement outcome distribution of $S^{\vec{m}}$ from that of $S^{\vec{n}}$. For instance, if we perform measurements in the eigenstate $\varrho = \frac{1}{2}(I + \vec{n} \cdot \vec{\sigma})$ then $\mathrm{tr}[\varrho S^{\vec{n}}(+)] = 1$ and the outcome of $S^{\vec{n}}$ is completely predictable, while $\mathrm{tr}[\varrho S^{\vec{m}}(+)] = \mathrm{tr}[\varrho S^{\vec{m}}(-)] = \frac{1}{2}$. △

It is possible to distinguish several different but related formulations of complementarity. We will discuss two variants, the first suitable for finite outcome observables and the second suitable for real outcome observables. For simplicity we restrict the discussion to sharp observables, although also some pairs of other types of observable can be argued to be complementary. More detailed overviews of different formulations can be found in [28], [30], [117].

Value complementarity

We expect complementary observables to give as different probability distributions as possible. Two maximally different probability distributions in a finite outcome set are, evidently, the Dirac probability distribution, supported at a single point, and the uniform probability distribution, supported equally at all points. This motivates the following definition.

Definition 3.40 Two sharp observables A and B with finite numbers (N and M, respectively) of outcomes are *value complementary* if the following implications are valid for any state ϱ and all outcomes $x \in \Omega_A$, $y \in \Omega_B$:

$$\text{tr}[\varrho A(x)] = 1 \quad \Rightarrow \quad \text{tr}[\varrho B(y)] = 1/M,$$
$$\text{tr}[\varrho B(y)] = 1 \quad \Rightarrow \quad \text{tr}[\varrho A(x)] = 1/N.$$

Exercise 3.41 In relation to Example 3.39, show that two spin component observables are value complementary exactly when they correspond to orthogonal directions.

Value-complementary observables exists in every finite dimension, as we demonstrate next. Let $\mathcal{H}$ be a d-dimensional Hilbert space and let $\{\varphi_i\}_{i=0}^{d-1}$ and $\{\psi_j\}_{j=0}^{d-1}$ be two orthonormal bases for $\mathcal{H}$. Each orthonormal basis defines an associated sharp observable (see Example 3.28). The value complementarity of the observables related to the bases $\{\varphi_i\}$ and $\{\psi_j\}$ means that

$$|\langle \varphi_i | \psi_j \rangle|^2 = \frac{1}{d} \qquad \forall i, j = 0, \ldots, d-1. \tag{3.36}$$

Two orthonormal bases $\{\varphi_i\}$ and $\{\psi_j\}$ satisfying (3.36) are called *mutually unbiased*

To construct mutually unbiased bases (MUBs) and hence value complementary observables, let $\{\varphi_i\}_{i=0}^{d-1}$ be an orthonormal basis for $\mathcal{H}$ and fix a complex number $\zeta \neq 1$ satisfying $\zeta^d = 1$. In other words, ζ is a dth root of unity and $\zeta = e^{2\pi i k/d}$ for some $k = 1, \ldots, d-1$. Another orthonormal basis, mutually unbiased with respect to the first, is obtained by setting

$$\psi_j = \frac{1}{\sqrt{d}} \sum_{i=0}^{d-1} \zeta^{ij} \varphi_i \tag{3.37}$$

for every $j = 0, \ldots, d-1$. Since $|\zeta| = 1$, the relation (3.36) follows.

Exercise 3.42 Verify that the vectors ψ_j defined in (3.37) form an orthonormal basis.

Let us return to Example 3.39. Instead of considering only two observables, we can choose three orthogonal unit vectors $\vec{n}, \vec{m}, \vec{k}$ in $\mathbb{R}^3$; then the related spin component observables $S^{\vec{n}}, S^{\vec{m}}$ and $S^{\vec{k}}$ are pairwise value complementary. There is no fourth orthogonal unit vector, and hence no possibility of extending this set by a fourth value-complementary observable.

Similarly, one can ask for the maximal number of orthonormal bases in dimension d such that they are all pairwise mutually unbiased. In spite of many efforts this question is still open. Wootters and Fields [147] observed that there cannot be more than $d+1$ mutually unbiased bases in a d-dimensional Hibert space. Namely, it follows from (3.36) that the operators $d|\varphi_i\rangle\langle\varphi_i|-I$ and $d|\psi_j\rangle\langle\psi_j|-I$ are orthogonal in the Hilbert–Schmidt inner product. Therefore, $d+1$ mutually unbiased bases lead to d^2 orthogonal operators, and d^2 is the dimension of the inner product space $\mathcal{T}(\mathcal{H})$. Ivanović [82] demonstrated that if the dimension d of the Hilbert space is a prime number then there exist $d+1$ mutually unbiased bases. This result has been extended to cases where d is a power of a prime number. However, a generalization to an arbitrary dimension or a counter-example is not currently known. We refer to the review article [56] for further details on this topic.

Probabilistic complementarity

Let us move to observables with outcome set $\mathbb{R}$. There is no uniform probability distribution on $\mathbb{R}$, and we therefore cannot apply value complementarity directly in these cases. The following is one possible modification of value complementarity.

Definition 3.43 Two sharp observables A and B with outcome space $(\mathbb{R}, \mathcal{B}(\mathbb{R}))$ are *probabilistically complementary* if the following implications are valid for any state ϱ and all bounded intervals $X, Y \subset \mathbb{R}$:

$$\text{tr}[\varrho A(X)] = 1 \quad \Rightarrow \quad 0 < \text{tr}[\varrho B(Y)] < 1,$$
$$\text{tr}[\varrho B(Y)] = 1 \quad \Rightarrow \quad 0 < \text{tr}[\varrho A(X)] < 1.$$

The prime example of probabilistically complementary observables is, of course, the canonical position and momentum observables.

Example 3.44 (*Complementarity of position and momentum*)
Let $\mathcal{H} = L^2(\mathbb{R})$ and Q and P be the canonical position and momentum observables, respectively. Their probability distributions for a pure state $\varrho = |\psi\rangle\langle\psi|$ are given by

$$\text{tr}[\varrho Q(X)] = \int_X |\psi(x)|^2 \, dx, \qquad \text{tr}[\varrho P(Y)] = \int_Y |\widehat{\psi}(y)|^2 \, dy.$$

Let us suppose that

$$\int_X |\psi(x)|^2 \, dx = 1$$

for some bounded interval $X \subset \mathbb{R}$. It follows that ψ must have compact support. By the Paley–Wiener theorem (a classical result in harmonic analysis), a function

$\psi \in L^2(\mathbb{R})$ has compact support if and only if its Fourier transform $\hat{\psi} \in L^2(\mathbb{R})$ extends to an entire function of exponential type (see e.g. [122]). The zeroes of an entire function are isolated, implying that its restriction to the real line cannot vanish on any open set. This implies that

$$0 < \int_Y |\widehat{\psi}(y)|^2 \, dy < 1$$

whenever Y is a bounded interval. We can switch the roles of $\widehat{\psi}$ and ψ and derive a similar conclusion from the assumption that $\widehat{\psi}$ has compact support. Therefore, Q and P are probabilistically complementary observables.   △

3.3 Informationally complete observables

Perhaps the most essential purpose of a measurement is to gain knowledge about the state of the system under investigation. Typically, the system is in an unknown state and one is trying to find out what this is. In the best case one can specify the state completely. This leads to the concept of informational completeness, introduced by Prugovečki [119].

3.3.1 Informational completeness

Let A be an observable and ϱ_1, ϱ_2 two different states. On the one hand, if

$$\Phi_A(\varrho_1) \neq \Phi_A(\varrho_2),$$

where $\Phi_A(\varrho)$ is the probability distribution of the outcomes of A in state ϱ, then a measurement of A makes a distinction between the states ϱ_1 and ϱ_2. Namely, as the probability distributions are different we see a difference in the measurement statistics corresponding to ϱ_1 and ϱ_2. On the other hand, if

$$\Phi_A(\varrho_1) = \Phi_A(\varrho_2)$$

then there is no difference in the measurement outcome statistics of ϱ_1 and ϱ_2. This motivates the following terminology.

Definition 3.45 A collection $\{A, B, \ldots\}$ of observables is *informationally complete* if, for every $\varrho_1, \varrho_2 \in \mathcal{S}(\mathcal{H})$,

$$\left. \begin{array}{l} \Phi_A(\varrho_1) = \Phi_A(\varrho_2), \\ \Phi_B(\varrho_1) = \Phi_B(\varrho_2), \\ \vdots \end{array} \right\} \quad \Rightarrow \quad \varrho_1 = \varrho_2.$$

In particular, a single observable A is *informationally complete* if the implication

$$\Phi_A(\varrho_1) = \Phi_A(\varrho_2) \qquad \Rightarrow \qquad \varrho_1 = \varrho_2 \tag{3.38}$$

holds for every $\varrho_1, \varrho_2 \in \mathcal{S}(\mathcal{H})$.

In other words, the informational completeness of a collection {A, B, ...} means that for two different states $\varrho_1 \neq \varrho_2$, at least one observable in the collection gives different probability distributions for ϱ_1 and ϱ_2. Therefore, every state can be uniquely determined from the measurement data.

It is not surprising that we can form informationally complete collections of observables. For instance, it follows from Proposition 1.21 that the set of all sharp observables consisting of one-dimensional projections is informationally complete. In practice, the task is to find a physically realizable collection of observables and then to minimize the number of observables or to optimize the collection with respect to some other criteria.

Before discussing some general results on informational completeness, we will discuss some standard examples.

Example 3.46 (*Informationally complete collection of qubit observables*)
Let us consider a qubit system. As we saw in Example 2.21, qubit states can be described by Bloch sphere vectors. Every unit vector $\vec{a}$ in the Bloch sphere determines a one-dimensional projection and hence a two-outcome sharp observable A, as explained in Example 3.33. For a state ϱ corresponding to a Bloch vector $\vec{r}$, we get $\mathrm{tr}[\varrho A(1)] = \frac{1}{2}(1 + \vec{r} \cdot \vec{a})$ and thus $\vec{r} \cdot \vec{a} = 2\mathrm{tr}[\varrho A(1)] - 1$.

Let A, B and C be three sharp observables, determined by the unit vectors $\vec{a}, \vec{b}, \vec{c} \in \mathbb{R}^3$, respectively. If the vectors $\vec{a}, \vec{b}$ and $\vec{c}$ are linearly independent then the collection {A, B, C} is informationally complete. Namely, in this case any Bloch vector $\vec{r}$ is uniquely determined from the inner products $\vec{r} \cdot \vec{a}, \vec{r} \cdot \vec{b}$ and $\vec{r} \cdot \vec{c}$. One should remember that the sharp observables A, B and C do not commute and hence require different measurement setups. △

Exercise 3.47 Continuing Example 3.46, prove that the set {A, B, C} is not informationally complete if the vectors $\vec{a}, \vec{b}, \vec{c}$ are linearly dependent. [Hint: You need to demonstrate that there are two Bloch vectors $\vec{r} \neq \vec{s}$ such that $\vec{r} \cdot \vec{a} = \vec{s} \cdot \vec{a}$, $\vec{r} \cdot \vec{b} = \vec{s} \cdot \vec{b}$ and $\vec{r} \cdot \vec{c} = \vec{s} \cdot \vec{c}$.]

Example 3.48 (*Pauli problem*)
Let $\mathcal{H} = L^2(\mathbb{R})$ and let Q and P be the canonical position and momentum observables (see Examples 3.32 and 3.35). Since

$$\langle \psi | Q(X)\psi \rangle = \int_X |\psi(x)|^2 \, dx, \qquad \langle \psi | P(Y)\psi \rangle = \int_Y |\widehat{\psi}(y)|^2 \, dy,$$

we see that knowing the measurement outcome distribution of Q (resp. P) in a pure state $|\psi\rangle\langle\psi|$ is the same as knowing the function $|\psi(\cdot)|$ (resp. $|\widehat{\psi}(\cdot)|$). The set $\{Q, P\}$ is not informationally complete, since the functions $|\psi(\cdot)|$ and $|\widehat{\psi}(\cdot)|$ do not determine the vector ψ uniquely up to a phase factor. To see this, let $\vartheta : \mathbb{R} \to \mathbb{R}$ be the rectangle function, defined as

$$\vartheta(x) = \begin{cases} 0 & \text{if } |x| > \frac{1}{2}, \\ \frac{1}{2} & \text{if } |x| = \frac{1}{2}, \\ 1 & \text{if } |x| < \frac{1}{2}. \end{cases}$$

We define

$$\psi_{\pm}(x) = \tfrac{1}{\sqrt{3}}\big(\pm\vartheta(x+1) + \vartheta(x) \mp \vartheta(x-1)\big). \tag{3.39}$$

Then $|\psi_+(\cdot)| = |\psi_-(\cdot)|$. Using the linearity of the Fourier transform and the fact that Fourier transformation converts translation into multiplication by a character, we see also that $|\widehat{\psi}_+(\cdot)| = |\widehat{\psi}_-(\cdot)|$. But ψ_+ and ψ_- are not constant multiples of one another, hence we have two different pure states that have the same position and the same momentum distributions.

The question of the informational completeness of the set $\{Q, P\}$ was first posed by Wolfgang Pauli and it is therefore called the Pauli problem. Nowadays the term 'Pauli problem' is also used to refer to several variants of the original problem, and some of these variants are still open. For instance, no exhaustive characterization of observables A seems to be known such that the collection $\{Q, P, A\}$ is informationally complete. We refer to [26], [48], [49], [63] for further details on this topic.   △

The existence of a single informationally complete observable is perhaps not evident from Definition 3.45. Before giving some examples of such observables, let us notice a simple necessary criterion.

Proposition 3.49 Let $\mathcal{H}$ be a Hilbert space of dimension $d < \infty$. If an observable A on $\mathcal{H}$ is informationally complete then its outcome set Ω contains at least d^2 points.

Proof Let A be an observable with an outcome space $\Omega = \{x_1, \ldots, x_n\}$, where $n < d^2$. Since the real inner product space $T_s(\mathcal{H})$ of selfadjoint operators is d^2-dimensional, there exists a selfadjoint operator $T \neq O$ such that $\mathrm{tr}\big[TA(x_j)\big] = 0$ for all $x_j \in \Omega$. Moreover, since $\sum_j A(x_j) = I$ we have $\mathrm{tr}[T] = 0$. We define an operator $\varrho_0 = d^{-1}(I + \|T\|^{-1} T)$. As ϱ_0 is positive (recall Exercise 1.32) and $\mathrm{tr}\big[\varrho_0\big] = 1$, it is a state. Now $\mathrm{tr}\big[\varrho_0 A(x_j)\big] = \mathrm{tr}\big[\frac{1}{d}IA(x_j)\big]$ for all $x_j \in \Omega$, meaning that A cannot distinguish between the total mixture $\frac{1}{d}I$ and the state ϱ_0. This shows that A cannot be informationally complete.   □

In the following example we demonstrate that a single observable can be informationally complete. An informationally complete observable A is called a *minimal informationally complete observable* if the outcome space Ω has the smallest possible number of elements, $|\Omega| = d^2$.

Example 3.50 (*Minimal informationally complete observable*)
Let $\mathcal{H}$ be a finite-dimensional Hilbert space with dimension d and write $\Omega = \{1, \ldots, d\}$. We follow [40] in the construction of an informationally complete observable A. First we fix an orthonormal basis $\{\varphi_j\}_{j=1}^d$ for $\mathcal{H}$. For every $j, k \in \Omega$, we denote by P_{jk} the following one-dimensional projections:

$$P_{jj} = |\varphi_j\rangle\langle\varphi_j|,$$

$$P_{jk} = \left|\frac{1}{\sqrt{2}}(\varphi_j + \varphi_k)\right\rangle\left\langle\frac{1}{\sqrt{2}}(\varphi_j + \varphi_k)\right| \qquad if \, j > k,$$

$$P_{jk} = \left|\frac{1}{\sqrt{2}}(\varphi_j + i\varphi_k)\right\rangle\left\langle\frac{1}{\sqrt{2}}(\varphi_j + i\varphi_k)\right| \qquad if \, j < k.$$

We write

$$T = \sum_{j,k=1}^d P_{jk}. \tag{3.40}$$

Since $T \geq \sum_{j=1}^d P_{jj} = I$, the operator T is positive and has square root $T^{1/2}$. Moreover, it follows from $T \geq I$ that $T^{1/2}$ is invertible. For every $j, k \in \Omega$, we then define

$$A(j, k) := T^{-1/2} P_{jk} T^{-1/2}. \tag{3.41}$$

Each operator $A(j, k)$ is positive, and the normalization condition $\sum_{j,k} A(j, k) = I$ is satisfied. Hence A is an observable with outcome set $\Omega \times \Omega$.

Suppose that ϱ_1 and ϱ_2 are two states such that the numbers $\mathrm{tr}[\varrho_1 P_{jk}]$ and $\mathrm{tr}[\varrho_2 P_{jk}]$ are the same for all indices $j, k \in \Omega$. This implies that $\langle\varphi_j|\varrho_1\varphi_k\rangle = \langle\varphi_j|\varrho_2\varphi_k\rangle$ for all $j, k \in \Omega$, and thus $\varrho_1 = \varrho_2$. This reasoning holds true also if the operators P_{jk} were replaced by the operators $A(j, k)$. Therefore, the observable A is informationally complete. Since $|\Omega \times \Omega| = d^2$, it is a minimal informationally complete observable. $\triangle$

Suppose that an observable A is informationally complete. Since the measurement statistics of A specifies the initial state ϱ uniquely, we expect that the measurement statistics of any observable can be calculated from the measurement statistics of A. The following proposition is a precise formulation of this intuitive idea.

Proposition 3.51 Suppose that dim $\mathcal{H} < \infty$. An observable A is informationally complete if and only if every selfadjoint operator can be written as a real linear combination of the elements belonging to the range of A.

Proof Let A be an observable on $\mathcal{H}$. We denote by Ā the collection of all real linear combinations of the elements in ran(A). This is a linear subspace in the real vector space $\mathcal{L}_s(\mathcal{H})$ of all selfadjoint operators. We denote by $A^\perp$ the orthogonal complement of Ā with respect to the Hilbert–Schmidt inner product, i.e.

$$A^\perp := \{B \in \mathcal{L}_s(\mathcal{H}) \ : \ \operatorname{tr}[BA(X)] = 0 \ \forall X\}.$$

Since $I \in \operatorname{ran}(A)$, we observe that $\operatorname{tr}[B] = 0$ for any $B \in A^\perp$. It follows directly from this definition that, for any two states ϱ_1 and ϱ_2, the condition $\Phi_A(\varrho_1) = \Phi_A(\varrho_2)$ is equivalent to $\varrho_1 - \varrho_2 \in A^\perp$.

Suppose that $A^\perp = \{0\}$. This implies that $\Phi_A(\varrho_1) = \Phi_A(\varrho_2)$ only when $\varrho_1 - \varrho_2 = 0$. Hence, A is informationally complete.

Now suppose that $A^\perp \neq \{0\}$, which means that there exists a nonzero selfadjoint operator $T \in A^\perp$. Split T into its positive and negative parts, $T = T^+ - T^-$. Since $\operatorname{tr}[T] = 0$, we have $\operatorname{tr}[T^+] = \operatorname{tr}[T^-] \equiv t$. We set $\varrho_1 = T^+/t$ and $\varrho_2 = T^-/t$, and clearly $\varrho_1 - \varrho_2 = T/t \in A^\perp$. This implies that $\Phi_A(\varrho_1) = \Phi_A(\varrho_2)$, hence A is not informationally complete. □

If $\mathcal{H}$ is infinite dimensional then Proposition 3.51 has to be slightly modified. Notice that the criterion for informational completeness can be written as $(A^\perp)^\perp = \mathcal{L}_s(\mathcal{H})$, using the notation of the proof. This is equivalent to informational completeness in the infinite-dimensional case also, but now $(A^\perp)^\perp$ has to be understood as the double annihilator of A with respect to the duality $\mathcal{T}(\mathcal{H})^* = \mathcal{L}(\mathcal{H})$. The reader is referred to [32], [131] for details.

Example 3.52 (*Informational completeness of the spin direction observable*)
Using Proposition 3.51 it is easy to see that the spin direction observable D is informationally complete. As we observed in Example 3.13, all the effects $\frac{1}{2}(I + \frac{1}{2}\vec{n} \cdot \vec{\sigma})$, $\vec{n} \in \mathbb{S}^2$, are in the range of D. Since the identity operator I is also in the range of D, we can write all the operators $\vec{n} \cdot \vec{\sigma}$, $\vec{n} \in \mathbb{S}^2$ as real linear combinations of the elements in ran(D). A general selfadjoint operator in $\mathbb{C}^2$ is a real linear combination of I and the Pauli operators $\sigma_x, \sigma_y, \sigma_z$. Thus, D is informationally complete.   △

While it may sometimes be hard to check whether a given observable is informationally complete, there are some simple *necessary* conditions for an observable to have this property. Two such criteria, first noted in [26], [32], are given in the next proposition.

Proposition 3.53 Let A be an informationally complete observable with outcome space $(\Omega, \mathcal{F})$, and let $X \in \mathcal{F}$.

(a) $A(X)$ does not have both eigenvalues 0 and 1.

(b) If $A(X)$ is not a scalar multiple of the identity operator I then there is at least one effect $A(Y)$ in the range of A that does not commute with $A(X)$.

Proof (a): Assume that $A(X)$ has eigenvalues 0 and 1. Let $\psi_1, \psi_0 \in \mathcal{H}$ be unit vectors such that $A(X)\psi_1 = \psi_1$ and $A(X)\psi_0 = 0$. The vectors ψ_1 and ψ_0 are orthogonal, since

$$\langle \psi_1 | \psi_0 \rangle = \langle A(X)\psi_1 | \psi_0 \rangle = \langle \psi_1 | A(X)\psi_0 \rangle = 0.$$

Moreover, we will set $\neg X \equiv \Omega \smallsetminus X$; then

$$A(\neg X)\psi_1 = (I - A(X))\,\psi_1 = 0,$$
$$A(\neg X)\psi_0 = (I - A(X))\,\psi_0 = \psi_0.$$

Set $\psi = \frac{1}{\sqrt{2}}(\psi_1 + \psi_0)$. For any $Y \in \mathcal{F}$ we then have

$$A(X \cap Y)\psi = \frac{1}{\sqrt{2}}A(X \cap Y)\psi_1, \quad A(\neg X \cap Y)\psi = \frac{1}{\sqrt{2}}A(\neg X \cap Y)\psi_0,$$

and therefore

$$\langle \psi | A(Y)\psi \rangle = \tfrac{1}{2}\langle \psi_1 | A(X \cap Y)\psi_1 \rangle + \tfrac{1}{2}\langle \psi_0 | A(\neg X \cap Y)\psi_0 \rangle.$$

Now set $\varrho = \frac{1}{2}|\psi_1\rangle\langle\psi_1| + \frac{1}{2}|\psi_0\rangle\langle\psi_0|$. Then

$$\mathrm{tr}\big[\varrho A(Y)\big] = \mathrm{tr}\big[\varrho A(X \cap Y)\big] + \mathrm{tr}\big[\varrho A(\neg X \cap Y)\big]$$
$$= \tfrac{1}{2}\langle \psi_1 | A(X \cap Y)\psi_1 \rangle + \tfrac{1}{2}\langle \psi_0 | A(\neg X \cap Y)\psi_0 \rangle.$$

This shows that $\Phi_A(|\psi\rangle\langle\psi|) = \Phi_A(\varrho)$, which contradicts the fact that A is informationally complete.

(b): Assume that $A(X)A(Y) = A(Y)A(X)$ for every $Y \in \mathcal{F}$. Let $U \equiv e^{iA(X)}$ be the unitary operator defined by $A(X)$ (see subsection 1.2.4). Since $A(X)$ is not a multiple of the identity operator I, there is a unit vector $\psi \in \mathcal{H}$ which is not an eigenvector of $A(X)$. Thus ψ is not an eigenvector of U either, and hence the vectors ψ and $U\psi$ are not parallel. Choose $\varrho_1 = |\psi\rangle\langle\psi|$ and $\varrho_2 = |U\psi\rangle\langle U\psi|$, in which case $\varrho_1 \neq \varrho_2$.

Let $Y \in \mathcal{F}$. As $A(X)$ commutes with $A(Y)$, U also commutes with $A(Y)$. We then obtain

$$\mathrm{tr}\big[\varrho_2 A(Y)\big] = \mathrm{tr}\big[|\psi\rangle\langle\psi|U^*A(Y)U\big] = \mathrm{tr}\big[|\psi\rangle\langle\psi|A(Y)\big] = \mathrm{tr}\big[\varrho_1 A(Y)\big],$$

and therefore $\Phi_A(\varrho_1) = \Phi_A(\varrho_2)$. This is in contradiction with the fact that A is informationally complete. $\square$

We saw earlier that any sharp observable in a finite d-dimensional Hilbert space has at most d nonzero outcomes (Proposition 3.31), which is less than the minimal number of outcomes for informationally complete observables (Proposition 3.49). Hence, a sharp observable is not informationally complete. Proposition 3.53 allows one to extend this conclusion to infinite-dimensional Hilbert spaces since no sharp observable satisfies the criteria (a) and (b) in Proposition 3.53. We conclude that *no sharp observable is informationally complete.*

3.3.2 Symmetric informationally complete observables

There are many different informationally complete observables. Can some be considered better than the others? In this subsection we introduce a subclass of informationally complete observables that can, with some justification, be called optimal.

Let $\mathcal{H}$ be a finite d-dimensional Hilbert space. If A is a discrete informationally complete observable then by Proposition 3.51 any state ϱ can be expressed as a linear combination $\varrho = \sum_x r_x A(x)$. This linear combination is unique if A is a minimal informationally complete observable, since in that case the effects $A(x)$ form a basis. The measurement outcome probabilities are

$$p_\varrho(x) = \text{tr}[\varrho A(x)] = \sum_y \text{tr}[A(x)A(y)] r_y.$$

The linear independence of $A(x)$ implies that the matrix $\mathfrak{A}_{xy} \equiv \text{tr}[A(x)A(y)]$ is invertible, hence we can write

$$r_y = \sum_x \mathfrak{A}_{yx}^{-1} p_\varrho(x).$$

This is the *reconstruction formula* for calculating the input state ϱ from the measurement statistics. To make the reconstruction formula as uncomplicated as possible, the matrix $\mathfrak{A}_{xy}$ should have a simple form. This motivates the following definition.

Definition 3.54 Let $\mathcal{H}$ be a finite d-dimensional Hilbert space. An observable A on $\mathcal{H}$ is *symmetric informationally complete (SIC)* if:

(i) A is minimal informationally complete (thus Ω_A has d^2 elements);
(ii) each effect $A(x)$ is a rank-1 operator;
(iii) $\text{tr}[A(x)] = $ constant for all $x \in \Omega_A$;
(iv) $\text{tr}[A(x)A(y)] = $ constant for all $x \neq y \in \Omega_A$.

The constants in conditions (iii) and (iv) are fixed by the dimension d of the Hilbert space $\mathcal{H}$. Namely, suppose that A is a SIC observable and that $\text{tr}[A(x)] = \alpha$

for all x and that $\text{tr}\left[A(x)A(y)\right] = \beta$ for all $x \neq y$. First, since $\sum_x A(x) = I$ we get $d^2\alpha = d$ and hence $\alpha = 1/d$. Moreover, since $A(x) = \alpha P(x)$ for some one-dimensional projection $P(x)$, we get $\text{tr}\left[A(x)^2\right] = \alpha^2 = 1/d^2$. Using this we obtain on the one hand

$$\sum_{x,y}\text{tr}\left[A(x)A(y)\right] = \sum_{x \neq y}\text{tr}\left[A(x)A(y)\right] + \sum_{x}\text{tr}\left[A(x)^2\right]$$
$$= \beta d^2(d^2 - 1) + d^2\alpha^2 = \beta d^2(d^2 - 1) + 1.$$

On the other hand,

$$\sum_{x,y}\text{tr}\left[A(x)A(y)\right] = \text{tr}[I] = d.$$

A comparison of these equations gives $\beta = 1/(d^2(d + 1))$.

Example 3.55 (*SIC qubit observable*)
A SIC qubit observable A has four outcomes. We can write these four effects in the form $A(j) = \frac{1}{4}(I + \vec{s}_j \cdot \vec{\sigma})$, where $\vec{s}_j$ are unit vectors in $\mathbb{R}^3$. The factor $\frac{1}{4}$ is determined by condition (iii). For $j \neq k$ we have the requirement that

$$\frac{1}{12} = \text{tr}\left[A(j)A(k)\right] = \frac{1}{8}(1 + \vec{s}_j \cdot \vec{s}_k). \tag{3.42}$$

This implies that $\vec{s}_j \cdot \vec{s}_k = -\frac{1}{3}$, which means that the vectors $\vec{s}_1, \vec{s}_2, \vec{s}_3$, and $\vec{s}_4$ point to the corners of a tetrahedron. A possible choice is

$$\vec{s}_1 = \tfrac{1}{\sqrt{3}}(1, 1, 1), \qquad \vec{s}_2 = \tfrac{1}{\sqrt{3}}(1, -1, -1),$$
$$\vec{s}_3 = \tfrac{1}{\sqrt{3}}(-1, 1, -1), \qquad \vec{s}_4 = \tfrac{1}{\sqrt{3}}(-1, -1, 1).$$

If we set $p_k = \text{tr}\left[\varrho A(k)\right]$ then the reconstruction formula reads

$$\varrho = \frac{1}{2}I + \frac{2}{3}\left(\sum_k p_k\vec{s}_k\right) \cdot \vec{\sigma}$$

and the Bloch vector of the reconstructed state ϱ is recovered from the formula

$$\vec{r} = \sqrt{3}\left(2(p_1 + p_2) - 1, \quad 2(p_1 + p_3) - 1, \quad 2(p_1 + p_4) - 1\right). \tag{3.43}$$

$\triangle$

Curiously, it is not known whether a SIC observable exists in every finite dimension, even though there have been several insightful studies on this topic. The existence of SIC observables has been verified in (at least) dimensions 2–15. Additionally, there have been numerical studies, up to dimension 67, giving evidence for the existence of SIC observables in all the dimensions investigated [128].

3.3.3 State estimation

The essential goal of any experiment is to obtain probability distributions for the
possible measurement outcomes. But wait: how can we measure probabilities?
What we actually observe is a finite sequence of measurement outcomes, and the
probability distribution is afterwards abstracted from this sequence. In this subsection we illustrate briefly the idea of data processing, which finally allows for the
experimental confirmation of theoretical models. We will introduce one specific
method for the extraction of a probability distribution from the observed outcome
frequencies. For more details and references we refer to [110].

Suppose that we perform a measurement of an n-valued informationally complete observable A for a system in an unknown input state ϱ. Once we agree that
a probability distribution p has been obtained, we can employ the reconstruction
formula $\varrho = \Phi_A^{-1}(p)$ to recover the state. Suppose that in N experimental runs we
observe an outcome sequence $j_1, \ldots, j_N \equiv \vec{j}$. From this data set we can calculate N_j, the number of the occurrences of each outcome j. The frequencies given
by the fractions $f(j) \equiv N_j/N$ form a probability distribution, i.e. $f(j) \geq 0$
and $\sum_j f(j) = 1$. The simplest thing would be to set $p(j) = f(j)$ and exploit
the reconstruction formula. Unfortunately, as we will demonstrate in the following
example, that kind of approach may fail to give a relevant answer.

Example 3.56 (*'Unphysical' result*)
Suppose we perform $N = 20$ measurement runs of the qubit SIC observable A
defined in Example 3.55. We would like to determine the unknown input state.
The possible measurement outcomes are $1, 2, 3, 4$, and the obtained measurement
data are

$$1, 4, 2, 3, 1, 2, 3, 1, 1, 1, 1, 2, 1, 1, 2, 3, 1, 3, 1, 1.$$

Thus, the frequencies of the outcomes are $f(1) = \frac{11}{20}$, $f(2) = f(3) = \frac{4}{20}$ and
$f(4) = \frac{1}{20}$. If we now set $p(j) = f(j)$ we should be able to identify a unique density operator ϱ satisfying $\mathrm{tr}[\varrho A(j)] = p(j)$. Employing the reconstruction formula
(3.43) we get the innocent-looking Bloch vector

$$\vec{r} = \sqrt{3}\left(\frac{1}{2}, \frac{1}{2}, \frac{1}{5}\right),$$

but $\|\vec{r}\|^2 = 3\left(\frac{1}{4} + \frac{1}{4} + \frac{1}{25}\right) > 1$. This means that the probability distribution p
does not correspond to any state. The small size of our data and its oversimplified
interpretation is the reason for this unphysical result. △

Exercise 3.57 Let us continue the discussion begun in Example 3.56, and demonstrate that having a small sample of data and setting $p = f$ can cause problems
for any observed frequency. For a given finite data set, there is a certain number of
possible frequency distributions and only some of them correspond to a valid state.

For example, for $N = 1$ there are four possible frequencies ($f(1) = 1$ or $f(2) = 1$ or $f(3) = 1$ or $f(4) = 1$); however, none of them describes a physically valid state because there is no state ϱ which would give $\text{tr}[\varrho A(j)] = 1$ for a single outcome j. For the given observable A find the smallest N for which the setting $p = f$ may give a valid quantum state. [Hint: Analyze the cases $N = 2, 3, \ldots$]

The previous discussion illustrates explicitly the fact that the probability distributions obtained in a fixed quantum measurement do not cover the whole space of potential probability distributions, i.e. the set $\Phi_A(\mathcal{S}(\mathcal{H}))$ is typically a proper subset of $Prob(\Omega)$. In any case, even if the frequency distribution lies inside $\Phi_A(\mathcal{S}(\mathcal{H}))$, our reconstruction is always only an estimate of the input state. We denote by $\hat{\varrho}$ the estimated state and by ϱ the 'true' input state. The general goal of the estimation procedure is to assign a unique and valid estimate $\hat{\varrho}$ for arbitrary experimental data.

One of the most popular estimation procedures is based on the concept of *likelihood*. Informally: given that the data is $\vec{j}$, what is the likelihood that the measurement was performed on a state ϱ?

Let us first recall that the conditional probability $p(\vec{j}|\varrho)$ for observing an outcome sequence $\vec{j} = j_1, \ldots, j_N$, under the condition that the input state is ϱ, is

$$p(\vec{j}|\varrho) = \prod_{k=1}^{N} p(j_k|\varrho) = \prod_{j=1}^{n} p(j|\varrho)^{N_j}. \tag{3.44}$$

Here N_j is the number of occurrences of outcome j and $p(j|\varrho) = \text{tr}[\varrho A(j)]$. We emphasize that the different runs of an experiment are assumed to be independent. The probability $p(\vec{j}|\varrho)$ does not depend on the order of the outcomes but only on their relative occurrences. The likelihood function ℓ is defined as

$$\ell(\varrho|\vec{j}) := \frac{1}{N} \log p(\vec{j}|\varrho) = \sum_{j=1}^{n} \frac{N_j}{N} \log p(j|\varrho),$$

and the so-called *maximum likelihood estimate* is

$$\hat{\varrho} = \arg \max_{\varrho \in \mathcal{S}(\mathcal{H})} \ell(\varrho|\vec{j}). \tag{3.45}$$

This formula means that $\hat{\varrho}$ is the element of $\mathcal{S}(\mathcal{H})$ that maximizes the likelihood function. Obviously, this makes sense only if a maximum exists and is unique.

A practical evaluation of the expression (3.45) can be difficult, and typically one has to use numerical optimization methods. The likelihood function ℓ is nonpositive, i.e. $\ell \leq 0$. Let us set $p_j = p(E_{\vec{j}}|\varrho)$, so that $\ell(\varrho|E_{\vec{j}}) = \sum_j f_j \log p_j$; it is closely related to the so-called *relative entropy*, which is defined via the formula $H(q||p) = \sum_j q(j)(\log q(j) - \log p(j))$, where q, p are probability distributions. Setting $q(j) = f(j)$ we obtain the identity

$$\ell(\varrho|\vec{j}) = -H(f) - H(f||p), \tag{3.46}$$

where $H(f) := -\sum_j f(j) \log f(j)$ is the Shannon entropy of the probability distribution f. Since $H(f)$ is a positive constant and $H(f\|p)$ is nonnegative (see for instance [90]), it follows that the (global) maximum is unique and is obtained when the relative entropy vanishes. This happens if and only if $f = p$. In other words, providing that the observed frequency is allowed, it maximizes the likelihood and gives directly a valid maximum likelihood estimate $\hat{\varrho}$. Let us note that, especially for small statistics, the likelihood ℓ may have several local maxima; this represents potential problems for numerical algorithms.

Example 3.58 In this example we will employ the maximum likelihood method to obtain a physically valid estimate from the data recorded in Example 3.56. The likelihood function reads

$$\ell(\varrho|\vec{j}) = \frac{11}{20} \log p(1|\varrho) + \frac{4}{20} \left(\log p(2|\varrho) + \log p(3|\varrho) \right) + \frac{1}{20} \log p(4|\varrho).$$

We write ϱ in the Bloch form $\varrho = \frac{1}{2}(I + \vec{r} \cdot \vec{\sigma})$ for a vector $\vec{r} \in \mathbb{R}^3$ with $\|\vec{r}\| \leq 1$. Then $p(j|\varrho) = \frac{1}{4}(1 + \vec{r} \cdot \vec{s}_j)$, and we can write the likelihood function as a function of the three parameters $\vec{r} = (x, y, z) \in \mathbb{R}^3$:

$$\ell(x, y, z) = \frac{11}{20} \log(\sqrt{3} + x + y + z) + \frac{4}{20} \log(\sqrt{3} + x - y - z)$$

$$+ \frac{4}{20} \log(\sqrt{3} - x + y - z) + \frac{1}{20} \log(\sqrt{3} - x - y + z)$$

$$- \log 4\sqrt{3}.$$

Under the condition $x^2 + y^2 + z^2 \leq 1$ this function has a unique maximum. Let us note that $\ell(x, y, z) = \ell(y, x, z)$; thus, owing to uniqueness the optimal Bloch vector has the form $\vec{r} = (x, x, z)$. Applying normalization we get $z = \pm\sqrt{1 - 2x^2}$ and $x \in [-1/\sqrt{2}, 1/\sqrt{2}]$. The likelihood function then reads

$$\ell(x) = \frac{11}{20} \log(\sqrt{3} + 2x \pm \sqrt{1 - 2x^2}) + \frac{2}{5} \log(\sqrt{3} \mp \sqrt{1 - 2x^2})$$

$$+ \frac{1}{20} \log(\sqrt{3} - 2x \pm \sqrt{1 - 2x^2}) - \log 4\sqrt{3}, \tag{3.47}$$

and achieves its maximum at approximately $x = 0.698$ (see Figure 3.9). In conclusion, maximum likelihood processing of the observed data produces the pure state $\hat{\varrho} = \frac{1}{2}[I + 0.698(\sigma_x + \sigma_y) + 0.16\sigma_z)]$. $\triangle$

3.4 Testing quantum systems

As we saw in Section 3.3, informationally complete observables provide a universal tool enabling us to make a decision on any property of a quantum system.

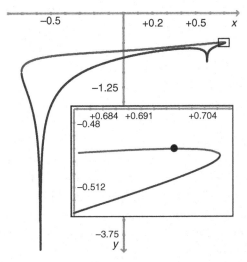

Figure 3.9 Plot of the likelihood function from (3.47). The inset shows the detail near the maximum value (solid circle) of the likelihood function.

However, there are at least two reasons why it is of interest to design experiments testing only particular properties of quantum systems. First, it may be difficult to implement informationally complete observables or to process the observed data, especially when the system of interest is large. Second, a complete description is much more than we usually need. The complete and precise knowledge of a quantum state is in reality quite unfeasible and, at the same time, is not usually needed.

3.4.1 Complete versus incomplete information

Any question about properties of a system can be formulated as follows. The task is to design an experiment that enables us to decide between different alternatives or hypotheses regarding the possible states of the system. These alternatives define a set of potential answers and the goal of an experiment is to choose the alternative that maximizes our success. Depending on whether the set of alternatives is finite or infinite, we have a discrimination or estimation problem, respectively. In the latter case it is impossible to give an error-free answer because it is impossible in practice to measure probability distributions precisely. In this section we will focus on discrimination problems.

Perhaps the simplest version of a quantum discrimination problem is the state discrimination problem, in which a system is known to be either in state ϱ_1 or in state ϱ_2. Hence, our task is to choose one of these two alternatives. For this purposes it is sufficient to design an observable A for which $\Phi_A(\varrho_1) \neq \Phi_A(\varrho_2)$, where

$\Phi_A(\varrho)$ is the probability of state ϱ. It follows from the definition of a quantum state that such an observable exists, because otherwise these states would not be considered to be different. In fact, informationally complete observables satisfy the above requirement for all pairs ϱ_1, ϱ_2. However, there are other suitable observables that are not necessarily informationally complete. Once the family of all suitable observables is characterized, the next question is which of them is optimal. In other words, we are interested in determining which observable induces the most readily distinguishable probability distributions $\Phi_A(\varrho_1)$ and $\Phi_A(\varrho_2)$. In order to find an answer we need to define some figure of merit for probability distributions.

Instead of doing so directly, let us change the problem slightly. Let us note that, although our task is to distinguish the probability distributions $\Phi_A(\varrho_1)$ and $\Phi_A(\varrho_2)$, the problem concerning the finiteness of the data partially remains. But do we actually need to reconstruct or estimate the probabilities? Is it not sufficient to base our reasoning directly on the observed sequence of events? In principle the task is to distinguish between two alternatives, hence a two-outcome observable could be sufficient! Even a single click experiment with a single copy of the system available can be used to draw nontrivial and meaningful conclusions on the particular state of the system.

To give a simple example, imagine that we toss a coin only once and observe the outcome *head*. What can we say about the coin? Without any additional assumption we can conclude only that the probability of getting head is nonzero. However, if we have been told that the coin is either a fair one or an unfair one having the same value *tail* on both sides, then our conclusion can be much stronger. In fact, in this restricted case we may with certainty confirm that the coin is fair.

This example illustrates that even in statistical theories there are situations in which certain inferences can be made from a finite number of experimental runs. Quantum theory is no exception. Under certain circumstances and with suitable a priori knowledge, quantum states may be uniquely identified even from individual measurement outcomes. This kind of task, which is not based on measurement statistics but on some finite number of measurement outcomes, is called a *nonstatistical discrimination problem*. By contrast, in a *statistical discrimination problem* we tend to measure the probability distribution. For a general reference on this type of problem we recommend the classic text [75] by Helstrom and the more recent review book [110].

3.4.2 Unambiguous discrimination of states

The goal of *unambiguous state discrimination* (or *error-free state discrimination*) is to identify an unknown state ϱ out of some set of possible states $\{\varrho_1, \ldots, \varrho_n\}$ without an error (see Figure 3.10). We assume that only a single copy of a system

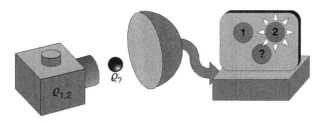

Figure 3.10 Unambiguous discrimination.

is available and that η_j is the a priori probability that the unknown state ϱ is ϱ_j. The main question is which sets of states can be discriminated in an unambiguous way. This problem was originally introduced by Ivanovic [83].

In the general scheme of unambiguous state discrimination we accept the possibility that no conclusion will be reached. However, when there is a conclusion it has to be correct. In contrast with unambiguous state discrimination, there are also probabilistic state discrimination schemes; these are discussed briefly in subsection 3.4.3. In the present subsection we focus on the unambiguous discrimination of two pure states. We refer to [42] and [21] for reviews of various state-discrimination strategies.

As a special type of unambiguous discrimination, we say that a set of states can be *perfectly discriminated* if the probability of getting a conclusive outcome is 1. Hence, a set of states $\{\varrho_1, \ldots, \varrho_n\}$ can be perfectly discriminated if and only if there exists an observable A with outcome space $\Omega = \{1, \ldots, n\}$ such that

$$\mathrm{tr}\big[\varrho_1\mathsf{A}(1)\big] = \mathrm{tr}\big[\varrho_2\mathsf{A}(2)\big] = \cdots = \mathrm{tr}\big[\varrho_n\mathsf{A}(n)\big] = 1. \qquad (3.48)$$

Indeed, in the perfect discrimination case one must be able to draw the correct conclusion in every possible measurement outcome. This is exactly the condition (3.48).

Exercise 3.59 Let $P_1, \ldots, P_n$ be a collection of orthogonal pure states. Show that the set $\{P_1, \ldots, P_n\}$ can be perfectly discriminated. [Hint: Recall Example 3.28.]

Proposition 3.60 Let P_1 and P_2 be two pure states. They can be perfectly discriminated if and only if they are orthogonal, i.e. $P_1 P_2 = O$.

Proof We have already seen in Exercise 3.59 that two orthogonal pure states can be perfectly discriminated. Therefore we will show the necessity of mutual orthogonality.

If P_1 and P_2 can be perfectly discriminated then there exists an observable A with outcome space $\Omega = \{1, 2\}$ such that $\mathrm{tr}\big[P_1\mathsf{A}(1)\big] = \mathrm{tr}\big[P_2\mathsf{A}(2)\big] = 1$. By

Proposition 2.34 this implies that

$$A(1)P_1 = P_1, \qquad A(2)P_2 = P_2.$$

Since $A(2) = I - A(1)$, the latter condition gives $A(1)P_2 = 0$. Therefore we obtain

$$\mathrm{tr}[P_1 P_2] = \mathrm{tr}\big[A(1)P_1 P_2\big] = \mathrm{tr}\big[P_1 P_2 A(1)\big] = \mathrm{tr}\big[P_1(A(1)P_2)^*\big] = 0,$$

which means that P_1 and P_2 are orthogonal. □

This relation between orthogonality and perfect discrimination provides an operational interpretation of the Hilbert space dimension. The number of pure states that can be perfectly discriminated cannot exceed the dimension of the Hilbert space.

Example 3.61 (*Impossibility of identifying the state of a single system*)
The concept of informational completeness refers to statistical information; to identify an unknown input state we need the total probability distribution of measurement outcomes. It should not come as a surprise, therefore, that one cannot identify the unknown state if only a single quantum system is available. A hypothetical observable performing this task would perfectly discriminate all states. Since not all pure states are orthogonal, we can see from Proposition 3.60 that this kind of observable cannot exist.

There is also a stronger trade-off between statistical information (informational completeness) and single-shot information (perfect discrimination). Suppose that an observable A can perfectly discriminate two states ϱ_1 and ϱ_2. This would mean that $\mathrm{tr}\big[\varrho_1 A(j_1)\big] = 1$ for some outcome j_1 and similarly $\mathrm{tr}\big[\varrho_2 A(j_2)\big] = 1$ for some other outcome j_2. But then $\mathrm{tr}\big[\varrho_2 A(j_1)\big] = 0$, and we would conclude that the effect $A(j_1)$ has both eigenvalues 0 and 1. This violates a necessary condition for informational completeness, derived in Proposition 3.53. Hence if an observable A can perfectly discriminate some states, it cannot be informationally complete. △

Suppose that a single quantum system in one of the two pure states ϱ_1, ϱ_2 is given and that these states are not orthogonal. As perfect discrimination is not possible, we turn to a more modest question: is it possible to identify the state ϱ_1 unambiguously? In comparison with the previous discussion, we are not interested in whether the unknown state is ϱ_2; only the conclusion $\varrho = \varrho_1$ matters. We say that a state ϱ_1 can be *unambiguously identified* within the set $\{\varrho_1, \varrho_2\}$ if and only if there exists an observable A with two outcomes 1 and ? such that

$$\mathrm{tr}\big[\varrho_1 A(1)\big] > 0, \qquad \mathrm{tr}\big[\varrho_2 A(1)\big] = 0. \tag{3.49}$$

The effect $A(?) = I - A(1)$ corresponds to an *inconclusive* outcome since no unambiguous conclusion is assigned to it. The success of the state identification is quantified by means of the probability to observe the conclusive outcome 1, i.e.

$$p_{\text{success}} = \eta \, \text{tr}[\varrho_1 A(1)], \tag{3.50}$$

where η is the a priori probability that the unknown state ϱ is ϱ_1.

Example 3.62 (*Unambiguous identification of a pure state*)
Let us consider the identification of two pure states P_1 and P_2. The no-error condition $\text{tr}[P_2 A(1)] = 0$ in (3.49) is equivalent to $\text{tr}[P_2 A(?)] = 1$. By Proposition 2.34 this implies that $P_2 \leq A(?)$ or, in other words, $A(1) \leq I - P_2$. Consequently, the probability $\text{tr}[P_1 A(1)]$ is maximal if we set $A(1) = I - P_2$. Hence, the best achievable probability of success is

$$p_{\text{success}} = \eta(1 - \text{tr}[P_1 P_2]). \tag{3.51}$$

Notice that $\text{tr}[P_1 P_2] = 1$ only when $P_1 = P_2$. Therefore, $p_{\text{success}} > 0$ whenever $P_1 \neq P_2$. We conclude that, for any pair of different pure states, it is possible to carry out the identification task with nonzero probability of success even if only a single copy of the system is available.   △

Example 3.63 (*Unambiguous identification of a mixed state*)
Let ϱ_1, ϱ_2 be two states that are not necessarily pure. The condition for identification of ϱ_1 from the set $\{\varrho_1, \varrho_2\}$ is still the set of constraints (3.49). However, the identification task cannot be performed for all pairs of states. Consider an effect E such that $\text{tr}[\varrho_1 E] > 0$ and write

$$S_E^\perp = \{\varrho \in \mathcal{S}(\mathcal{H}) : \text{tr}[\varrho E] = 0\}.$$

It follows that the two-outcome observable A, defined as $A(1) = E, A(?) = I - E$, can be employed to identify the state ϱ_1 from the set $\{\varrho_1, \varrho_2\}$ provided that $\varrho_2 \in S_E^\perp$. However, if for all effects E the inequality $\text{tr}[\varrho_1 E] > 0$ implies that $\text{tr}[\varrho_2 E] > 0$ then the state ϱ_1 cannot be identified from the set $\{\varrho_1, \varrho_2\}$. It follows that while for pure states the success probability given in (3.51) is symmetric with respect to the exchange of ϱ_1 and ϱ_2, it is not necessarily symmetric for mixed states. For example, if $\mathcal{H}$ is finite dimensional and ϱ_2 is the total mixture $\frac{1}{d}I$, then identification is impossible for any state ϱ_1. Thus, no state can be identified from the total mixture. However, if we choose $\varrho_1 = \frac{1}{d}I$ then we see that the total mixture can be identified with nonzero success probability provided that the support of ϱ_2 is not the whole Hilbert space $\mathcal{H}$.   △

Let us return to the unambiguous discrimination of two pure states $P_1 = |\psi_1\rangle\langle\psi_1|$ and $P_2 = |\psi_2\rangle\langle\psi_2|$ (assuming that $P_1 \neq P_2$). If the states are not orthogonal then, according to Proposition 3.60, they cannot be perfectly discriminated. However, as we have seen previously, each pure state can be unambiguously distinguished from any other pure state. We denote by A and B the observables identifying the states P_1 and P_2 in the optimal way, respectively. Hence,

$A(1) = I - P_2$ and $B(1) = I - P_1$. Let us then fix a number $0 < q < 1$ and define an observable C with outcome space $\{1, 2, ?\}$ in the following way:

$$C(1) = q(I - P_2), \quad C(2) = (1 - q)(I - P_1), \quad C(?) = q P_2 + (1 - q) P_1.$$

The observable C is a mixture of A and B (compare with (3.20) in subsection 3.1.5). The inconclusive outcomes of A and B are identified as a single inconclusive outcome '?'. As C is a mixture of A and B and these are the optimal observables for identifying the states P_1 and P_2, we expect that C will be able to discriminate these states.

The observable C leads to the following list of probabilities:

$$p_{P_1}^C(1) = q(1 - \text{tr}[P_1 P_2]) \neq 0, \qquad p_{P_2}^C(1) = 0,$$
$$p_{P_1}^C(2) = 0, \qquad p_{P_2}^C(2) = (1 - q)(1 - \text{tr}[P_1 P_2]) \neq 0,$$
$$p_{P_1}^C(?) = 1 - q + q\text{tr}[P_1 P_2] \neq 0, \qquad p_{P_2}^C(?) = q + (1 - q)\text{tr}[P_1 P_2] \neq 0.$$

Therefore if we get the outcome 1, we can draw the conclusion that $\varrho = P_1$. In a similar way, the outcome 2 leads to the conclusion that $\varrho = P_2$. If the outcome ? is obtained then no conclusion can be made.

We conclude that, although perfect discrimination is not possible, there exists an observable discriminating between the nonorthogonal pure states P_1 and P_2 provided that an inconclusive result is allowed. As before, the success probability p_{success} is quantified as the probability of getting a conclusive result. In the above scheme we get

$$p_{\text{success}} = \eta \, \text{tr}[C(1) P_1] + (1 - \eta) \, \text{tr}[C(2) P_2]$$
$$= q\eta \, \text{tr}[P_2(I - P_1)] + (1 - q)(1 - \eta) \, \text{tr}[P_1(I - P_2)]$$
$$= (q\eta + (1 - q)(1 - \eta))(1 - \text{tr}[P_1 P_2]).$$

For an equal a priori distribution of the states P_1 and P_2 (i.e. $\eta = \frac{1}{2}$), unambiguous discrimination by C is successful with probability

$$p_{\text{success}} = \tfrac{1}{2}(1 - \text{tr}[P_1 P_2]) = \tfrac{1}{2}(1 - |\langle \psi_1 | \psi_2 \rangle|^2). \tag{3.52}$$

Does the above procedure give the best achievable success probability? To find an optimal solution, we are looking for an observable D with three outcomes $1, 2, ?$ such that

$$\text{tr}[P_1 D(2)] = 0, \qquad \text{tr}[P_2 D(1)] = 0 \tag{3.53}$$

and the success probability

$$p_{\text{success}} = \eta \, \text{tr}[P_1 D(1)] + (1 - \eta) \, \text{tr}[P_2 D(2)]$$

is as large as possible.

Before we come to the optimal value of p_{success}, we will derive an upper bound on the success probability; this was originally given by Feng *et al.* in [59]. Before we can formulate this bound we need the following lemma.

Lemma 3.64 Let $T \in \mathcal{T}(\mathcal{H})$. Then

$$\sup_{U \in \mathcal{U}(\mathcal{H})} |\text{tr}[UT]| = \|T\|_{\text{tr}} .$$

Proof For each $A \in \mathcal{L}(\mathcal{H})$ with $\|A\| \leq 1$, we have on the one hand

$$|\text{tr}[AT]| \leq \|A\| \, \|T\|_{\text{tr}} \leq \|T\|_{\text{tr}} .$$

On the other hand, if $T = V|T|$ is the polar decomposition of T then

$$\text{tr}[V^*T] = \text{tr}[|T|] = \|T\|_{\text{tr}}$$

and $\|V^*\| = \|V\| \leq 1$. We conclude that

$$\sup_{A \in \mathcal{L}(\mathcal{H}), \|A\| \leq 1} |\text{tr}[AT]| = \|T\|_{\text{tr}} . \tag{3.54}$$

If $\dim \mathcal{H} < \infty$, the operator V in the polar decomposition of T is unitary and the claim therefore follows from the previous calculation. In the general case, we recall that the Russo–Dye theorem (Corollary 1 in [125]) states that

$$\sup_{A \in \mathcal{L}(\mathcal{H}), \|A\| \leq 1} |\text{tr}[AT]| = \sup_{U \in \mathcal{U}(\mathcal{H})} |\text{tr}[UT]|. \tag{3.55}$$

A comparison of (3.54) and (3.55) yields the claim. $\qquad\square$

Proposition 3.65 Let ϱ_1 and ϱ_2 be two states occurring with a priori probabilities η and $1 - \eta$. The probability of success of their unambiguous discrimination has the following upper bound:

$$p_{\text{success}} \leq 1 - 2\sqrt{\eta(1-\eta)} \, \text{tr}\big[|\sqrt{\varrho_1}\sqrt{\varrho_2}|\big] . \tag{3.56}$$

Proof The probability of success can be written in the form $p_{\text{success}} = 1 - p_{\text{fail}}$, where $p_{\text{fail}} = \text{tr}\big[\mathsf{D}(?)(\eta\varrho_1 + (1-\eta)\varrho_2)\big]$ is the failure probability. We then have

$$\begin{aligned}
p_{\text{fail}}^2 &= \eta^2(\text{tr}\big[\mathsf{D}(?)\varrho_1\big])^2 + (1-\eta)^2(\text{tr}\big[\mathsf{D}(?)\varrho_2\big])^2 \\
&\quad + 2\eta(1-\eta) \, \text{tr}\big[\mathsf{D}(?)\varrho_1\big] \text{tr}\big[\mathsf{D}(?)\varrho_2\big] \\
&\geq 4\eta(1-\eta)\text{tr}\big[\mathsf{D}(?)\varrho_1\big] \text{tr}\big[\mathsf{D}(?)\varrho_2\big] ,
\end{aligned}$$

where we have used the inequality $a^2 + b^2 \geq 2ab$ for $a = \eta \, \text{tr}\big[\mathsf{D}(?)\varrho_1\big]$ and $b = (1-\eta) \, \text{tr}\big[\mathsf{D}(?)\varrho_2\big]$. Using the Cauchy–Schwartz inequality we obtain further that, for any unitary operator U,

$$\text{tr}\big[D(?)\varrho_1\big]\,\text{tr}\big[D(?)\varrho_2\big] = \text{tr}\Big[U\sqrt{\varrho_1}\sqrt{D(?)}\sqrt{D(?)}\sqrt{\varrho_1}U^\dagger\Big]$$
$$\times\text{tr}\Big[\sqrt{\varrho_2}\sqrt{D(?)}\sqrt{D(?)}\sqrt{\varrho_2}\Big]$$
$$\geq \big(\text{tr}\big[U\sqrt{\varrho_1}D(?)\sqrt{\varrho_2}\big]\big)^2,$$

Since $D(?) = I - D(1) - D(2)$ and, by the conditions for unambiguous discrimination, $D(1)\varrho_2 = D(2)\varrho_1 = O$ it follows that $\sqrt{\varrho_1}D(?)\sqrt{\varrho_2} = \sqrt{\varrho_1}\sqrt{\varrho_2}$. Thus,

$$p_{\text{fail}} \geq 2\sqrt{\eta(1-\eta)}\,\text{tr}\big[U\sqrt{\varrho_1}\sqrt{\varrho_2}\big] \tag{3.57}$$

for all unitary operators U. According to Lemma 3.64 this leads to the formula

$$p_{\text{fail}} \geq 2\sqrt{\eta(1-\eta)}\,\text{tr}\big[|\sqrt{\varrho_1}\sqrt{\varrho_2}|\big]. \tag{3.58}$$

$\square$

For two pure states distributed with equal a priori probabilities, Proposition 3.65 gives

$$p_{\text{success}} \leq 1 - |\langle\psi_1|\psi_2\rangle|. \tag{3.59}$$

From (3.52) it is straightforward to see that observable C does not reach this upper bound. Namely, the equation

$$\frac{1}{2}\left(1 - |\langle\psi_1|\psi_2\rangle|^2\right) = 1 - |\langle\psi_1|\psi_2\rangle|$$

holds only if $|\langle\psi_1|\psi_2\rangle| = 1$, but this would mean that $\varrho_1 = \varrho_2$. Is there an observable reaching this bound? The following example gives a positive answer.

Example 3.66 (*Optimal observable for unambiguous discrimination of pure states*)

Consider a pair of pure states $P_1 = |\psi_1\rangle\langle\psi_1|$, $P_2 = |\psi_2\rangle\langle\psi_2|$ appearing with equal a priori probabilities. The optimal observable D_{opt} for unambiguous discrimination of these pure states is given by

$$D_{\text{opt}}(1) = \frac{1}{1 + |\langle\psi_1|\psi_2\rangle|}(P - |\psi_2\rangle\langle\psi_2|), \tag{3.60}$$

$$D_{\text{opt}}(2) = \frac{1}{1 + |\langle\psi_1|\psi_2\rangle|}(P - |\psi_1\rangle\langle\psi_1|), \tag{3.61}$$

$$D_{\text{opt}}(?) = I - D_{\text{opt}}(1) - D_{\text{opt}}(2), \tag{3.62}$$

where P is the projection onto the two-dimensional subspace of $\mathcal{H}$ spanned by the vectors ψ_1, ψ_2, i.e.

$$P = \frac{1}{N}(P_1 + P_2 - P_1 P_2 - P_2 P_1),$$

with $N = 1-\text{tr}[P_1 P_2]$. Since $P\psi_j = \psi_j$ it follows that $(P - P_j)P_j = O$ and hence the no-error conditions $\text{tr}[P_1 \mathsf{D}_{\text{opt}}(2)] = \text{tr}[P_2 \mathsf{D}_{\text{opt}}(1)] = 0$ are satisfied. Moreover,

$$(P_1 + P_2)P = \frac{1}{N}(P_1 + P_2 - P_1 P_2 P_1 - P_2 P_1 P_2)$$

$$= \frac{1}{N}(P_1 + P_2)(1 - |\langle\psi_1|\psi_2\rangle|^2) = P_1 + P_2. \qquad (3.63)$$

Using all these relations it is straightforward to verify that the success probability reaches the upper bound (for the case $\eta_1 = \eta_2 = \frac{1}{2}$)

$$p_{\text{success}} = \frac{1}{2}(\text{tr}[P_1 \mathsf{D}_{\text{opt}}(1)] + \text{tr}[P_2 \mathsf{D}_{\text{opt}}(2)])$$

$$= \frac{1}{2(1 + |\langle\psi_1|\psi_2\rangle|)}(\text{tr}[(P_1 + P_2)P] - 2\,\text{tr}[P_1 P_2])$$

$$= \frac{1}{1 + |\langle\psi_1|\psi_2\rangle|}(1 - \text{tr}[P_1 P_2])$$

$$= \frac{(1 - |\langle\psi_1|\psi_2\rangle|)(1 + |\langle\psi_1|\psi_2\rangle|)}{1 + |\langle\psi_1|\psi_2\rangle|} = 1 - |\langle\psi_1|\psi_2\rangle|,$$

which proves that the considered observable is indeed optimal. $\triangle$

For a general pair of mixed states the optimal observable for their unambiguous discrimination is not known. However, it is known that, unlike in the case of pure states, not every pair of mixed states can be unambiguously discriminated, which means that the upper bound is not attainable. In what follows we will provide a necessary condition for the unambiguous discrimination of mixed states and describe a cryptographic application based on the unambiguous discrimination problem.

Let us recall that the support $\text{supp}(T)$ of an operator T consists of all vectors that are orthogonal to the vectors in the kernel of T.

Proposition 3.67 Let ϱ_1 and ϱ_2 be two states that can be unambiguously discriminated. Then neither $\text{supp}(\varrho_1) \subseteq \text{supp}(\varrho_2)$ nor $\text{supp}(\varrho_2) \subseteq \text{supp}(\varrho_1)$ holds.

Proof Suppose that, on the contrary, $\text{supp}(\varrho_1) \subseteq \text{supp}(\varrho_2)$. This is equivalent to the opposed inclusion of the kernels, $\ker(\varrho_2) \subseteq \ker(\varrho_1)$. However, for a state ϱ and an effect E, the condition $\text{tr}[\varrho E] = 0$ is equivalent to the requirement that $E^{1/2}\psi \in \ker(\varrho)$ for every $\psi \in \mathcal{H}$. Thus for any effect E the condition $\text{tr}[\varrho_2 E] = 0$ implies that $\text{tr}[\varrho_1 E] = 0$. It follows that we can never unambiguously conclude that the unknown state is ϱ_1. Similarly, in the case $\text{supp}(\varrho_2) \subseteq \text{supp}(\varrho_1)$ we can never conclude that the unknown state is ϱ_2. Therefore, the condition on the supports is a necessary condition that ϱ_1 and ϱ_2 can be unambiguously discriminated. $\square$

The following result is a simple consequence of Proposition 3.67.

Proposition 3.68 Let $\mathcal{H}$ be a two-dimensional Hilbert space. Two different states $\varrho_1, \varrho_2 \in \mathcal{S}(\mathcal{H})$ can be unambiguously discriminated if and only if they are both pure.

Proof We saw earlier that two different pure states can be unambiguously discriminated. To prove the other implication, let us assume that one state, say ϱ_1, is mixed. As we saw in subsection 2.1.3, the state ϱ_1 can be written in the form $\varrho_1 = \frac{1}{2}(I + \vec{r} \cdot \vec{\sigma})$, where $\|\vec{r}\| < 1$. Let us define a pure state $\omega = \frac{1}{2}(I + \vec{n} \cdot \vec{\sigma})$ with $\vec{n} = \vec{r}/\|\vec{r}\|$. Then

$$\varrho_1 = (1 - \|\vec{r}\|)\frac{1}{2}I + \|\vec{r}\|\,\omega,$$

which shows that ϱ_1 can be written as a convex combination of the total mixture and the pure state ω. Since $\mathrm{supp}(I) = \mathcal{H}$, it follows that $\mathrm{supp}(\varrho_1) = \mathcal{H}$. Therefore, according to Proposition 3.67 it is not possible to unambiguously discriminate ϱ_1 and ϱ_2. $\qquad\square$

Example 3.69 *(Key distribution protocol B92)*
A nice application of the unambiguous state discrimination scheme was presented by Charles Bennett in [15]; this was a quantum key distribution protocol nowadays known as B92. The goal of the key distribution is to establish a secret key represented by strings of perfectly correlated bits shared solely by two communicating parties, Alice and Bob. (See subsection 5.2.4 below for more details on quantum key distributions.)

The key distribution protocol B92 works as follows. Alice randomly prepares a system in one of two pure states P_0, P_1 and sends the system to Bob. Bob then performs a measurement of the observable $\mathsf{D}_{\mathrm{opt}}$, unambiguously discriminating between the given pair of states. Repeating the experiment n times, Alice's preparations define a string of n random bits $\vec{x} = (x_1, \ldots, x_n)$ such that $x_j \in \{0, 1\}$. Bob's measurements define a string $\vec{y} = (y_1, \ldots, y_n)$ with $y_j \in \{0, 1, ?\}$, where ? is associated with an inconclusive outcome. Bob's string matches Alice's string whenever $y_j \neq ?$. Therefore, Bob announces the positions in which he found inconclusive outcomes and both Alice and Bob simply erase these entries from their strings. After that Alice's and Bob's reduced strings match perfectly and represent a secret key shared between Alice and Bob.

The security of the B92 protocol relies on the nonorthogonality of the states P_0, P_1. For orthogonal states the correlations between the strings are perfect without any public communication and the whole procedure can be eavesdropped by a third party without its being detected. By reducing the orthogonality we also reduce the ability to eavesdrop and the anonymity of the eavesdropper. The security of the

key can be verified if part of the shared key is released and compared. This veri-
fication phase of the protocol is based on the same principles as that discussed in
subsection 5.2.4 below.

Let us note that this scheme, although elegant, is not very practical owing to
its fragility with respect to noise. In fact, even for an arbitrarily small amount of
noise the reliability of the conclusion is lost, since the conditions $\text{tr}[P_1 D_{\text{opt}}(0)] =$
$\text{tr}[P_0 D_{\text{opt}}(1)] = 0$ no longer hold. In physical terms, noise transforms the pure
states P_0, P_1 into mixed states ϱ_0', ϱ_1'. For a qubit system, Proposition 3.68 implies
that ϱ_0' and ϱ_1' cannot be unambiguously discriminated. △

3.4.3 How distinct are two states?

Although the set of states is naturally endowed with various types of distance function, the question is which of these (if any) expresses the similarity of two states.
Operationally, the distance of two preparators (preparation procedures) should be
related to our ability to design an experimental procedure allowing us to distinguish between them. The information acquired in measurements comes from the
observed measurement outcome distributions. Therefore, it is natural to use these
probability distributions to introduce the notion of the distance between two states.
Loosely speaking, the idea is that two states are considered to be close to each
other if, for all observables, it is very difficult to distinguish the associated probability distributions. We can use both individual effects and observables to define
distances between a pair of quantum states. In the present subsection we exploit the
success probability of single-shot state-discrimination tasks to quantify the distance
between quantum states.

Minimum-error state discrimination

Suppose we know a priori that an unknown state $\varrho_?$ is either ϱ_1 or ϱ_2. For simplicity
let us assume that both these options are equally likely. We measure an observable
A once and obtain a single outcome x_j. If we conclude that the state was ϱ_1 then the
probability $p_j = \text{tr}[\varrho_1 A(x_j)]$ determines our success rate in guessing before the
measurement that the unknown state $\varrho_?$ was actually ϱ_1, whereas the probability
$q_j = \text{tr}[\varrho_2 A(x_j)]$ quantifies our error. These probabilities interchange their roles
if our conclusion is that $\varrho_? = \varrho_2$. Our goal is to find an observable for which the
error (averaged over all outcomes) is the smallest.

Grouping the outcomes according to the success, in each measurement, of a
specific fixed guess, for example that the particle was in the state ϱ_1, we end up
with a two-outcome observable described by the effects C_1 and $C_2 = I - C_1$.
These effects are associated with conclusions $\varrho_? = \varrho_1$ and $\varrho_? = \varrho_2$, respectively.
For such a binary observable the average error probability is given by the formula

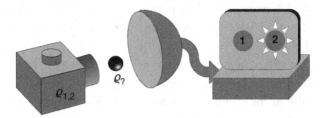

Figure 3.11 Illustration of minimum-error discrimination. The screen on the right is registering that the unknown state was ϱ_2.

$$P_{\text{error}} = \tfrac{1}{2} \text{tr}\big[C_1\varrho_2 + C_2\varrho_1\big] = \tfrac{1}{2}\left(1 + \text{tr}\big[C_1(\varrho_2 - \varrho_1)\big]\right). \qquad (3.64)$$

Naturally, we are interested in finding the observable that minimizes this quantity. We call this task *minimum-error state discrimination* (see Figure 3.11).

The probability of error is minimized when $\text{tr}\big[C_1(\varrho_2 - \varrho_1)\big]$ is as small as possible. If we write the selfadjoint operator $\varrho_2 - \varrho_1$ in the spectral decomposition form $\varrho_2 - \varrho_1 = \sum_j \lambda_j |\varphi_j\rangle\langle\varphi_j|$ then

$$\text{tr}\big[C_1(\varrho_2 - \varrho_1)\big] = \sum_j \lambda_j \langle\varphi_j|C_1\varphi_j\rangle \geq \sum_{j:\lambda_j<0} \lambda_j.$$

The minimum value of $\sum_{j:\lambda_j<0} \lambda_j$ can be reached by choosing C_1 as the projection onto the eigenvectors of the operator $\varrho_2 - \varrho_1$ associated with the negative eigenvalues, i.e.

$$C_1 = \sum_{j:\lambda_j<0} |\varphi_j\rangle\langle\varphi_j|.$$

Since $\text{tr}\big[\varrho_2 - \varrho_1\big] = \text{tr}\big[\varrho_2\big] - \text{tr}\big[\varrho_1\big] = 0$ it follows that for the selfadjoint operator $\varrho_2 - \varrho_1$ the sum of its positive eigenvalues equals the sum of its negative eigenvalues. This implies that the optimal error probability is symmetric with respect to the exchange of ϱ_1, ϱ_2, and we can write

$$\text{tr}\big[C_1(\varrho_2 - \varrho_1)\big] = -\tfrac{1}{2}\sum_j |\lambda_j| = -\tfrac{1}{2}\|\varrho_2 - \varrho_1\|_{\text{tr}}. \qquad (3.65)$$

Let us also note that

$$\text{tr}\big[C_2(\varrho_2 - \varrho_1)\big] = \text{tr}\big[(I - C_1)(\varrho_2 - \varrho_1)\big] = \tfrac{1}{2}\|\varrho_2 - \varrho_1\|_{\text{tr}}. \qquad (3.66)$$

In summary, for the optimal error probability for discrimination of the states ϱ_1, ϱ_2 we find

$$P_{\text{error}} = \tfrac{1}{2}(1 - \tfrac{1}{2}\|\varrho_1 - \varrho_2\|_{\text{tr}}). \qquad (3.67)$$

This formula gives a clear operational meaning for the trace norm and the associated *trace distance*

$$D(\varrho_1, \varrho_2) := \tfrac{1}{2} \|\varrho_1 - \varrho_2\|_{\text{tr}}. \tag{3.68}$$

The larger is the trace distance between the states, the smaller is the error probability. If $p_{\text{error}} = \tfrac{1}{2}$ then the states are necessarily the same, because then $D(\varrho_1, \varrho_2) = 0$.

Fidelity

An alternative state-discrimination task is one in which either the conclusions are error-free or no conclusion is made. This is known as the unambiguous discrimination problem and was discussed in subsection 3.4.2. In this case failure is quantified by the probability of an inconclusive result (associated with the effect $E_?$),

$$p_{\text{fail}} = \tfrac{1}{2} \text{tr}[E_?(\varrho_1 + \varrho_2)], \tag{3.69}$$

where ϱ_1, ϱ_2 are known states and $E_? = I - E_1 - E_2$ is such that $\text{tr}[E_1 \varrho_2] = \text{tr}[E_2 \varrho_1] = 0$. Again, we assume that states are a priori distributed with the same probability. A general solution is not known, but for pure states ϱ_1, ϱ_2 we found in subsection 3.4.2 that the optimal value is

$$p_{\text{fail}} = |\langle \psi_1 | \psi_2 \rangle|, \tag{3.70}$$

where $\psi_1, \psi_2 \in \mathcal{H}$ are the vectors corresponding to the states ϱ_1, ϱ_2, respectively.
According to Proposition 3.65 the quantity

$$F(\varrho_1, \varrho_2) := \text{tr}\left[\left| \sqrt{\varrho_1} \sqrt{\varrho_2} \right| \right] = \text{tr}\left[\sqrt{\sqrt{\varrho_1} \varrho_2 \sqrt{\varrho_1}} \right], \tag{3.71}$$

provides an upper bound on the optimal success probability for the unambiguous discrimination of a pair of states ϱ_1, ϱ_2. This function is called the *fidelity* and represents another common way of measuring the difference between quantum states.

Exercise 3.70 (*Basic properties of the fidelity*)
Prove the following properties of the fidelity. For all states ϱ_1, ϱ_2:

1. $F(\varrho_1, \varrho_2) = 1$ if and only if $\varrho_1 = \varrho_2$;
2. $0 \le F(\varrho_1, \varrho_2) \le 1$;
3. the fidelity is invariant under unitary conjugation: $F(U\varrho_1 U^*, U\varrho_2 U^*) = F(\varrho_1, \varrho_2)$ for all unitary operators U.

3.5 Relations between observables

In earlier sections we discussed properties and qualities of observables. A supplementary point of view is obtained when we compare different observables. We will concentrate on three relations between observables. These relations can be used to make statements like 'A is better than B' precise. We will formulate such relations in terms of statistical mappings, which were introduced in subsection 3.1.2 as an alternative mathematical description of observables. (Recall that Φ_A denotes the statistical mapping corresponding to an observable A.)

3.5.1 State distinction and state determination

Informational completeness is an absolute, which may or may not be possessed by an observable. In this subsection we generalize this binary division into a finer picture, by introducing two preorderings.

Definition 3.71 Let A and B be observables. If, for all states $\varrho_1, \varrho_2 \in S(\mathcal{H})$,

$$\Phi_A(\varrho_1) = \Phi_A(\varrho_2) \qquad \Rightarrow \qquad \Phi_B(\varrho_1) = \Phi_B(\varrho_2) \qquad (3.72)$$

then we write $B \preccurlyeq_i A$ and say that the *state-distinction power* of A is greater than or equal to that of B. If $B \preccurlyeq_i A \preccurlyeq_i B$, we say that A and B are *informationally equivalent*, and write $A \sim_i B$.

Exercise 3.72 Confirm that $\preccurlyeq_i$ is a preorder and that $\sim_i$ is an equivalence relation. [Hint: To confirm that $\preccurlyeq_i$ is a preorder, you need to verify that $A \preccurlyeq_i A$ for all observables A (reflexivity) and that $A \preccurlyeq_i B \preccurlyeq_i C$ implies $A \preccurlyeq_i C$ for all observables A, B and C (transitivity).]

Comparing the above definition with the definition of an informationally complete observable (subsection 3.3.1), we notice immediately that if A is informationally complete then $B \preccurlyeq_i A$ for any observable B. Moreover, if A is informationally complete and $A \sim_i B$ then B is also informationally complete. The state-distinction power is therefore a generalization of the concept of informational completeness.

If the equivalence class of all informationally complete observables corresponds to the best state-distinction power, what is the worst equivalence class?

Example 3.73 (*Trivial observable*)
An observable A is *trivial* if it does not distinguish any pair of states, meaning that

$$\Phi_A(\varrho_1) = \Phi_A(\varrho_2) \qquad (3.73)$$

for all $\varrho_1, \varrho_2 \in S(\mathcal{H})$. Condition (3.73) is equivalent to the fact that each effect $A(X)$ is a multiple of the identity operator I. For instance, coin-tossing observables (see Example 3.4) are trivial observables.

If A is a trivial observable then A $\preceq_i$ B for any other observable B. Moreover, if B $\preceq_i$ A then B is also a trivial observable. △

By definition, an informationally complete observable gives a unique probability distribution for each state. If an observable A is not informationally complete, it can still give a unique probability distribution for some state ϱ_1. This means that for all states $\varrho \in \mathcal{S}(\mathcal{H})$, we have

$$\Phi_A(\varrho) = \Phi_A(\varrho_1) \quad \Rightarrow \quad \varrho = \varrho_1.$$

In this case, we say that the state ϱ_1 is *determined* by A and we denote by $\mathcal{D}_A$ the set of states determined by A.

Definition 3.74 Let A and B be observables. If $\mathcal{D}_B \subseteq \mathcal{D}_A$ then we write B $\preceq_d$ A and say that the *state-determination* power of A is greater than or equal to B. If B $\preceq_d$ A $\preceq_d$ B, we denote A $\sim_d$ B.

As in the case of the state-distinction power, we see that if A is informationally complete then B $\preceq_d$ A for any observable B and that A $\sim_d$ B only if B is informationally complete.

Exercise 3.75 Confirm that $\preceq_d$ is a preorder and that $\sim_d$ is an equivalence relation.

The two preorderings $\preceq_i$ and $\preceq_d$ are obviously related, and we make the following simple observation.

Proposition 3.76 For two observables A and B, the condition B $\preceq_i$ A implies that B $\preceq_d$ A.

Proof Assume that B $\preceq_i$ A and that $\varrho_1 \in \mathcal{D}_B$. For every $\varrho \in \mathcal{S}(\mathcal{H})$, we then have

$$\Phi_A(\varrho_1) = \Phi_A(\varrho) \quad \Rightarrow \quad \Phi_B(\varrho_1) = \Phi_B(\varrho) \quad \Rightarrow \quad \varrho_1 = \varrho.$$

This shows that $\varrho_1 \in \mathcal{D}_A$ and therefore that $\mathcal{D}_B \subseteq \mathcal{D}_A$. □

The converse of Proposition 3.76 is not true; B $\preceq_d$ A does not imply that B $\preceq_i$ A. This becomes clear in the following.

Proposition 3.77 Let A be a binary sharp observable with outcome set $\Omega = \{x_1, x_2\}$. If neither $A(x_1)$ nor $A(x_2)$ is a one-dimensional projection then $\mathcal{D}_A = \emptyset$.

Proof For each state ϱ, the probability distribution $\Phi_A(\varrho)$ is uniquely characterized by the number $\text{tr}[\varrho A(x_1)]$, since $\text{tr}[\varrho A(x_2)] = 1 - \text{tr}[\varrho A(x_1)]$. Hence, to prove the statement, it is enough to show that for each number $0 \leq p \leq 1$ there are two different states ϱ and ϱ' such that $p = \text{tr}[\varrho A(x_1)] = \text{tr}[\varrho' A(x_1)]$.

If $A(x_1) = O$ or $A(x_2) = O$ then A is a trivial observable and $\mathcal{D}_A = \emptyset$. Hence, we will assume that $A(x_1)$ and $A(x_2)$ are both nonzero and not one-dimensional projections. We can then write them as sums, $A(x_1) = \sum_{k=1}^{r} P_k$ and $A(x_2) = \sum_{k=1}^{s} Q_k$, of orthogonal one-dimensional projections, where $r, s \geq 2$ (see the end of subsection 1.2.3). Also, any two of one-dimensional projections from different sums are orthogonal since

$$P_k + Q_k \leq A(x_1) + A(x_2) = I.$$

For any number $0 \leq p \leq 1$, we write $\varrho = pP_1 + (1-p)Q_1$ and $\varrho' = pP_2 + (1-p)Q_2$. Then $\mathrm{tr}[\varrho A(x_1)] = p$ and $\mathrm{tr}[\varrho' A(x_1)] = p$. To confirm that ϱ and ϱ' are different operators, we multiply them by the projection $P_1 + Q_1$, obtaining

$$(P_1 + Q_1)\varrho = pP_1 + (1-p)Q_1, \qquad (P_1 + Q_1)\varrho' = O.$$

This shows that $(P_1 + Q_1)\varrho \neq (P_1 + Q_1)\varrho'$, hence $\varrho \neq \varrho'$. $\qquad\square$

Proposition 3.77 shows, in particular, that if A is a binary sharp observable not containing a one-dimensional projection then $A \sim_d B$ for any trivial observable B. But we have seen earlier that every nontrivial observable has strictly better state distinction power than a trivial observable (see also Exercise 3.78 below). This demonstrates that the preorderings $\preccurlyeq_i$ and $\preccurlyeq_d$ are different.

Exercise 3.78 Let A be a binary sharp observable with outcome set $\Omega = \{x_1, x_2\}$ and assume that $A(x_1)$ and $A(x_2)$ are both nonzero operators. Find two states ϱ and ϱ' such that $\Phi_A(\varrho) \neq \Phi_A(\varrho')$. Conclude that A has strictly better state-distinction power than any trivial observable.

3.5.2 Coarse-graining

Coarse-graining means, generally speaking, a reduction in the quality of the statistical description of a system. The statistical information related to an observable A is most directly encoded in the statistical map Φ_A, and we therefore use statistical maps in the following definition. Recall that $Prob(\Omega)$ denotes the convex set of all probability measures on Ω.

Definition 3.79 Let A and B be observables. We say that B is a *coarse-graining* of A, and write $B \preccurlyeq_c A$, if there exists an affine mapping $V : Prob(\Omega_A) \to Prob(\Omega_B)$ such that

$$\Phi_B = V \circ \Phi_A, \tag{3.74}$$

where the circle symbol denotes a composition of mappings. If $B \preccurlyeq_c A \preccurlyeq_c B$ then we write $A \sim_c B$.

Exercise 3.80 Show that $\preccurlyeq_c$ is a preorder and $\sim_c$ is an equivalence relation.

The idea behind the coarse-graining relation is that the additional mapping V reduces the quality of the statistical description. We have already encountered some examples of coarse-graining where this idea can be seen in practice. The discretization procedure, explained in Example 3.13, leads to a coarse-graining of the original observable. The reduction in the statistical description is also manifest in the following simple result.

Proposition 3.81 If $B \preccurlyeq_c A$ then $B \preccurlyeq_i A$.

Proof Assume that $B \preccurlyeq_c A$ and that ϱ_1, ϱ_2 are states for which $\Phi_A(\varrho_1) = \Phi_A(\varrho_2)$. Then

$$\Phi_B(\varrho_1) = V \circ \Phi_A(\varrho_1) = V \circ \Phi_A(\varrho_2) = \Phi_B(\varrho_2).$$

This shows that $B \preccurlyeq_i A$. $\qquad\qquad\qquad\qquad\qquad\qquad\qquad\qquad\qquad\qquad\quad\square$

The converse of Proposition 3.81 is not true; $B \preccurlyeq_i A$ does not imply that $B \preccurlyeq_c A$. This fact will be demonstrated later in subsection 3.6.2.

There are types of coarse-graining other than discretization. One specific class of coarse-grainings arises from the idea that additional noise in a measurement setting can blur events. Let us recall that a *fuzzy set* on Ω is a function $\widetilde{X}$ from Ω to the interval $[0, 1]$. The value $\widetilde{X}(x)$ is interpreted as the membership degree of a point x in $\widetilde{X}$. Any ordinary set X can be identified with a special type of fuzzy set, a *crisp set*, taking only the values 0 $(x \notin X)$ and 1 $(x \in X)$. Values $\widetilde{X}(x)$ intermediate between 0 and 1 indicate that a point x has partial membership in $\widetilde{X}$.

Let $(\Omega, \mathcal{F})$ be an outcome space and suppose that v is a mapping from the product set $\Omega \times \mathcal{F}$ to the interval $[0, 1]$ such that each mapping $X \mapsto v(x, X)$ is a probability measure. We interpret this kind of mapping as transforming an event X into a fuzzy event $\widetilde{X}(x) = v(x, X)$. We can now define a mapping V on $Prob(\Omega)$ by the formula

$$[Vp](X) = \int v(x, X)\, dp(x). \qquad\qquad\qquad (3.75)$$

The requirement that all mappings $X \mapsto v(x, X)$ are probability measures guarantees that the image Vp is a probability measure. This leads to the following definition.

Definition 3.82 Let A and B be two observables defined on an outcome space $(\Omega, \mathcal{F})$. We say that B is an *unsharp* or *fuzzy version* of A if there exists an affine mapping V of the form (3.75) and if $\Phi_B = V \circ \Phi_A$.

In other words, unsharp versions are a special kind of coarse-graining that is produced by fuzzification of the outcome space. Here is an example.

Example 3.83 (*Unsharp position observable*)
Each probability density function on $\mathbb{R}$ gives rise to an unsharp position observable. Namely, let f be a probability density function on $\mathbb{R}$, i.e. a nonnegative integrable function with normalization $\int_{-\infty}^{\infty} f(x)\,dx = 1$. For any $x \in \mathbb{R}$ and $X \in \mathcal{B}(\mathbb{R})$, we write

$$v_f(x, X) = \int_X f(x - y)\,dy = (f * \chi_X)(x).$$

Here $f * \chi_X$ is the convolution of f and the characteristic function χ_X. The mapping v_f on $\mathbb{R} \times \mathcal{B}(\mathbb{R})$ has the property that each $X \mapsto v(x, X)$ is a probability measure. If we apply v_f to make a coarse-graining of the position observable Q, we get an *unsharp position observable* Q_f. It can be written in the form

$$\langle \psi | \mathsf{Q}_f(X) \psi \rangle = \int (f * \chi_X)(x) |\psi(x)|^2 \, dx.$$

We observe that $\langle \psi | \mathsf{Q}_f(X) \psi \rangle$ is like $\langle \psi | \mathsf{Q}(X) \psi \rangle$ except that the crisp set X has been transformed into the fuzzy set $f * \chi_X$. This type of fuzzy position variable was introduced in [5]. $\triangle$

In the rest of this subsection we will concentrate on finite outcome spaces. Let $\Omega_\mathsf{A} = \{a_1, \ldots, a_k\}$ and $\Omega_\mathsf{B} = \{b_1, \ldots, b_\ell\}$ be finite sets, and let V be an affine mapping from $Prob(\Omega_\mathsf{A})$ to $Prob(\Omega_\mathsf{B})$. Every probability measure on Ω_A is a convex combination of the point measures $\delta_{a_1}, \ldots, \delta_{a_k}$, and similarly for Ω_B. Thus, we can write

$$V(\delta_{a_1}) = v_{11}\delta_{b_1} + \cdots + v_{\ell 1}\delta_{b_\ell},$$

$$\vdots \tag{3.76}$$

$$V(\delta_{a_k}) = v_{1k}\delta_{b_1} + \cdots + v_{\ell k}\delta_{b_\ell},$$

for some numbers $v_{11}, \ldots, v_{\ell k}$. The properties of V imply that $0 \le v_{ij} \le 1$ and $\sum_i v_{ij} = 1$. In other words, $[v_{ij}]$ is a *stochastic matrix*. Condition (3.74) can now be written in the form

$$\mathsf{B}(b_i) = \sum_{j=1}^{k} v_{ij} \mathsf{A}(a_j). \tag{3.77}$$

Notice that if $\Omega_\mathsf{A} = \Omega_\mathsf{B}$ then B in (3.77) is an unsharp version of A. Let us also remark that this kind of matrix construction can be written down if Ω_A and Ω_B are countable (but not necessarily finite) sets, such as $\mathbb{N}$ or $\mathbb{Z}$.

Example 3.84 (*Binary qubit observables*)

In Example 3.33 we discussed a sharp qubit observable A that is defined by a unit vector $\vec{a} \in \mathbb{R}^3$. The projections corresponding to this observable are given by $\mathsf{A}(\pm 1) = \frac{1}{2}(I \pm \vec{a} \cdot \vec{\sigma})$. A stochastic 2×2 matrix $[v_{ij}]$ is completely determined by one of its rows and has the form

$$\begin{pmatrix} v_{11} & v_{12} \\ 1 - v_{11} & 1 - v_{12} \end{pmatrix}.$$

We conclude that the most general binary qubit observable B that is a coarse-graining of A is given by

$$\mathsf{B}(1) = v_{11}\mathsf{A}(1) + v_{12}\mathsf{A}(-1) = \frac{1}{2}\left((v_{11} + v_{12})I + (v_{11} - v_{12})\vec{a} \cdot \vec{\sigma}\right).$$

We can write $\mathsf{B}(1)$ in the form

$$\mathsf{B}(1) = \frac{1}{2}\left(\beta I + \vec{b} \cdot \vec{\sigma}\right), \tag{3.78}$$

where $\beta = v_{11} + v_{12}$ and $\vec{b} = (v_{11} - v_{12})\vec{a}$. From $0 \leq v_{ij} \leq 1$ follows that

$$\|\vec{b}\| \leq \beta \leq 2 - \|\vec{b}\|. \tag{3.79}$$

However, we saw in Example 2.33 that every qubit effect has the form (3.78) for some parameters β and $\vec{b}$ satisfying (3.79). This leads to the conclusion that every binary qubit observable is an unsharp version of some sharp qubit observable.

The previous conclusion on binary qubit observables is a very specific instance of a general result that *every commutative observable is a coarse-graining of some sharp observable*. We refer to [3], [85] for further details. △

If the sets Ω_A and Ω_B have equally many elements then the stochastic matrix $[v_{ij}]$ corresponding to coarse-graining is a square matrix. A special class of square stochastic matrices consists of *doubly stochastic matrices*; these satisfy an extra requirement that $\sum_j v_{ij} = 1$. Thus, not only the columns but also the rows sum to 1. Doubly stochastic matrices correspond to a specific kind of coarse-graining. The *Birkhoff–von Neumann theorem* states that every doubly stochastic matrix is a convex combination of permutation matrices. (A permutation matrix is a matrix that has exactly one entry 1 in each row and each column; the other entries are all 0.) We can easily verify the Birkhoff–von Neumann theorem for 2×2 matrices. A doubly stochastic 2×2 matrix $[v_{ij}]$ is determined by one of its elements (say v_{11}) and hence we can write it in the form

$$\begin{pmatrix} v_{11} & 1 - v_{11} \\ 1 - v_{11} & v_{11} \end{pmatrix} = v_{11}\begin{pmatrix} 1 & 0 \\ 0 & 1 \end{pmatrix} + (1 - v_{11})\begin{pmatrix} 0 & 1 \\ 1 & 0 \end{pmatrix}.$$

Permutation matrices correspond to permutations of the measurement outcomes, while their convex combination leads to a mixture of observables in the sense defined in subsection 3.1.5.

Example 3.85 (*Unbiased binary qubit observables*)
We saw in Example 3.84 that each binary qubit observable can be obtained from some sharp qubit observable by coarse-graining. Which binary qubit observables correspond to doubly stochastic matrices?

Suppose that the stochastic matrix $[\nu_{ij}]$ in Example 3.84 is doubly stochastic. This means that $\nu_{11} + \nu_{12} = 1$ and hence $\beta = 1$. The coarse-grained observable B is given by

$$\mathsf{B}(\pm 1) = \tfrac{1}{2}\left(I \pm \vec{b} \cdot \vec{\sigma}\right) \tag{3.80}$$

and the parameter restrictions (3.79) reduce to $\|\vec{b}\| \le 1$. Binary qubit observables having the form (3.80) are *unbiased*; otherwise they are *biased*. The unbiasedness condition is equivalent to the requirement that the operators B(1) and B(−1) have the same eigenvalues. In this sense, an unbiased qubit observable is more similar to sharp qubit observables than a biased one. △

3.6 Example: photon-counting observables

In this section we demonstrate some concepts studied earlier in a quantum optical context. At the beginning we give a very brief explanation of the quantum optical description of a single-mode electromagnetic field. Only basic concepts are needed, but a reader who is not familiar with this topic can find further explanations in the books by Gerry and Knight [64] and Walls and Milburn [139].

3.6.1 Single-mode electromagnetic field

In the usual quantum optical description, a photon is an excitation of a single mode of the electromagnetic field. Let $\mathcal{H}$ be an infinite-dimensional Hilbert space associated with one fixed mode. We choose an orthonormal basis for $\mathcal{H}$ and denote the basis vectors by Dirac kets $|0\rangle$, $|1\rangle$, ... We also write $\zeta_k = |k\rangle\langle k|$. The state ζ_0 represents the vacuum state of the electromagnetic field, while the state ζ_k is taken to represent an excited state of the field containing k photons. The states $\zeta_0, \zeta_1, \ldots$ are called *number states*.

The *number observable* N is the sharp observable associated with the orthonormal basis $\{|0\rangle, |1\rangle, \ldots\}$, and it has $\mathbb{N} = \{0, 1, \ldots\}$ as its outcome space. Hence, the effects of N are defined as $\mathsf{N}(n) = |n\rangle\langle n|$. In the above quantum optical setting N describes the ideal photon-counting observable. Namely, for a number state ζ_k, the related probability distribution in a N-measurement is

$$\text{tr}\big[\zeta_k N(n)\big] = |\langle k|n\rangle|^2 = \delta_{kn}.$$

The *number operator* N is the (unbounded) selfadjoint operator corresponding to N, that is,

$$N = \sum_{n=0}^{\infty} n N(n).$$

Thus, $\text{tr}\big[\varrho N\big]$ gives the expectation value $\langle N\rangle_\varrho$ of N for a system in the state ϱ.

To illustrate this framework further, we recall that for every $z \in \mathbb{C}$ a *coherent state* $|z\rangle$ is defined as

$$|z\rangle = e^{-|z|^2/2} \sum_{n=0}^{\infty} \frac{z^n}{\sqrt{n!}} |n\rangle.$$

Coherent states describe the electromagnetic field generated by a laser. The number $|z|$ is proportional to the energy of the field. Hence, we expect that a larger $|z|$ value implies more photons. Indeed, the expectation or average value of N in the state $|z\rangle$ is

$$\langle N\rangle_{|z\rangle} = \langle z|N|z\rangle = |z|^2.$$

Exercise 3.86 Calculate the probability distribution of N in a coherent state $|z\rangle$. Confirm that it is a Poisson distribution.

3.6.2 Nonideal photon-counting observables

A nonideal photodetector does not detect all the photons hitting the detector. Let us assume that a photon hitting the detector is detected with probability ϵ, $0 \leq \epsilon \leq 1$. Therefore, if the electromagnetic field is in the number state ζ_k then we expect that the probability $p(n|k)$ that the detector clicks n times is

$$\begin{aligned}
p(n|k) &= \binom{k}{n}\epsilon^n (1-\epsilon)^{k-n} \qquad &\text{if } n \leq k, \\
p(n|k) &= 0 \qquad &\text{if } n > k.
\end{aligned} \tag{3.81}$$

With the usual conventions that $0^0 = 1$ and $0^n = 0$ for $n \geq 1$ these probabilities make sense also in the extreme cases $\epsilon = 0$ and $\epsilon = 1$.

We want to describe this kind of nonideal photodetector as a coarse-graining of the number observable. For the number state ζ_k, the number observable N gives the point probability measure δ_k. Hence, a coarse-graining mapping V maps δ_k to the probability distribution $p(\cdot|k)$. As explained in subsection 3.5.2, V can be written as a matrix $[v_{kn}]$ with respect to the point measures. Comparison with (3.81) shows that

$$
v_{kn} = \begin{cases} 0 & \text{if } k < n, \\ \binom{k}{n}\epsilon^n(1-\epsilon)^{k-n} & \text{if } k \geq n. \end{cases}
$$

Exercise 3.87 Confirm that $[v_{kn}]$ is a stochastic $\mathbb{N} \times \mathbb{N}$ matrix.

We conclude that a photodetector with efficiency ϵ can be described by the observable N^ϵ defined by

$$
\mathsf{N}^\epsilon(n) = \sum_{k=0}^\infty v_{kn}\mathsf{N}(k) = \sum_{k=n}^\infty \binom{k}{n}\epsilon^n(1-\epsilon)^{k-n}|k\rangle\langle k|, \qquad n \in \mathbb{N}. \tag{3.82}
$$

We call N^ϵ a *photon-counting observable*. On the one hand, the number observable N is thus the photon-counting observable N^1 with ideal efficiency, $\epsilon = 1$. On the other hand, the photon-counting observable N^0 with the worst efficiency, $\epsilon = 0$, is a trivial observable.

The photon-counting observable N^ϵ should be seen as an unsharp or imprecise version of N. Its form does not say anything about the mechanism or source of the imprecision, and N^ϵ can arise in many different ways. One possible setting that produces this kind of observable is an ideal photodetector that has a beam splitter in front of it. If the transparency of the beam splitter is ϵ then the observable representing the whole setup is exactly N^ϵ. We refer to Chapter VII.3 in [34] for a derivation of this result.

By our construction N^ϵ is a coarse-graining of N. As one could expect, the converse is not true. Actually, the set of photon-counting observables is totally ordered by the coarse-graining relation and the ordering corresponds to the ordering of the efficiencies. The following two results were proved in [73].

Proposition 3.88 Let $\epsilon_1, \epsilon_2 \in [0, 1]$ and let N^{ϵ_1} and N^{ϵ_2} be the corresponding photon-counting observables. The relation $\mathsf{N}^{\epsilon_1} \preccurlyeq_c \mathsf{N}^{\epsilon_2}$ holds if and only if $\epsilon_1 \leq \epsilon_2$.

Proof Let us first assume that $\mathsf{N}^{\epsilon_1} \preccurlyeq_c \mathsf{N}^{\epsilon_2}$. This means that there exists a stochastic matrix μ such that

$$
\mathsf{N}^{\epsilon_1}(n) = \sum_{k=0}^\infty \mu_{kn}\mathsf{N}^{\epsilon_2}(k), \qquad n \in \mathbb{N}.
$$

For every $m, n \in \mathbb{N}$, we get

$$
\langle m|\mathsf{N}^{\epsilon_1}(n)|m\rangle = \sum_{k=0}^\infty \mu_{kn}\langle m|\mathsf{N}^{\epsilon_2}(k)|m\rangle. \tag{3.83}
$$

Substituting (3.82) into both sides of (3.83) we obtain the identity

$$\binom{m}{n}\epsilon_1^n(1-\epsilon_1)^{m-n} = \sum_{k=0}^{m}\mu_{kn}\binom{m}{k}\epsilon_2^k(1-\epsilon_2)^{m-k}. \tag{3.84}$$

Setting $m = n = 0$ we obtain $1 = \mu_{00}$. Since $\sum_n \mu_{kn} = 1$ and $\mu_{kn} \geq 0$ it follows that $\mu_{0n} = 0$ for all $n \geq 1$. Now let us set $m = n = 1$. We obtain $\epsilon_1 = \mu_{01}(1 - \epsilon_2) + \mu_{11}\epsilon_2$. The previous setting implies $\mu_{01} = 0$, and thus we obtain $\mu_{11} = \epsilon_1\epsilon_2^{-1}$. As μ is a stochastic matrix, we have $\mu_{11} \leq 1$. This can hold only if $\epsilon_1 \leq \epsilon_2$.

Let us therefore assume that $\epsilon_1 \leq \epsilon_2$. Define

$$\mu_{kn} = \begin{cases} 0 & \text{if } k < n, \\ \binom{k}{n}\epsilon_1^n\epsilon_2^{-k}(\epsilon_2 - \epsilon_1)^{k-n} & \text{if } k \geq n. \end{cases}$$

Then μ is a stochastic matrix and, for each n, we have

$$\sum_{k=0}^{\infty}\mu_{kn}\mathsf{N}^{\epsilon_2}(k) = \sum_{k=n}^{\infty}\sum_{m=k}^{\infty}\binom{k}{n}\binom{m}{k}\epsilon_1^n(\epsilon_2 - \epsilon_1)^{k-n}(1-\epsilon_2)^{m-k}|m\rangle\langle m|$$

$$= \sum_{m=n}^{\infty}\epsilon_1^n\left(\sum_{k=n}^{m}\binom{k}{n}\binom{m}{k}(\epsilon_2 - \epsilon_1)^{k-n}(1-\epsilon_2)^{m-k}\right)|m\rangle\langle m|$$

$$= \sum_{m=n}^{\infty}\binom{m}{n}\epsilon_1^n(1-\epsilon_1)^{m-n}|m\rangle\langle m| = \mathsf{N}^{\epsilon_1}(n).$$

Thus we conclude that $\mathsf{N}^{\epsilon_1} \preceq_c \mathsf{N}^{\epsilon_2}$. $\qquad\square$

Proposition 3.89 Let $\epsilon \neq 0$. Then $\mathsf{N}^{\epsilon} \sim_i \mathsf{N}$.

Proof As the claim is trivial in the case $\epsilon = 1$, we will assume that $0 < \epsilon < 1$. Moreover, since $\mathsf{N}^{\epsilon} \preceq_c \mathsf{N}$ we have $\mathsf{N}^{\epsilon} \preceq_i \mathsf{N}$ by Proposition 3.81. To prove that $\mathsf{N} \preceq_i \mathsf{N}^{\epsilon}$, let $\varrho_1, \varrho_2 \in \mathcal{S}(\mathcal{H})$ and assume that $\Phi_{\mathsf{N}^{\epsilon}}(\varrho_1) = \Phi_{\mathsf{N}^{\epsilon}}(\varrho_2)$. By (3.82) this means that, for every $n \in \mathbb{N}$,

$$\sum_{k=n}^{\infty}\binom{k}{n}\epsilon^n(1-\epsilon)^k\langle k|\varrho_1|k\rangle = \sum_{k=n}^{\infty}\binom{k}{n}\epsilon^n(1-\epsilon)^k\langle k|\varrho_2|k\rangle. \tag{3.85}$$

We can eliminate ϵ^n from both sides and move all the terms to one side. Hence, (3.85) is equivalent to the condition

$$\sum_{k=n}^{\infty}\binom{k}{n}(1-\epsilon)^k\langle k|\varrho_1 - \varrho_2|k\rangle = 0. \tag{3.86}$$

Write $a_k = (1 - \epsilon)^k \langle k | \varrho_1 - \varrho_2 | k \rangle$ for every $k \in \mathbb{N}$. Since $|a_k| \le (1 - \epsilon)^k$, the formula

$$f(z) := \sum_{k=0}^{\infty} a_k z^k$$

defines an analytic function in the region $|z| < (1 - \epsilon)^{-1}$. The nth derivative of f is

$$f^{(n)}(z) = \sum_{k=n}^{\infty} k(k-1) \cdots (k-n+1) a_k z^{k-n} = \frac{1}{n!} \sum_{k=n}^{\infty} \binom{k}{n} a_k z^{k-n},$$

and thus (3.86) implies that $f^{(n)}(1) = 0$ for every $n \in \mathbb{N}$. Therefore $f \equiv 0$ and hence $a_k = 0$ for every $k \in \mathbb{N}$. This means that $\langle k | \varrho_1 | k \rangle = \langle k | \varrho_2 | k \rangle$ for every $k \in \mathbb{N}$. We conclude that $\Phi_{\mathsf{N}}(\varrho_1) = \Phi_{\mathsf{N}}(\varrho_2)$ and therefore $\mathsf{N} \preccurlyeq_i \mathsf{N}^\epsilon$. □

It is perhaps surprising that ideal and nonideal photon-counting observables are informationally equivalent. By Proposition 3.76 this also implies that they have the same state-determination power. However, an important difference between these observables is that the induced probability distributions possess different statistical properties. Roughly speaking, the difference is in the statistical distance between the induced probabilities. In a sense, the probabilities induced by a sharp photon-counting observable are more distinguishable than the probabilities induced by nonideal observables. This is the result of the coarse-graining procedure – a coarse-grained probability distribution is fuzzier and statistically noisier than the original.

Exercise 3.90 The functions

$$d(\vec{p}, \vec{q}) = \frac{1}{2} \sum_j |p_j - q_j| \qquad \text{and} \qquad f(\vec{p}, \vec{q}) = \sum_j \sqrt{p_j q_j}$$

quantify the distance of two probability distributions. Calculate the probability distributions of N and N^ϵ ($0 < \epsilon < 1$) for number states ζ_1 and ζ_2. Evaluate the functions d and f for the calculated probability distributions and conclude that N leads to better-separated probability distributions than N^ϵ.

4

Operations and channels

In Chapters 2 and 3 we assumed a model according to which an experiment is split into two parts – preparation and measurement. This led us to the associated concepts of states and observables, respectively. On the basis of this picture we can think about two types of apparatus, which can be characterized abstractly as follows.

- *Preparators* are devices producing quantum states. No input is required, but a quantum output is produced.
- *Measurements* are input–output devices that accept a quantum system as their input and produce a classical output in the form of measurement outcome distributions.

This setting is summarized in Figure 2.2. Furthermore, in subsection 3.5.2 we discussed coarse-graining relations; coarse-graining can be depicted as a box with a classical input and a classical output. This line of thought indicates that at least one type of input–output box is so far missing from our discussion – a device taking a quantum input and producing a quantum output. Hence, we say that

- *quantum channels* are input–output devices transforming quantum states into quantum states.

Each quantum channel can be placed between arbitrary preparation and measurement devices (Figure 4.1). Channels, and a slightly more general concept, that of *operations*, are the topics of this chapter.

4.1 Transforming quantum systems

In this section we discuss the mathematical definition of operations and channels. The notion of an operation was introduced by Haag and Kastler [70] within the

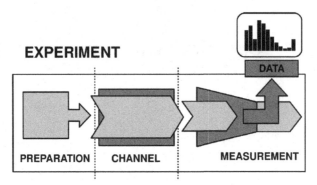

Figure 4.1 The basic statistical framework of an experiment including a quantum channel.

framework of algebraic quantum field theory. Later Kraus [88], Lindblad [93] and Davies [51] discussed the requirement for operations to be completely positive.

4.1.1 Operations and complete positivity

By an *operation* we mean the most general transformation that can be performed on a given ensemble of quantum systems. It is assumed that every operation is applicable to any state. The passage of a photon through an optical fibre or a polarizer is a paradigmatic example of an operation. From the point of view of the observed probabilities (defined for a given state and observable), an operation can be understood either as part of the preparations or as part of the measurement. Mathematically this means that an operation can be formulated equivalently either as a mapping on states (*Schrödinger picture*) or as a mappings on effects (*Heisenberg picture*). We will use mostly the Schrödinger picture.

It may happen that an operation destroys some fraction of the systems in the initial ensemble. For this reason, in order to formulate a mathematical description for quantum operations it is convenient to introduce a set of *subnormalized states*. The central notion of quantum theory is the probability rule, i.e. the prescription determining the probability distribution for each pair consisting of a state and an observable. For subnormalized states the predicted probabilities sum up to a number less than or equal to 1. In this case we can extend the outcome space of the system's transformers (devices affecting the system) by adding an empty outcome, which formally describes the loss of systems in an experiment. The set of subnormalized states $\tilde{S}(\mathcal{H})$ consists of positive trace class operators ϱ with $\mathrm{tr}[\varrho] \leq 1$; thus

$$\tilde{S}(\mathcal{H}) = \{\varrho \in T(\mathcal{H}) : \varrho \geq O, 0 \leq \mathrm{tr}[\varrho] \leq 1\},$$

where $T(\mathcal{H})$ is the set of trace class operators.

In the Schrödinger picture a quantum operation $\mathcal{N}$ is defined as a mapping transforming the set of states $S(\mathcal{H})$ into the set of subnormalized states $\tilde{S}(\mathcal{H})$. The statistical indistinguishability of different convex decompositions of the same state ϱ cannot be changed by any operation. Hence, an operation $\mathcal{N}$ must preserve convex combinations, meaning that

$$\mathcal{N}\left(\sum_j \lambda_j \varrho_j\right) = \sum_j \lambda_j \mathcal{N}(\varrho_j) \tag{4.1}$$

for all $\varrho_j \in S(\mathcal{H})$ and $0 \leq \lambda_j \leq 1$, $\sum_j \lambda_j = 1$. A mapping $\mathcal{N}$ satisfying (4.1) has a unique linear extension to the linear space of trace class operators $T(\mathcal{H})$. (This can be proved in a similar way to Proposition 2.30.) Therefore, we consider operations to be linear mappings on $T(\mathcal{H})$.

Example 4.1 *(No-cloning theorem – weak version)*
A cloning machine is a hypothetical device that produces a clone (a duplicate) of any unknown state. It takes one state as an input and gives back two of the same kind. The second is a duplicate of the first in the sense that no experiment would discover any difference between them. Hence, the cloning machine acts as $\varrho \otimes \omega_0 \mapsto \varrho \otimes \varrho$ for all states ϱ, where ω_0 is a fixed 'blank state' of the system. However, owing to its violation of linearity this kind of transformation is not an operation. Namely, let $\sum_j \lambda_j \varrho_j$ be a mixture of some states. Then

$$\left(\sum_j \lambda_j \varrho_j\right) \otimes \omega_0 \mapsto \left(\sum_j \lambda_j \varrho_j\right) \otimes \left(\sum_j \lambda_j \varrho_j\right) = \sum_{k,j} \lambda_k \lambda_j \varrho_k \otimes \varrho_j, \tag{4.2}$$

but on assuming the linearity of the cloning transformation we get

$$\left(\sum_j \lambda_j \varrho_j\right) \otimes \omega_0 = \sum_j \lambda_j (\varrho_j \otimes \omega_0) \mapsto \sum_j \lambda_j \varrho_j \otimes \varrho_j. \tag{4.3}$$

These two expressions do not generally coincide. For example, choose $\{\varrho_j\}$ to be a collection of orthogonal pure states. Then the numbers $\lambda_k \lambda_j$ and λ_j are exactly the eigenvalues of the final states in (4.2) and (4.3), respectively. But $\lambda_k \lambda_j < \lambda_j$, and the two final states are therefore different.

The fact that the linearity of quantum theory forbids the duplication of an unknown state was observed by several scientists independently at the beginning of the 1980s. This impossibility statement is nowadays known as the *no-cloning*

theorem. We refer to the review articles [126] and [41] for the history of this theorem and for further references. Later, in Example 4.17, we will discuss a more
subtle version of the no-cloning theorem.   △

Not every linear mapping $\mathcal{N}$ on $\mathcal{T}(\mathcal{H})$ describes an operation. Since a valid
operation maps states into subnormalized states, $\mathcal{N}$ has to satisfy

- $\text{tr}\big[\mathcal{N}(\varrho)\big] \leq 1$,
- $\mathcal{N}(\varrho) \geq O$,

for all states $\varrho \in \mathcal{S}(\mathcal{H})$.

Every positive trace class operator can be expressed as a scalar multiple of a
state. Therefore, the first requirement implies that $\text{tr}[\mathcal{N}(T)] \leq \text{tr}[T]$ for all positive
operators $T \in \mathcal{T}(\mathcal{H})$, and we say that $\mathcal{N}$ is *trace nonincreasing*. If $\text{tr}\big[\mathcal{N}(\varrho)\big] =$
$\text{tr}\big[\varrho\big]$ for all states $\varrho \in \mathcal{S}(\mathcal{H})$, we say that $\mathcal{N}$ is *trace preserving*. Since every trace
class operator can be expressed as a linear combination of states, we conclude
that $\mathcal{N}$ is trace preserving if and only if $\text{tr}[\mathcal{N}(T)] = \text{tr}[T]$ for all $T \in \mathcal{T}(\mathcal{H})$.
Similarly, the second requirement implies that $\mathcal{N}(T) \geq O$ for all positive operators
$T \in \mathcal{T}(\mathcal{H})$, and we say that $\mathcal{N}$ is *positive*. At first sight these two conditions should
be enough to define a valid quantum operation. However, there is an additional
requirement, which becomes visible only when we consider composite systems.

Consider a state of a composite system $A + B$ consisting of two subsystems
A and B. The latter subsystem can be assumed to be finite dimensional. We can
extend any operation $\mathcal{N}_A$ acting on subsystem A to a mapping $\mathcal{N}_A \otimes \mathcal{I}_B$ acting on
the composite system, where $\mathcal{I}_B$ is the identity mapping defined on the subsystem
B. This should not make any difference as $\mathcal{I}_B$ corresponds to a situation in which
we do nothing. For instance, if we act on a state $\varrho_{AB} = \varrho_A \otimes \varrho_B$, then

$$(\mathcal{N}_A \otimes \mathcal{I}_B)(\varrho_A \otimes \varrho_B) = \mathcal{N}_A(\varrho_A) \otimes \varrho_B.$$

In spite of this innocent-looking action, it is a nontrivial requirement that $\mathcal{N}_A \otimes \mathcal{I}_B$
transforms states of the composite system $A+B$ into subnormalized states. Namely,
the positivity of $\mathcal{N}_A$ does not guarantee the positivity of the extended mapping
$\mathcal{N}_A \otimes \mathcal{I}_B$. If it is positive for all extensions $\mathcal{I}_B$ then we say that the map $\mathcal{N}_A$ is
completely positive. Otherwise the map $\mathcal{N}_A$ leads to negative probabilities in some
extension, which makes it unphysical. Thus, to avoid such situations we need the
following concept.

Definition 4.2 A linear mapping $\mathcal{N}_A : \mathcal{T}(\mathcal{H}_A) \to \mathcal{T}(\mathcal{H}_A)$ is *completely positive*
if the mapping $\mathcal{N}_A \otimes \mathcal{I}_B$ on $\mathcal{T}(\mathcal{H}_A \otimes \mathcal{H}_B)$ is positive for all finite dimensional
extensions $\mathcal{H}_B$.

There are two possible objections to the physical content of this definition. Is
it enough to restrict the definition to finite-dimensional extensions $\mathcal{H}_B$? And is it

enough to concentrate only on the identity mapping $\mathcal{I}_B$? We will see later that the answer to both these questions is yes. After we have characterized all completely positive mappings, it is easy to see that a combination $\mathcal{N}_A \otimes \mathcal{N}_B$ of two completely positive mappings is completely positive.

In the following example we demonstrate that there are positive mappings which are not completely positive.

Example 4.3 *(Partial transposition)*
Let $\{\varphi_j\}_{j=1}^d$ be an orthonormal basis for a finite-dimensional Hilbert space $\mathcal{H}$. We write $e_{jk} := |\varphi_j\rangle\langle\varphi_k|$ for every $j, k = 1, \ldots, d$. *Transposition* (related to the basis $\{\varphi_j\}_{j=1}^d$) is a linear mapping $\tau : T(\mathcal{H}) \rightarrow T(\mathcal{H})$ acting on the operators e_{jk} as $\tau(e_{jk}) = e_{kj}$. Hence, the transposition mapping is a specific permutation of the operators $\{e_{jk}\}$. If an operator $M \in T(\mathcal{H})$ is written as a matrix $[M_{jk}]$ in the orthonormal basis $\{\varphi_j\}_{j=1}^d$ then the matrix of $\tau(M)$ is, of course, the transposed matrix $[M_{jk}]^T = [M_{kj}]$. Since the transposed matrix $[M_{jk}]^T$ has the same eigenvalues as $[M_{jk}]$, we conclude that the transposition τ is a positive mapping.

Let us now consider the unit vector

$$\psi_+ \equiv \frac{1}{\sqrt{d}} \sum_{j=1}^d \varphi_j \otimes \varphi_j$$

in the tensor product space $\mathcal{H} \otimes \mathcal{H}$. Surprisingly, applying $\tau_A \otimes \mathcal{I}_B$ to the state $|\psi_+\rangle\langle\psi_+|$ gives an operator which is not positive. We obtain

$$\tau_A \otimes \mathcal{I}_B(|\psi_+\rangle\langle\psi_+|) = \frac{1}{d} \sum_{j,k} \tau_A \otimes \mathcal{I}_B(e_{jk} \otimes e_{jk}) = \frac{1}{d} \sum_{j,k} e_{kj} \otimes e_{jk},$$

and this operator is not positive since

$$\left(\frac{1}{d} \sum_{j,k} e_{kj} \otimes e_{jk} \right) (\varphi_1 \otimes \varphi_2 - \varphi_2 \otimes \varphi_1) = -(\varphi_1 \otimes \varphi_2 - \varphi_2 \otimes \varphi_1).$$

As a consequence we see that the transposition τ is an example of a positive but not completely positive map. The mapping $\tau_A \otimes \mathcal{I}_B$ is commonly called *partial transposition* (of the subsystem A). △

Exercise 4.4 Show that the transposition map τ introduced in Example 4.3 is trace preserving. Hence, the only reason why it is not a valid operation is its failure on complete positivity.

It is now time to summarize all these requirements on quantum operations and channels.

Definition 4.5 A mapping $\mathcal{N}$ on $\mathcal{T}(\mathcal{H})$ is an *operation* (a *channel*) if it is

1. linear,
2. completely positive,
3. trace nonincreasing (trace preserving).

We denote by $\mathcal{O}$ the set of all operations and by $\mathcal{O}_c$ the set of all channels.

Some examples are now in order.

Example 4.6 (*Unitary channel*)
As we saw in Section 2.3, state automorphisms can be represented as linear mappings on $\mathcal{T}(\mathcal{H})$ having the form $\sigma_U(T) = UTU^*$ for some unitary or antiunitary operator U. If U is unitary then σ_U is a channel. To see this, let T be a positive trace class operator on a composite system $\mathcal{H}_A \otimes \mathcal{H}_B$. Then, for every vector $\psi \in \mathcal{H}_A \otimes \mathcal{H}_B$, we obtain

$$\langle \psi | (\sigma_U \otimes \mathcal{I}_B) T \psi \rangle = \langle \psi | (U \otimes I) T (U^* \otimes I) \psi \rangle = \left\langle \tilde{\psi}_U \middle| T \tilde{\psi}_U \right\rangle \geq 0,$$

where we have written $\tilde{\psi}_U = (U^* \otimes I)\psi$. The positivity of T implies that $(\sigma_U \otimes \mathcal{I}_B) T$ is positive also, hence the mapping σ_U is completely positive. Moreover, $\mathrm{tr}[\sigma_U(T)] = \mathrm{tr}[T]$ for all trace class operators T. We will show in subsection 4.2.1 that antiunitary operators do not define channels.   △

Example 4.7 (*Simple operations*)
Let $S \in \mathcal{L}(\mathcal{H})$ and define a mapping $\mathcal{N}_S$ on $\mathcal{T}(\mathcal{H})$ by $\mathcal{N}_S(T) = STS^*$. Then $\mathcal{N}$ is linear and also positive, since $STS^* \geq 0$ whenever $T \geq 0$. To see that $\mathcal{N}$ is completely positive, we repeat the derivation given in Example 4.6. Further, we have

$$\mathrm{tr}[\mathcal{N}_S(T)] = \mathrm{tr}\left[STS^*\right] = \mathrm{tr}\left[S^*ST\right]. \tag{4.4}$$

If $T \geq O$, this shows that $\mathrm{tr}[\mathcal{N}_S(T)] \leq \mathrm{tr}[T]$ whenever $S^*S \leq I$. The requirement $S^*S \leq I$ is equivalent to $\|S\| \leq 1$ (see subsection 1.2.1). In conclusion, we see that any bounded operator S satisfying $\|S\| \leq 1$ defines an operation $\mathcal{N}_S$. It is clear from (4.4) that $\mathcal{N}_S$ is a channel exactly when $S^*S = I$.

We will see in subsection 4.2.3 that all operations are sums of operations having the form $\mathcal{N}_S$. In this sense, the latter are the simplest building blocks, and we call them simple operations.   △

Quantum operations are interpreted as the most general actions that we can implement on quantum systems. While channels are deterministic transformations, an operation $\mathcal{N}$ can be interpreted as a probabilistic transformation that occurs with probability $\mathrm{tr}\left[\mathcal{N}(\varrho)\right]$. This aspect of an operation will be discussed in more detail in Chapter 5. In the rest of this chapter we give our attention mostly to quantum channels describing deterministic quantum processes between preparation and measurement.

Exercise 4.8 Let $P \in \mathcal{P}(\mathcal{H})$ be a projection, $O \neq P \neq I$. Recall from subsection 1.2.3 that $P^{\perp} = I - P$ is also a projection. We denote by $\mathcal{N}_P$ and $\mathcal{N}_{P^{\perp}}$ the related simple operations, as defined in Example 4.7. Show that the mapping $\mathcal{N}$ defined as

$$\mathcal{N}(\varrho) = \mathcal{N}_P(\varrho) + \mathcal{N}_{P^{\perp}}(\varrho)$$

is a channel but that it is not of the form $\mathcal{N} = \sigma_U$ for any unitary operator U. [Hint: σ_U maps all pure states into pure states. Does $\mathcal{N}$ do the same?]

4.1.2 Schrödinger versus Heisenberg picture

As explained in Chapter 2, states and effects are dual objects. Consequently, each operation defined in the Schrödinger picture as a linear mapping $\mathcal{N}$ on $\mathcal{T}(\mathcal{H})$ induces a linear mapping $\mathcal{N}^*$ on the dual space $\mathcal{L}(\mathcal{H})$. The connection between $\mathcal{N}$ and $\mathcal{N}^*$ is given by the formula

$$\mathrm{tr}[\mathcal{N}(T)E] = \mathrm{tr}\big[T\mathcal{N}^*(E)\big] \tag{4.5}$$

holding for all trace class operators $T \in \mathcal{T}(\mathcal{H})$ and all bounded operators $E \in \mathcal{L}(\mathcal{H})$. The mapping $\mathcal{N}^*$ describes the same transformation of the system as $\mathcal{N}$ but in the Heisenberg picture. In the Heisenberg picture the effects rather than the states are transformed. Since $(\mathcal{N} \otimes \mathcal{I})^* = \mathcal{N}^* \otimes \mathcal{I}$, the complete positivity of $\mathcal{N}$ is equivalent to the complete positivity of $\mathcal{N}^*$ defined on $\mathcal{L}(\mathcal{H})$. The identity

$$\mathrm{tr}[\mathcal{N}(T)] = \mathrm{tr}[\mathcal{N}(T)I] = \mathrm{tr}\big[T\mathcal{N}^*(I)\big]$$

shows that $\mathcal{N}$ is trace nonincreasing if and only if the dual operation $\mathcal{N}^*$ satisfies $\mathcal{N}^*(I) \leq I$. In the case of channels, the trace-preserving property of $\mathcal{N}$ corresponds to the *unitality* of $\mathcal{N}^*$, which means that $\mathcal{N}^*(I) = I$. In conclusion, we see that a channel $\mathcal{N} : \mathcal{T}(\mathcal{H}) \to \mathcal{T}(\mathcal{H})$ defines a linear mapping $\mathcal{N}^* : \mathcal{L}(\mathcal{H}) \to \mathcal{L}(\mathcal{H})$ which is completely positive and unital.

Exercise 4.9 Let $S \in \mathcal{L}(\mathcal{H})$ be a bounded operator satisfying $\|S\| \leq 1$ and let $\mathcal{N}_S$ be the corresponding operation (see Example 4.7). Show that $(\mathcal{N}_S)^* = \mathcal{N}_{S^*}$.

A technical remark has to be mentioned regarding the case of infinite-dimensional Hilbert spaces. As we have seen, channels can be defined directly in the Heisenberg picture in terms of linear mappings on $\mathcal{L}(\mathcal{H})$. In this case, in addition to complete positivity and unitality there is an extra requirement, namely, *normality* (see Theorem 1.71). This is a kind of continuity requirement which guarantees that there is also a valid description in the Schrödinger picture.

Example 4.10 (*Complete state-space contraction*)

We fix a positive trace class operator $F \neq O$ and define a linear mapping $\mathcal{E}_F$ on $\mathcal{T}(\mathcal{H})$ by the formula

$$\mathcal{E}_F(T) = \frac{\text{tr}[T]}{\text{tr}[F]} F.$$

Let us now specify the action of $\mathcal{E}_F$ in the Heisenberg picture. By definition

$$\text{tr}\big[T\mathcal{E}_F^*(A)\big] = \text{tr}[\mathcal{E}_F(T)A] = \frac{\text{tr}[T]}{\text{tr}[F]} \text{tr}[FA],$$

and this is required to hold for all $T \in \mathcal{T}(\mathcal{H})$ and all $A \in \mathcal{L}(\mathcal{H})$. It follows that

$$\mathcal{E}_F^*(A) = \frac{\text{tr}[FA]}{\text{tr}[F]} I.$$

In summary, in the Schrödinger picture the channel $\mathcal{E}_F$ maps all trace class operators into the one-dimensional linear subspace spanned by the positive trace class operator F, whereas in the Heisenberg picture all operators are mapped into the one-dimensional subspace spanned by the identity operator I. If ϱ is a state then $\mathcal{E}_F(\varrho) = F/\text{tr}[F]$. In other words, the whole state space is contracted into a single point represented by the state $F/\text{tr}[F]$. For this reason we say that $\mathcal{E}_F$ is a complete state-space contraction. △

So far, we have defined channels only in the case where their domain and codomain are identical. However, we can also have channels which map $\mathcal{T}(\mathcal{H})$ into some other space $\mathcal{T}(\mathcal{K})$. The defining conditions (linearity, complete positivity, trace preservation) are just the same. In the following we will give demonstrations of this type of channel.

Example 4.11 (*Addition of an uncorrelated ancilla*)

Let $\mathcal{H}_A$, $\mathcal{H}_{\text{anc}}$ be two Hilbert spaces and let ξ be a fixed state of the ancillary system (ancilla) $\mathcal{H}_{\text{anc}}$. We define a linear mapping $\mathcal{P}_\xi$ from $\mathcal{T}(\mathcal{H}_A)$ into $\mathcal{T}(\mathcal{H}_A \otimes \mathcal{H}_{\text{anc}})$ by $T \mapsto T \otimes \xi$. This mapping is trace preserving, since

$$\text{tr}\big[\mathcal{P}_\xi(T)\big] = \text{tr}[T \otimes \xi] = \text{tr}[T]\,\text{tr}[\xi] = \text{tr}[T].$$

If $T \geq O$ then $\mathcal{P}_\xi(T) = T \otimes \xi \geq O$. Therefore, $\mathcal{P}_\xi$ is positive. An extension $\mathcal{P}_\xi \otimes \mathcal{I}_B$ acts on $\Omega_{AB} \in \mathcal{T}(\mathcal{H}_A \otimes \mathcal{H}_B)$ according to $(\mathcal{P}_\xi \otimes \mathcal{I}_B)(\Omega_{AB}) = \Omega_{AB} \otimes \xi_{\text{anc}}$, and this implies that $\mathcal{P}_\xi$ is completely positive. In conclusion, the addition of an ancillary system yields a proper channel. The form of $\mathcal{P}_\xi$ in the Heisenberg picture is determined by the relation

$$\text{tr}\big[\mathcal{P}_\xi(T)A\big] = \text{tr}\big[T\mathcal{P}_\xi^*(A)\big],$$

which is required to hold for all $T \in \mathcal{T}(\mathcal{H}_A)$ and $A \in \mathcal{L}(\mathcal{H}_A \otimes \mathcal{H}_{\mathrm{anc}})$. Since

$$\mathrm{tr}\big[\mathcal{P}_\xi(T)A\big] = \mathrm{tr}[(T \otimes \xi)A] = \mathrm{tr}[(T \otimes I)((I \otimes \xi)A)]$$
$$= \mathrm{tr}[T(\mathrm{tr}_B[(I \otimes \xi)A])]$$

it follows that

$$\mathcal{P}_\xi^*(A) = \mathrm{tr}_B[(I \otimes \xi)A].$$

One can now check that $\mathcal{P}_\xi^*$ is unital. △

Example 4.12 (*Partial trace in the Heisenberg picture*)
The partial-trace mapping $\mathrm{tr}_B : \mathcal{T}(\mathcal{H}_A \otimes \mathcal{H}_B) \rightarrow \mathcal{T}(\mathcal{H}_A)$ is linear and transforms states into states. In the Heisenberg picture we have a mapping $(\mathrm{tr}_B)^*$ from $\mathcal{L}(\mathcal{H}_A)$ to $\mathcal{L}(\mathcal{H}_A \otimes \mathcal{H}_B)$. By the definition of tr_B, we obtain

$$\mathrm{tr}[\mathrm{tr}_B[T]A] = \mathrm{tr}[TA \otimes I]$$

for all $T \in \mathcal{T}(\mathcal{H}_A \otimes \mathcal{H}_B)$ and for all $A \in \mathcal{L}(\mathcal{H}_A)$. Therefore, we conclude that

$$(\mathrm{tr}_B)^*(A) = A \otimes I$$

for all $A \in \mathcal{L}(\mathcal{H}_A)$. It is now clear that $(\mathrm{tr}_B)^*$ (and hence also tr_B) is completely positive. △

We see that in Example 4.11 the Schrödinger picture looks simpler than the Heisenberg picture, while in Example 4.12 the opposite is true. It is therefore recommended that one should have in mind both descriptions, and switch from one to the other whenever it is convenient to do so.

4.2 Physical model of quantum channels

We introduced quantum channels as transformations of quantum states that satisfy certain mathematical properties in order to preserve the basic framework of quantum statistics introduced in Chapter 2. No additional assumption was made. In this section we reintroduce quantum channels, starting from a slightly different point of view.

4.2.1 Isolated versus open systems

For the purposes of dynamics it is useful to distinguish two different types of system, *isolated* and *open*. A possible definition of this distinction is to say that a system is isolated if all its changes are reversible; otherwise the system is open. As was shown in Section 2.3, the reversibility condition requires that transformations of the system are either unitary or antiunitary. Although isolated systems are only mathematical idealizations of physical reality, it is a common paradigm in physics that every open system is embedded in a larger isolated system; this additional part is conventionally called the *environment*. This implies that, in principle, any evolution is expected to be reversible if a sufficiently large environment is taken into account.

Let us denote by $\mathcal{H}_S$ and $\mathcal{H}_E$ the Hilbert spaces of the system and the environment, respectively. Consider a general input state $\omega \in \mathcal{S}(\mathcal{H}_S \otimes \mathcal{H}_E)$ and some unitary operator $U : \mathcal{H}_S \otimes \mathcal{H}_E \to \mathcal{H}_S \otimes \mathcal{H}_E$ acting on the system plus environment and mapping the state ω into $\omega' = U\omega U^*$. An observer having access to the system S finds the following state change:

$$\mathrm{tr}_E[\omega] \equiv \varrho \mapsto \varrho' \equiv \mathrm{tr}_E[U\omega U^*]. \tag{4.6}$$

Therefore, once the assignment $\varrho \mapsto \omega$ is fixed for all states ϱ, we can calculate the (open) system's evolution.

We say that a map $\mathcal{P} : \mathcal{S}(\mathcal{H}_S) \to \mathcal{S}(\mathcal{H}_S \otimes \mathcal{H}_E)$ is a *preparation map* if it satisfies the compatibility relation $\mathrm{tr}_E[\mathcal{P}(\varrho)] = \varrho$ for all $\varrho \in \mathcal{S}(\mathcal{H}_S)$. A specific class of preparation maps is formed by independent preparation maps $\mathcal{P}_\xi$ acting via the relation

$$\mathcal{P}_\xi(\varrho) = \varrho \otimes \xi, \tag{4.7}$$

where $\xi \in \mathcal{S}(\mathcal{H}_E)$ is a fixed state. In fact, these mappings are channels and were introduced in Example 4.11. They describe the situation where the system and the environment are initially statistically independent, meaning that the initial state of the environment is independent of the channel input.

Proposition 4.13 Let $\mathcal{H}_E$ be the Hilbert space describing the environment, U a unitary operator on $\mathcal{H} \otimes \mathcal{H}_E$ and ξ a fixed state of the environment. Then the induced mapping

$$\mathcal{E} : \mathcal{S}(\mathcal{H}_S) \to \mathcal{S}(\mathcal{H}_S), \qquad \mathcal{E}(\varrho) - \mathrm{tr}_E[U\varrho \otimes \xi U^*] \tag{4.8}$$

is a channel.

Proof Formula (4.8) can be written as the composite mapping $\mathcal{E} = \mathrm{tr}_E \circ \sigma_U \circ \mathcal{P}_\xi$. Linearity, complete positivity and trace preservation follow from the properties of the mappings $\mathcal{P}_\xi$, U and tr_E. Each is a channel, as we saw earlier. $\square$

In Proposition 4.13 we assumed that U is a unitary operator. Since antiunitary operators are also state isomorphisms, could we prove a similar result using antiunitary operators instead of unitary operators? The answer to this question is negative, as shown in the following result.

Proposition 4.14 Let A be an antiunitary operator acting on $\mathcal{H}$. The mapping σ_A is positive, but not completely positive.

Proof As shown in Proposition 1.52, the antiunitary operator A can be written as a composition of a unitary operator U and the complex conjugate operator J with respect to some basis $\{\varphi_j\}$ of the Hilbert space $\mathcal{H}$. For every unit vector $\psi \in \mathcal{H}$, we obtain

$$P_{J\psi} = \tau(P_\psi),$$

where τ is the transposition mapping with respect to the orthonormal basis $\{\varphi_j\}$, as defined in Example 4.3. Consequently, the complex conjugation of vectors is associated with the transposition of pure states. Since the transposition mapping is positive but not completely positive, it follows that σ_A is also positive but not completely positive. $\qquad\square$

Our conclusion from Proposition 4.14 is that if σ_A is applied to a subsystem then the total system can become unphysical, i.e. described by a nonpositive operator. Therefore, we can restrict ourselves to unitary mappings when describing the evolution of the closed system $\mathcal{H}_S \otimes \mathcal{H}_E$.

Example 4.15 (*SWAP gate*)
In this example we introduce one of the simplest interactions between a system and its environment – the so-called SWAP gate. Suppose that $\mathcal{H}_E$ and $\mathcal{H}_S$ are isomorphic Hilbert spaces and $I_0 : \mathcal{H}_S \to \mathcal{H}_E$ is an isomorphism. We define the *SWAP operator* by the formula

$$V_{\text{SWAP}}(\varphi \otimes \psi) = I_0^{-1}\psi \otimes I_0\varphi. \tag{4.9}$$

In what follows we will not use the isomorphisms I_0, I_0^{-1} explicitly in the formulas. We will simply write $V_{\text{SWAP}}(\varphi \otimes \psi) = \psi \otimes \varphi$. This operator is selfadjoint and unitary. Indeed, the identities

$$\langle \varphi' \otimes \psi' | V_{\text{SWAP}}(\varphi \otimes \psi) \rangle = \langle V_{\text{SWAP}}(\varphi' \otimes \psi') | \varphi \otimes \psi \rangle,$$
$$V_{\text{SWAP}}^2(\varphi \otimes \psi) = \varphi \otimes \psi,$$

hold for all $\varphi \otimes \psi, \varphi' \otimes \psi' \in \mathcal{H}_S \otimes \mathcal{H}_E$ and, by linearity, are valid for all vectors in $\mathcal{H}_S \otimes \mathcal{H}_E$. The action of the associated unitary channel is best described by the following relation:

$$\sigma_{\text{SWAP}}(T \otimes S) = V_{\text{SWAP}}(T \otimes S)V_{\text{SWAP}} = S \otimes T.$$

(Here the last form is actually $I_0^{-1}SI_0 \otimes I_0T I_0^{-1}$, but we again omit the explicit form of the isomorphisms.) $\qquad\triangle$

Exercise 4.16 Since eigenvalues of unitary operators are complex numbers with modulus 1 and eigenvalues of selfadjoint operators are real numbers, it follows that the possible eigenvalues of V_{SWAP} are ± 1. Find some eigenvectors corresponding to 1 and -1, hence verify that both these numbers are eigenvalues of V_{SWAP}.

Example 4.17 (*No-cloning theorem for pure states – strong version*)
In Example 4.1 we showed that the cloning transformation $\varrho \mapsto \varrho \otimes \varrho$ is not linear, hence no channel can implement such a transformation. The essential point is that the input state ϱ is unknown. In this example we show that an even stronger statement holds; a pair of nonorthogonal pure states cannot be cloned. This 'no-go' theorem can be seen as a starting point for quantum cryptography, as cloning would be an ideal strategy for an eavesdropper.

Consider a pair of unit vectors $\psi, \phi \in \mathcal{H}$ and corresponding pure states P_ψ and P_ϕ. According to our quantum model, a cloning machine acts as a unitary channel σ_U on a composite system consisting of the system itself (associated with $\mathcal{H}$) and some ancillary system described by a Hilbert space $\mathcal{H}_{anc} = \mathcal{H} \otimes \mathcal{H}'$. Without loss of generality we may assume that the initial state of the ancillary system is pure, $\xi_{anc} = P_\varphi$. (This fact follows from the purification procedure as described in subsection 2.4.2.) Consequently, the cloning transformation reads

$$P_\psi \otimes P_\varphi \mapsto P_\psi \otimes P_\psi \otimes P_{\psi'},$$
$$P_\phi \otimes P_\varphi \mapsto P_\phi \otimes P_\phi \otimes P_{\phi'},$$

where the purity of the states $P_{\psi'}$ and $P_{\phi'}$ is guaranteed by the unitarity of the channel σ_U. As unitary channels satisfy $\mathrm{tr}[\sigma_U[\varrho_1]\sigma_U[\varrho_2]] = \mathrm{tr}[\varrho_1\varrho_2]$ for all states ϱ_1, ϱ_2, it follows that the identity

$$\mathrm{tr}[P_\psi P_\phi] = (\mathrm{tr}[P_\psi P_\phi])^2 |\langle \phi'|\psi'\rangle|^2 \tag{4.10}$$

must hold for some $\phi', \psi' \in \mathcal{H}'$.

Suppose now that the states P_ψ and P_ϕ are different but not orthogonal, so that $0 < \mathrm{tr}[P_\psi P_\phi] < 1$. Condition (4.10) now gives

$$1 = \mathrm{tr}[P_\psi P_\phi] |\langle \phi'|\psi'\rangle|^2. \tag{4.11}$$

However, since $\mathrm{tr}[P_\psi P_\phi] < 1$ and $|\langle \phi'|\psi'\rangle|^2 \leq 1$, the above condition cannot be satisfied. In summary, we have proved that the cloning of two nonorthogonal pure states is impossible. △

Let us return to the discussion relating to Proposition 4.13. It is not too surprising that a combination of the mappings (in fact, channels) $\mathcal{P}_\xi$, σ_U and tr_E gives a channel. However, could there be something more general? What if $\mathcal{P}_\xi$ were replaced by a more general preparation map? Perhaps our construction gives only some special channels and to get other channels we need a more general construction. The final answer to this question is given in subsection 4.2.2. However, to increase our understanding it is instructive to take a look at a generalized construction.

Generally, having defined a preparation mapping $\mathcal{P}$ we can express the evolution $\mathcal{E}$ of the system S as

$$\mathcal{E} = \operatorname{tr}_E \circ \sigma_U \circ \mathcal{P}. \tag{4.12}$$

Let us suppose that the system and the environment are described by identical Hilbert spaces, i.e. $\mathcal{H}_E = \mathcal{H}_S$. We define a preparation mapping

$$\mathcal{P}_f(\varrho) = \varrho \otimes f(\varrho),$$

where $f : \mathcal{S}(\mathcal{H}_S) \to \mathcal{S}(\mathcal{H}_S)$ is an arbitrary function. If we choose $U = V_{\text{SWAP}}$ then the evolution formula (4.12) gives

$$\varrho \overset{\mathcal{P}}{\longmapsto} \varrho \otimes f(\varrho) \overset{\text{SWAP}}{\longmapsto} f(\varrho) \otimes \varrho \overset{\operatorname{tr}_E}{\longmapsto} f(\varrho). \tag{4.13}$$

Hence, if this kind of preparation mapping $\mathcal{P}_f$ is allowed, we could produce arbitrary state transformation $\varrho \mapsto f(\varrho)$.

The problem is, again, linearity. Unless the preparation map $\mathcal{P}$ is linear, it enables us to distinguish different convex decompositions of the same state. But this would necessitate a redefinition of the mathematical representation of states and effects.

In summary, the adopted model for the quantum evolution of open systems assumes that the preparation map is linear. This implies (see [36]) that it is necessarily of the form (4.7); thus, the preparator and the channel (modeled by the environment interacting with the system) are initially uncorrelated. Under such an assumption the evolution of an open system is described by a quantum channel, and the triple $\langle \mathcal{H}_E, U, \xi \rangle$ is called a *dilation* of the associated channel $\mathcal{E}$.

4.2.2 Stinespring's dilation theorem

It is a fundamental fact in the theory of open systems that Proposition 4.13 has a counterpart: every channel can be written as a concatenation of a preparation map, a unitary channel and a partial trace. In other words, every channel can be viewed as an effective description of a unitary channel describing the interaction between the system and its environment. The underlying mathematical result was originally proved by Stinespring [133]. Before stating his theorem we will introduce a relevant family of completely positive mappings.

Suppose that $\alpha : \mathcal{L}(\mathcal{H}) \to \mathcal{L}(\mathcal{K})$ is a *unital $*$-homomorphism*, i.e. α is a linear map such that $\alpha(AB) = \alpha(A)\alpha(B)$, $\alpha(A^*) = \alpha(A)^*$ and $\alpha(I_{\mathcal{H}}) = I_{\mathcal{K}}$. Every $*$-homomorphism is positive. Namely, we recall from subsection 1.2.2 that every positive operator $T \in \mathcal{L}(\mathcal{H})$ can be written in the form $T = A^*A$ for some operator $A \in \mathcal{L}(\mathcal{H})$ and any product of that form is positive. We obtain

$$\alpha(T) = \alpha(A^*A) = \alpha(A^*)\alpha(A) = \alpha(A)^*\alpha(A) \geq 0;$$

hence α is positive. For any Hilbert space $\mathcal{H}'$ the extension $\alpha \otimes \mathcal{I}_{\mathcal{H}'}$ from $\mathcal{L}(\mathcal{H} \otimes \mathcal{H}')$ to $\mathcal{L}(\mathcal{K} \otimes \mathcal{H}')$ is also a $*$-homomorphism. Therefore, α is a completely positive map. An example of a unital $*$-homomorphism is the map $(\mathrm{tr}_E)^*$ introduced in Example 4.12. In this case $\mathcal{K} = \mathcal{H} \otimes \mathcal{H}_E$ and $(\mathrm{tr}_E)^*(A) = A \otimes I_E$.

Now we are ready to formulate Stinespring's dilation theorem. For its proof we refer to the original work [133] and Paulsen's book [112].

Theorem 4.18 *(Stinespring's dilation theorem)*
If a linear map $\mathcal{E}^* : \mathcal{L}(\mathcal{H}) \to \mathcal{L}(\mathcal{H})$ is completely positive then there exists a Hilbert space $\mathcal{K}$, a bounded operator $V : \mathcal{H} \to \mathcal{K}$ and a unital $*$-homomorphism $\alpha : \mathcal{L}(\mathcal{H}) \to \mathcal{L}(\mathcal{K})$ such that

$$\mathcal{E}^*(T) = V^*\alpha(T)V \tag{4.14}$$

for all $T \in \mathcal{L}(\mathcal{H})$. If $\mathcal{E}^*$ is unital then $V^*V = I$.

Stinespring's original proof in [133] treated linear maps from an abstract C^*-algebra $\mathcal{A}$ into $\mathcal{L}(\mathcal{H})$. In our formulation we set $\mathcal{A} = \mathcal{L}(\mathcal{H})$. Then a unital $*$-homomorphism describes a channel in the Heisenberg picture if and only if it is unitarily equivalent to the map $(\mathrm{tr}_E)^*$. (In mathematical terms, every *normal* unital $*$-homomorphism is unitarily equivalent to the map $(\mathrm{tr}_E)^*$. A proof can be found in e.g. [51], p. 139.) Hence, assuming that $\mathcal{E}$ is a channel, we can set $\mathcal{K} = \mathcal{H} \otimes \mathcal{H}_E$ and (4.14) takes the form

$$\mathcal{E}^*(A) = V^*(A \otimes I_E)V$$

for some bounded operator V from $\mathcal{H}$ to $\mathcal{H} \otimes \mathcal{H}_E$ satisfying $V^*V = I$.

We fix a unit vector $\psi \in \mathcal{H}_E$ and write $U(\varphi \otimes \psi) = V\varphi$ for all vectors φ. It follows from $V^*V = I$ that

$$\langle U(\varphi' \otimes \psi) | U(\varphi \otimes \psi) \rangle = \langle V\varphi' | V\varphi \rangle = \langle \varphi' | V^*V\varphi \rangle = \langle \varphi' | \varphi \rangle$$

for all $\varphi, \varphi' \in \mathcal{H}$ and that U has an extension to a unitary operator on $\mathcal{H} \otimes \mathcal{H}_E$. We fix an orthonormal basis $\{\varphi_j\}$ for $\mathcal{H}$. Then, for all $\varrho \in \mathcal{S}(\mathcal{H})$, $T \in \mathcal{L}(\mathcal{H})$, we obtain

$$\mathrm{tr}\big[\mathcal{E}(\varrho)T\big] = \mathrm{tr}\big[\varrho\mathcal{E}^*(T)\big] = \mathrm{tr}\big[\varrho V^*(T\otimes I)V\big]$$

$$= \sum_j \langle V\varrho\varphi_j|(T\otimes I)V\varphi_j\rangle$$

$$= \sum_j \langle U(\varrho\varphi_j\otimes\psi)|(T\otimes I)U(\varphi_j\otimes\psi)\rangle$$

$$= \sum_j \langle \varphi_j\otimes\psi|(\varrho\otimes I)U^*(T\otimes I)U(\varphi_j\otimes\psi)\rangle$$

$$= \mathrm{tr}\big[(\varrho\otimes I)U^*(T\otimes I)U(I\otimes P_\psi)\big]$$

$$= \mathrm{tr}\big[T\,\mathrm{tr}_{\mathcal{K}}[U(\varrho\otimes P_\psi)U^*]\big].$$

We will now obtain the counterpart of Proposition 4.13.

Corollary 4.19 If $\mathcal{E} : \mathcal{T}(\mathcal{H}) \to \mathcal{T}(\mathcal{H})$ is a channel then there exist a Hilbert space $\mathcal{H}_E$, a pure state $\xi \in \mathcal{S}(\mathcal{H}_E)$ and a unitary operator U on $\mathcal{H}\otimes\mathcal{H}_E$ such that

$$\mathcal{E}(\varrho) = \mathrm{tr}_E[U(\varrho\otimes\xi)U^*] \tag{4.15}$$

for all $\varrho \in \mathcal{S}(\mathcal{H})$.

It is important to notice that different dilations $\langle\mathcal{H}_E, U, \xi\rangle$ and $\langle\mathcal{H}'_E, U', \xi'\rangle$ can determine the same channel $\mathcal{E}$ through formula (4.15). For instance, if V_E is a unitary operator on $\mathcal{H}_E$ then $\langle\mathcal{H}_E, U, \xi\rangle$ and $\langle\mathcal{H}_E, (I\otimes V_E)U, \xi\rangle$ determine the same channel. It may also happen that, for a fixed unitary operator U, different states ξ determine the same channel. We say that two dilations $\langle\mathcal{H}_E, U, \xi\rangle$ and $\langle\mathcal{H}'_E, U', \xi'\rangle$ are *equivalent* if they determine the same channel.

Example 4.20 (*Nonuniqueness of dilations*)
Suppose $\dim\mathcal{H} = 2$. If we set $F = I$ in Example 4.10, then we have the channel

$$T \mapsto \tfrac{1}{2}\mathrm{tr}[T]\,I \equiv \mathcal{A}_0(T). \tag{4.16}$$

To obtain a dilation of $\mathcal{A}_0$, let us first consider the unitary operator

$$U = \sum_{j=0}^{3} \sigma_j\otimes|\varphi_j\rangle\langle\varphi_j|,$$

where σ_j are the Pauli operators and the vectors φ_j form an orthonormal basis of a four-dimensional Hilbert space $\mathcal{H}_E$. Any state ξ satisfying $\langle\varphi_j|\xi\varphi_j\rangle = \tfrac{1}{4}$ for $j = 0, \ldots, 3$ induces the channel $\mathcal{A}_0$ (for instance, consider $\xi = \tfrac{1}{4}I_E$). Namely, expressing an operator $X \in \mathcal{T}(\mathcal{H})$ in the Bloch form $X = x_0 I + \vec{x}\cdot\vec{\sigma}$ gives

$$\mathrm{tr}_E[U(X\otimes\tfrac{1}{4}I)U^*] = \tfrac{1}{4}\sum_j \sigma_j X\sigma_j = x_0 I + \tfrac{1}{4}\sum_{j,k=1}^{3} x_k\sigma_j\sigma_k\sigma_j.$$

Using the identities $\sigma_j\sigma_k\sigma_j = -\sigma_k$ if $j \neq k$ (for $j, k = 1, 2, 3$) we see that $\sum_{j=0}^{3}\sigma_j\sigma_k\sigma_j = 2\sigma_k - 2\sigma_k = O$. Thus,

$$\text{tr}_E[U(X \otimes \tfrac{1}{4}I_E)U^*] = x_0 I = \mathcal{A}_0(X). \tag{4.17}$$

To have a look at another kind of dilation, we choose a two-dimensional Hilbert space $\mathcal{H}_E$, fix $U = V_{\text{SWAP}}$ and set $\xi = \tfrac{1}{2}I_E$. We obtain

$$\text{tr}_E[V_{\text{SWAP}}(X \otimes \tfrac{1}{2}I_E)V_{\text{SWAP}}] = \tfrac{1}{2}\text{tr}[X]\, I = \mathcal{A}_0(X). \tag{4.18}$$

In conclusion, $\langle \mathcal{H}_4, U, \tfrac{1}{4}I \rangle$ and $\langle \mathcal{H}_2, V_{\text{SWAP}}, \tfrac{1}{2}I \rangle$ are two different dilations of the same qubit channel $\mathcal{A}_0$.   △

Obviously, one reason for the nonuniqueness of dilations is that we have not limited the dimension of the ancillary Hilbert space $\mathcal{H}_E$. The dilation $\langle \mathcal{H}_E, U, \xi \rangle$ of a channel $\mathcal{E}$ is called *minimal* if no other dilation $\langle \mathcal{H}'_E, U', \xi' \rangle$ of $\mathcal{E}$ with $\dim \mathcal{H}'_E < \dim \mathcal{H}_E$ can be found. Even if we restricted ourselves to the minimal dilations of $\mathcal{E}$, we would still have the freedom that two dilations $\langle \mathcal{H}_E, U, \xi \rangle$ and $\langle \mathcal{H}_E, (I \otimes V_E)U, \xi \rangle$ are equivalent if V_E is a unitary operator on $\mathcal{H}_E$.

4.2.3 Operator-sum form of channels

In this subsection we introduce a very convenient alternative mathematical description of channels. Let us first recall that each bounded operator $S \in \mathcal{L}(\mathcal{H})$ with $S^*S \leq I$ determines an operation $\mathcal{N}_S$ by the formula $\mathcal{N}_S(\varrho) = S\varrho S^*$ (see Example 4.7). A sum $\mathcal{N}_{S_1} + \cdots + \mathcal{N}_{S_n}$ of n such operations is still an operation if $S_1^*S_1 + \cdots + S_n^*S_n \leq I$. It is a channel if $S_1^*S_1 + \cdots + S_n^*S_n = I$.

All operations are either finite or countable sums of the previous type. The following proposition is a consequence of Stinespring's dilation theorem.

Proposition 4.21 A linear mapping $\mathcal{E} : T(\mathcal{H}) \to T(\mathcal{K})$ is a channel if and only if there exists a (finite or infinite) sequence of bounded operators $A_1, A_2, \ldots$ such that

$$\mathcal{E}(T) = \sum_k A_k T A_k^*, \qquad \sum_k A_k^* A_k = I. \tag{4.19}$$

If $\dim \mathcal{H} < \infty$ then it is possible to choose $(\dim \mathcal{H})^2$ or fewer operators A_k in (4.19).

Proof We saw earlier that a sequence of bounded operators determines a channel. To prove the other direction, assume that $\mathcal{E}$ is a channel. By Corollary 4.19 it has a dilation $\langle \mathcal{H}_E, U, \xi \rangle$. We write $\xi = |\varphi_1\rangle\langle\varphi_1|$, where $\varphi_1 \in \mathcal{H}_E$ is a unit vector. Fix an orthonormal basis $\{\varphi_k\}_{k=1}^d$ for $\mathcal{H}_E$ (hence containing φ_1), and for each k define an operator A_k via the identity

$$\langle \psi | A_k \phi \rangle = \langle \psi \otimes \varphi_k | U \phi \otimes \varphi_1 \rangle,$$

which is required to hold for all $\psi, \phi \in \mathcal{H}$. We observe that

$$|\langle \psi | A_k \phi \rangle| = |\langle \psi \otimes \varphi_k | U \phi \otimes \varphi_1 \rangle| \le \|\psi\| \, \|\phi\| \, \|U\| \, ;$$

hence A_k is a bounded operator.

It follows that for all $\psi, \phi \in \mathcal{H}$ and for every pure state $|\eta\rangle\langle\eta| \in \mathcal{S}(\mathcal{H})$, we have

$$\left\langle \psi \middle| \mathcal{E}(|\eta\rangle\langle\eta|)\phi \right\rangle = \left\langle \psi \middle| \mathrm{tr}_E \left[U(|\eta\rangle\langle\eta| \otimes |\varphi_1\rangle\langle\varphi_1|)U^* \right] \phi \right\rangle$$

$$= \sum_k \left\langle \psi \otimes \varphi_k \middle| U(|\eta\rangle\langle\eta| \otimes |\varphi_1\rangle\langle\varphi_1|)U^*(\phi \otimes \varphi_k) \right\rangle$$

$$= \sum_k \left\langle \psi \otimes \varphi_k | U(\eta \otimes \varphi_1)\right\rangle \left\langle \eta \otimes \varphi_1 | U^*(\phi \otimes \varphi_k) \right\rangle$$

$$= \sum_k \left\langle \psi | A_k \eta \right\rangle \left\langle A_k \eta | \phi \right\rangle,$$

and thus $\mathcal{E}(|\eta\rangle\langle\eta|) = \sum_k A_k |\eta\rangle\langle\eta| A_k^*$. Since every state $\varrho \in \mathcal{S}(\mathcal{H})$ has a canonical convex decomposition into pure states, we conclude that $\mathcal{E}(\varrho) = \sum_k A_k \varrho A_k^*$.

For the last claim, we refer to [43]. $\qquad\qquad\square$

The form (4.19) is called an *operator-sum form*, or *Kraus form*, of the channel $\mathcal{E}$. The operators A_k in (4.19) are called *Kraus operators*.

Exercise 4.22 Prove the following: if a channel $\mathcal{E}$ has an operator-sum form $\mathcal{E}(T) = \sum_k A_k T A_k^*$ then $\mathcal{E}^*(T) = \sum_k A_k^* T A_k$ for every $T \in \mathcal{L}(\mathcal{H})$.

Example 4.23 (*Kraus form of a unitary channel*)
A unitary channel σ_U is defined as $\sigma_U(\varrho) = U\varrho U^*$, which is clearly a Kraus form. We can write, for instance, $\sigma_U(\varrho) = \frac{1}{2}U\varrho U^* + \frac{1}{2}U\varrho U^*$, and this corresponds to the Kraus operators $A_1 = A_2 = \frac{1}{\sqrt{2}}U$. All possible Kraus operators are scalar multiples of U. Namely, for a pure state $|\psi\rangle\langle\psi|$ we have $U|\psi\rangle\langle\psi|U^* = \sum_k A_k|\psi\rangle\langle\psi|A_k^*$, and hence $A_k|\psi\rangle\langle\psi|A_k^* = a_k U|\psi\rangle\langle\psi|U^*$ for some number $0 \le a_k \le 1$ (see Proposition 1.63). Since this is true for any unit vector $\psi \in \mathcal{H}$, we have $A_k = \sqrt{a_k}U$. $\qquad\qquad\triangle$

Example 4.24 (*Kraus form of contraction into the total mixture*)
Suppose that $\mathcal{H}$ is a finite d-dimensional Hilbert space. Let us consider again the channel $\mathcal{A}_0$ mapping the whole state space into the maximally mixed state $\frac{1}{d}I$. In Example 4.20 we introduced two specific dilations for this channel in the case $d = 2$. We will now derive two operator-sum forms of the channel $\mathcal{A}_0$.

Fix an orthonormal basis $\{\varphi_j\}_{j=1}^d$ of $\mathcal{H}$ and define operators $E_{jk} = |\varphi_j\rangle\langle\varphi_k|$ for every $j, k = 1, \ldots, d$. Since

$$\sum_{j,k} E_{jk}^* E_{jk} = \sum_{j,k} \langle\varphi_j|\varphi_j\rangle |\varphi_k\rangle\langle\varphi_k| = dI$$

it follows that $T \mapsto \frac{1}{d} E_{jk} T E_{jk}^*$ is a channel. We obtain

$$\mathcal{A}_0(T) = \frac{1}{d} \sum_{j,k} E_{jk} T E_{jk}^* = \operatorname{tr}[T] \frac{1}{d} I.$$

Therefore, the set $\{E_{jk}\}$ contains the Kraus operators for $\mathcal{A}_0$.

Next, suppose that $U_1, \ldots, U_{d^2}$ are unitary operators forming an orthogonal operator basis of $\mathcal{T}(\mathcal{H})$, i.e. $\operatorname{tr}[U_j^* U_k] = 0$ for all $j \neq k$ (see Example 1.64). The identity channel can be expressed as $T \mapsto \frac{1}{d^2} \sum_j U_j \operatorname{tr}[U_j^* T]$. Then for all $\psi, \phi \in \mathcal{H}$ we obtain

$$\operatorname{tr}[P_\psi P_\phi] = \frac{1}{d^2} \sum_j \operatorname{tr}[P_\psi U_j \operatorname{tr}[U_j^* P_\phi]] = \frac{1}{d^2} \sum_j \operatorname{tr}[P_\psi U_j] \operatorname{tr}[U_j^* P_\phi]$$

$$= \left\langle \psi \left| \left(\frac{1}{d^2} \sum_j |U_j\psi\rangle\langle U_j\phi| \right) \phi \right\rangle.$$

Since $\operatorname{tr}[P_\psi P_\phi] = \langle\psi|\operatorname{tr}[|\psi\rangle\langle\phi|]\phi\rangle$ it follows that

$$\frac{1}{d^2} \sum_j U_j |\psi\rangle\langle\phi| U_j^* = \operatorname{tr}[|\psi\rangle\langle\phi|] I. \tag{4.20}$$

A general operator T can be written as a linear combination of rank-1 operators; therefore, we can write

$$\frac{1}{d^2} \sum_j U_j T U_j^* = \operatorname{tr}[T] I = \mathcal{A}_0(T). \tag{4.21}$$

In other words the operators U_j/d are also Kraus operators for the channel $\mathcal{A}_0$. △

As should already be clear, the operator-sum form is not unique. The following result characterizes the freedom in the choice of Kraus operator in the description of a given operation.

Proposition 4.25 Two finite sets $\{A_1, \ldots, A_n\}$ and $\{B_1, \ldots, B_m\}$ of bounded operators define the same operation via the Kraus form if and only if

$$A_j = \sum_{k=1}^{m} u_{jk} B_k, \tag{4.22}$$

with complex numbers u_{jk} satisfying $\sum_j u_{jk} \bar{u}_{jl} = \delta_{kl}$.

Proof If (4.22) holds then we obtain

$$\sum_{j=1}^{n} A_j T A_j^* = \sum_{j=1}^{n} \sum_{k,l=1}^{m} u_{jk} \bar{u}_{jl} B_k T B_l^* = \sum_{k=1}^{m} B_k T B_k^*$$

for all $T \in \mathcal{T}(\mathcal{H})$, hence the two sets determine the same map.

Let us then suppose that $\sum_j A_j T A_j^* = \sum_k B_k T B_k^*$ for all $T \in \mathcal{T}(\mathcal{H})$. If we set $T = |\varphi\rangle\langle\varphi|$ for a unit vector $\varphi \in \mathcal{H}$ then

$$\sum_j |A_j\varphi\rangle\langle A_j\varphi| = \sum_k |B_k\varphi\rangle\langle B_k\varphi|,$$

where $\{A_j\varphi\}$, $\{B_k\varphi\}$ are collections of subnormalized vectors. Then Proposition 2.17 implies that $A_j\varphi = \sum_k u_{jk} B_k\varphi$. Since this is true for all unit vectors $\varphi \in \mathcal{H}$, we conclude that $A_j = \sum_k u_{jk} B_k$. $\square$

Exercise 4.26 For $d = 2$ find the explicit numbers u_{jk} relating the two Kraus forms presented in Example 4.24. [Hint: Set $U_j = \sigma_j$, where σ_j are the Pauli operators.]

4.3 Elementary properties of quantum channels

In this section we introduce some basic properties and concepts used for the characterization of quantum channels. Let us recall the notation: $\mathcal{M}_{cp}$ is the set of completely positive linear mappings, $\mathcal{O}$ is the set of quantum operations and $\mathcal{O}_c$ stands for the set of quantum channels.

4.3.1 Mixtures of channels

The experimental possibility of randomly switching between different apparatuses of the same type (preparators, observables, channels etc.) is reflected by the convex structures of their mathematical representatives. Indeed, it is straightforward to see that the sets $\mathcal{O}_c$ and $\mathcal{O}$ are convex, i.e.

$$\mathcal{E} = \lambda\mathcal{E}_1 + (1 - \lambda)\mathcal{E}_2 \tag{4.23}$$

is a channel (operation) for all $0 < \lambda < 1$ providing that $\mathcal{E}_1, \mathcal{E}_2$ are channels (operations).

As in the case of states and observables we can speak about *extremal channels*, i.e. those that cannot be written as a nontrivial convex mixture (nontrivial means that $\mathcal{E}_1 \neq \mathcal{E}_2$ in (4.23)). Before we formulate a criterion for a channel to be extremal, let us introduce two important families of extremal channels. Their extremality follows from the extremality of pure states in the set of all states.

- *Contractions into a pure state P.* This kind of channel $\mathcal{E}_P$ transforms the whole state space into a fixed pure state P (see Example 4.10). The extremality of $\mathcal{E}_P$ follows from the extremality of pure states in the set of all states (see subsection 2.1.2). In particular, let us assume that $\mathcal{E}_P = \lambda \mathcal{E}_1 + (1 - \lambda)\mathcal{E}_2$ for some $0 < \lambda < 1$. Then $P = \lambda \mathcal{E}_1(\varrho) + (1 - \lambda)\mathcal{E}_2(\varrho)$ for all states ϱ. However, since the pure state P is extremal in the set of states, it follows that $\mathcal{E}_1(\varrho) = \mathcal{E}_2(\varrho) = P$ for all ϱ and thus $\mathcal{E}_1 = \mathcal{E}_2 = \mathcal{E}_P$.
- *Unitary channels.* Let U be a unitary operator and σ_U the corresponding unitary channel. Assume that $\sigma_U = \lambda \mathcal{E}_1 + (1 - \lambda)\mathcal{E}_2$ for some $0 < \lambda < 1$. Then for every pure state P we obtain

$$P = U^* \sigma_U(P)U = \lambda U^* \mathcal{E}_1(P)U + (1 - \lambda)U^* \mathcal{E}_2(P)U.$$

It follows that $U^* \mathcal{E}_1(P)U = U^* \mathcal{E}_2(P)U = P$ for all pure states P and thus $\mathcal{E}_1 = \mathcal{E}_2 = \sigma_U$.

On a finite-dimensional Hilbert space we have the following complete characterization of extremal channels, first proved by Choi [43].

Proposition 4.27 Suppose that $\mathcal{H}$ is finite dimensional. A channel $\mathcal{E}$ is extremal if and only if it admits an operator-sum form such that the set $\{A_j^* A_k\}$ (induced by the associated Kraus operators) is linearly independent.

Proof Let $\mathcal{E}$ be a channel determined by the Kraus operators $\{A_j\}$. We assume that $\mathcal{E} = \frac{1}{2}(\mathcal{E}_1 + \mathcal{E}_2)$ for channels $\mathcal{E}_1(T) = \sum_s B_s T B_s^*$ and $\mathcal{E}_2(T) = \sum_r Z_r T Z_r$. Normalization implies that

$$\sum_j A_j^* A_j = \sum_s B_s^* B_s = \sum_r Z_r^* Z_r = I.$$

Since $\mathcal{E}(T) = \sum_j A_j T A_j^* = \frac{1}{2}\sum_s B_s T B_s^* + \frac{1}{2}\sum_r Z_r T Z_r$, it follows that $B_s = \sum_j \beta_{sj} A_j$ and $Z_r = \sum_j \zeta_{rj} A_j$. We thus obtain $\sum_j A_j^* A_j = \sum_s B_s^* B_s = \sum_s \overline{\beta}_{sj} \beta_{sj'} A_j^* A_{j'}$. Because of the linear independence of the operators $A_j^* A_{j'}$ it follows that $\sum_s \overline{\beta}_{sj} \beta_{sj'} = \delta_{jj'}$. Proposition 4.25 gives $\mathcal{E}_1 = \mathcal{E}$. Using the same line of argument we also find that $\mathcal{E}_2 = \mathcal{E}$, which proves the extremality of $\mathcal{E}$ and thus the sufficiency of the condition on the linear independence of the operators $A_j^* A_k$.

In order to prove the necessity of the linear independence condition, let us assume that $\mathcal{E}$ is an extremal channel expressed via the Kraus operators A_j. We want to show that the condition $\sum_{jk} \alpha_{jk} A_j^* A_k = 0$ implies that $\alpha_{jk} = 0$.

Let us first note that we can assume the α_{jk} form a hermitian matrix $[\alpha]$, because $\sum_{jk} \alpha_{jk} A_j^* A_k = 0$ is equivalent to $\sum_{jk} \overline{\alpha}_{kj} A_j^* A_k = 0$ and hence also to $\sum_{jk} (\alpha_{jk} \pm \overline{\alpha}_{kj}) A_j^* A_k = 0$. Moreover, if $\sum_{jk} \alpha_{jk} A_j^* A_k = 0$ then $\sum_{jk} \alpha'_{jk} A_j^* A_k = 0$ for $\alpha'_{jk} = c\alpha_{jk}$ and we may choose the constant c such that $-I \leq [\alpha] \leq I$. Define linear maps $\mathcal{F}_{\pm}(T) = \mathcal{E}(T) \pm \sum_{jk} \alpha_{jk} A_j T A_k^*$, i.e. $\mathcal{E} = \frac{1}{2}(\mathcal{F}_+ + \mathcal{F}_-)$. Define operators $B_r^{\pm} = \sum_j \beta_{rj}^{\pm} A_j$ such that $\sum_r \beta_{rj}^{\pm} \overline{\beta}_{rk}^{\pm} = \delta_{jk} \pm \alpha_{jk}$. Then $\mathcal{F}_{\pm}(T) = \sum B_r^{\pm} T (B_r^{\pm})^*$ and $\sum_r (B_r^{\pm})^* B_r^{\pm} = \sum_{r,j,k} \beta_{rk}^{\pm} \overline{\beta}_{rj}^{\pm} A_j^* A_k = \sum_{j,k} (\delta_{jk} \pm \alpha_{jk}) A_j^* A_k = \sum_j A_j^* A_j = I$. Hence $\mathcal{F}_{\pm}$ are channels. Using the extremality of $\mathcal{E}$ it follows that $\mathcal{F}_{\pm} = \mathcal{E}$ and $\beta_{rk}^{\pm}$ determine an isometry. That is, $\delta_{jk} = \sum_r \beta_{rj}^{\pm} \overline{\beta}_{rk}^{\pm} = \delta_{jk} \pm \alpha_{jk}$, meaning that $\alpha_{jk} = 0$ as required by the theorem. $\qquad \square$

Example 4.28 (*Amplitude-damping channel*)
An amplitude-damping channel $\mathcal{E}_\theta$ on a two-dimensional Hilbert space $\mathcal{H}$ is defined by two Kraus operators

$$A_\theta = |\varphi\rangle\langle\varphi| + \sqrt{1-\theta}|\varphi_\perp\rangle\langle\varphi_\perp|, \quad B_\theta = \sqrt{\theta}|\varphi\rangle\langle\varphi_\perp|, \qquad (4.24)$$

i.e. $\mathcal{E}_\theta(T) = A_\theta T A_\theta^* + B_\theta T B_\theta^*$. Here $\varphi, \varphi_\perp$ are orthogonal unit vectors and θ is a real number between 0 and 1. Since the operators

$$A_\theta^* A_\theta = |\varphi\rangle\langle\varphi| + (1-\theta)|\varphi_\perp\rangle\langle\varphi_\perp|, \quad B_\theta^* B_\theta = \theta|\varphi_\perp\rangle\langle\varphi_\perp|,$$
$$A_\theta^* B_\theta = \sqrt{\theta}|\varphi\rangle\langle\varphi_\perp|, \quad B_\theta^* A_\theta = \sqrt{\theta}|\varphi_\perp\rangle\langle\varphi| \qquad (4.25)$$

are linearly independent, Proposition 4.27 implies that all amplitude-damping channels are extremal. $\qquad \triangle$

In the case of a finite-dimensional system we identified also the other extreme relating to the convex structure of the state space – the maximally mixed state (see Example 2.28). Before we proceed with the question of a *maximally mixed channel* let us introduce an average unitary channel.

Example 4.29 (*Average unitary channel*)
The set of unitary channels is uncountable, so an average over all unitary channels has to be defined through an integral. Suppose that $d = \dim \mathcal{H} < \infty$. Let us consider the map

$$\mathcal{A}(T) := \int_{\mathcal{U}(\mathcal{H})} U T U^* \, dU,$$

where dU is the invariant Haar measure on $\mathcal{U}(\mathcal{H})$. The invariance implies that $[\mathcal{A}(T), U] = 0$ for all unitary operators U and, applying the Schur lemma, we can conclude that $\mathcal{A}(T)$ is proportional to the identity operator, i.e. $\mathcal{A}(T) = c(T)I$, where $c(\cdot)$ is a linear functional. Since $\mathcal{A}$ is trace preserving, it follows that

$$\text{tr}[T] = \text{tr}[\mathcal{A}(T)] = c(T)\,\text{tr}[I] = c(T)d;$$

thus $c(T) = \text{tr}[T]/d$. In summary, we see that

$$\mathcal{A}(T) = \frac{\text{tr}[T]}{d} I \equiv \mathcal{A}_0(T), \tag{4.26}$$

where $\mathcal{A}_0$ is the channel describing the contraction of the state space into the maximally mixed state. It has been discussed already, in Examples 4.10 and 4.20. $\triangle$

To identify a maximally mixed channel we will use an analogy with the concept of a maximally mixed state introduced near the end of subsection 2.1.3. We quantify a channel's mixedness through the output states. In particular, we introduce the *maximal output purity* $\mathcal{P}_{\max}(\mathcal{E}) := \max_\varrho \mathcal{P}(\mathcal{E}(\varrho))$ and *minimal output entropy* $S_{\min}(\mathcal{E}) := \min_\varrho S(\mathcal{E}(\varrho))$. We say that a channel $\mathcal{E}$ is *maximally mixed* if it minimizes the maximal output purity and maximizes the minimal output entropy. Let us note that, for any channel, $\mathcal{P}_{\max}(\mathcal{E}) \geq \mathcal{P}(\frac{1}{d}I)$ and $S_{\min}(\mathcal{E}) \leq S(\frac{1}{d}I)$. However, for the contraction into the maximally mixed state we have $\mathcal{P}_{\max}(\mathcal{A}_0) = \mathcal{P}(\frac{1}{d}I)$ and $S_{\min}(\mathcal{A}_0) = S(\frac{1}{d}I)$; hence, $\mathcal{A}_0$ is a maximally mixed channel. Its uniqueness follows from the uniqueness of the maximally mixed state. For instance, the relations $\mathcal{P}(\frac{1}{d}I) \leq \mathcal{P}(\mathcal{E}(\varrho)) \leq \mathcal{P}_{\max}(\mathcal{A}_0(\varrho)) = \mathcal{P}(\frac{1}{d}I)$ imply that $\mathcal{P}(\mathcal{E}(\varrho)) = \mathcal{P}(\frac{1}{d}I)$ for all states ϱ, but $\varrho = \frac{1}{d}I$ is the only state achieving the minimal value $\mathcal{P}(\frac{1}{d}I) = \frac{1}{d}$.

The maximally mixed channel $\mathcal{A}_0$ shares many features of the maximally mixed state. For example, it commutes with all unitary channels, i.e. we have $U\mathcal{A}_0(\varrho)U^* = \mathcal{A}_0(U\varrho U^*)$ for all $U \in \mathcal{U}(\mathcal{H})$ and $\varrho \in \mathcal{S}(\mathcal{H})$. However, for channels this feature is not unique since the identity channel also commutes with all unitary channels.

4.3.2 Concatenating channels

Let us consider channels acting on a fixed space $\mathcal{T}(\mathcal{H})$. Channels are functions and for this reason the set of channels is endowed with the binary operation of composition. Namely, a composition $\mathcal{E}_1 \circ \mathcal{E}_2$ of two channels $\mathcal{E}_1$ and $\mathcal{E}_2$ is also a channel. Physically this corresponds to the sequential implementation of the processes described by $\mathcal{E}_1$ and $\mathcal{E}_2$. To emphasize this physical point of view, we say that $\mathcal{E}_1 \circ \mathcal{E}_2$ is the *concatenation* of $\mathcal{E}_1$ and $\mathcal{E}_2$. Since concatenation is an associative binary operation, this means that the set of channels forms a semigroup.

Exercise 4.30 Show that the semigroup of channels is noncommutative, that is, give an example of two channels $\mathcal{E}_1$ and $\mathcal{E}_2$ such that $\mathcal{E}_1 \circ \mathcal{E}_2 \neq \mathcal{E}_2 \circ \mathcal{E}_1$. [Hint: Try complete state-space contractions.]

The identity channel $\mathcal{I}$ plays the role of unity. A channel $\mathcal{E}_1$ is the *inverse* of another channel $\mathcal{E}_2$ if

$$\mathcal{E}_1 \circ \mathcal{E}_2 = \mathcal{E}_2 \circ \mathcal{E}_1 = \mathcal{I}.$$

It should be clear from our earlier examples that not every channel has an inverse channel. For instance, the contraction channels discussed in Example 4.10 are not surjective functions and therefore cannot have an inverse function. However, suppose that a channel $\mathcal{E}$ has an inverse channel. Then $\mathcal{E}$ is a bijection on the set $S(\mathcal{H})$ and so it is actually a state automorphism (recall subsection 2.3.2). It follows that $\mathcal{E}$ is a unitary channel (combine Theorem 2.63 and Proposition 4.14). We thus conclude the following.

Proposition 4.31 A channel $\mathcal{E}$ has an inverse channel if and only if $\mathcal{E}$ is a unitary channel.

The subset of unitary channels is a group, since a concatenation of two unitary channels is again unitary and every unitary channel σ_U has its inverse unitary channel σ_{U^*}. This group is isomorphic to the quotient group $\mathcal{U}(\mathcal{H})/\mathbb{T}$, as explained in Section 2.3.

Decomposition of unitary channels

One task in the area of quantum computation is to run quantum algorithms, which are specified as unitary channels σ_U acting on a large number of qubits. It turns out that the implementation of a desired unitary channel is a challenging experimental problem. In the *circuit model* of quantum computation the desired multi-qubit unitary channel (called a quantum circuit) is decomposed into simpler unitary channels (called quantum gates) acting on a few qubits only. It is a key result in the theory of quantum computation that any unitary channel acting on arbitrarily many qubits can be realized as a sequence of *elementary quantum gates*. The set of elementary quantum gates consists of all one-qubit unitary channels that are enriched with only a single two-qubit unitary gate. This result was proved by Barenco *et al.* [9]. For more details on the decomposition of unitary channels for the purposes of quantum algorithms we refer to textbooks on quantum computation, for instance [99] or [104].

Example 4.32 (*Controlled-NOT gate*)
So far we have introduced two quantum gates – the Hadamard gate in Example 2.48 and the SWAP gate in Example 4.15. While the Hadamard gate is a special single-qubit gate, the SWAP gate acts on two d-dimensional systems. In the case $d = 2$ it determines a two-qubit unitary gate generated by the unitary operator V_{SWAP}. Any concatenation of SWAP gates and single-qubit gates transforms vector states $\psi_1 \otimes \cdots \otimes \psi_n$ into vector states $\psi_1' \otimes \cdots \otimes \psi_n'$. Therefore, for example, the vector state $\varphi \otimes \varphi$ cannot be transformed into a vector state $\frac{1}{\sqrt{2}}(\varphi \otimes \varphi + \varphi_\perp \otimes \varphi_\perp)$, where $\varphi_\perp$ is orthogonal to φ. (The fact that this vector state is not of the form $\psi_1' \otimes \psi_2'$ is proved later, in subsection 6.1.1.) We conclude that there are unitary channels that cannot be written as a composition of SWAP gates and one-qubit unitary gates.

The controlled-NOT gate is induced by the two-qubit unitary operator

$$V_{\text{CNOT}} = |\varphi_0\rangle\langle\varphi_0| \otimes I + |\varphi_1\rangle\langle\varphi_1| \otimes \sigma_x, \tag{4.27}$$

where $\varphi_j \in \mathcal{H}_2$ labels the so-called *computational basis* encoding the bit values 0 and 1, i.e. $\langle\varphi_0|\varphi_1\rangle = 0$. A direct calculation gives that $V_{\text{CNOT}}(V_H \otimes I)(\varphi_0 \otimes \varphi_0) = \frac{1}{\sqrt{2}}(\varphi_0 \otimes \varphi_0 + \varphi_1 \otimes \varphi_1)$, where V_H stands for the single-qubit Hadamard gate transforming φ_0 to $\frac{1}{\sqrt{2}}(\varphi_0 + \varphi_1)$. Of course, this is not a proof that compositions of the controlled-NOT gate and one-qubit unitary gates give all possible unitary channels, but this is indeed the case. For details we refer to the literature mentioned above.   △

Exercise 4.33 Find a decomposition of the SWAP gate into single-qubit gates and controlled-NOT gates. [Hint: Three controlled-NOT gates are needed.]

Example 4.34 (*Quantum discrete Fourier transform*)
The discrete Fourier transform maps a complex vector $(x_0, \ldots, x_{N-1})$ into another vector $(y_0, \ldots, y_{N-1})$ such that $\sqrt{N}\, y_k = \sum_{j=0}^{N-1} x_j e^{2i\pi jk/N}$. The operation of 'quantization' translates this mapping into a transformation of Hilbert space vectors,

$$\psi = \sum_{j=0}^{N-1} x_j \phi_j \qquad \mapsto \qquad \psi' = \frac{1}{\sqrt{N}} \sum_{j,k=0}^{N-1} x_j e^{2i\pi jk/N} \phi_k, \tag{4.28}$$

where $\{\phi_k\}$ is a fixed orthonormal basis of a d-dimensional Hilbert space. We can thus identify the unitary operator (verify its unitarity!)

$$V_{\text{Fourier}} = \frac{1}{\sqrt{N}} \sum_{j,k=0}^{N-1} e^{2i\pi jk/N} |\phi_k\rangle\langle\phi_j|. \tag{4.29}$$

Let us express the integer j in its binary representation as $j = j_1 \cdots j_n$, meaning that $j = \sum_{r=1}^{n} j_r 2^{n-r}$, where $j_r \in \{0, 1\}$ and the condition $n = \log_2 N$ fixes the number of bits. Applying V_{Fourier} to the vector state ϕ_j encoded into n qubits as $\varphi_{j_1} \otimes \cdots \otimes \varphi_{j_n}$ results in the following vector state transformation:

$$\phi_j \mapsto \frac{1}{2^{n/2}} \sum_{k_1=0}^{1} \cdots \sum_{k_n=0}^{1} e^{2i\pi j(\sum_{r=1}^{n} k_r 2^{-r})} \varphi_{k_1} \otimes \cdots \otimes \varphi_{k_n}$$

$$= \frac{1}{2^{n/2}} \sum_{k_1=0}^{1} \cdots \sum_{k_n=0}^{1} \bigotimes_{r=1}^{n} e^{2i\pi jk_r 2^{-r}} \varphi_{k_r} = \frac{1}{2^{n/2}} \bigotimes_{r=1}^{n} \left(\varphi_0 + e^{2i\pi j2^{-r}} \varphi_1\right)$$

$$= \frac{1}{2^{n/2}} (\varphi_0 + e^{2i\pi 0.j_n} \varphi_1) \otimes \cdots \otimes (\varphi_0 + e^{2i\pi 0.j_1 \cdots j_n} \varphi_1), \tag{4.30}$$

where $0.j_r \ldots j_m$ represents the binary fraction $j_r/2 + \cdots + j_m/2^{m-r+1}$. How many elementary gates do we need to implement a quantum Fourier gate?

Consider a two-qubit (controlled) unitary operator $V_{j,k} = I \otimes |\varphi_0\rangle\langle\varphi_0| + R_{1+k-j} \otimes |\varphi_1\rangle\langle\varphi_1|$ applied to the jth and kth qubit, where $R_t = |\varphi_0\rangle\langle\varphi_0| + e^{2i\pi/2^t}|\varphi_1\rangle\langle\varphi_1|$ is a single-qubit unitary operator. Applying the Hadamard gate to the first qubit maps ϕ_j into $\frac{1}{\sqrt{2}}(\varphi_0 + e^{2i\pi 0.j_1}\varphi_1) \otimes \varphi_{j_2} \otimes \cdots \otimes \varphi_{j_n}$. Further, applying a sequence of two-qubit gates $V_{1,n} \cdots V_{1,2}$, for each of which the first qubit is the target, we obtain the state

$$\frac{1}{\sqrt{2}}(\varphi_0 + e^{2i\pi 0.j_1 j_2 \ldots j_n}\varphi_1) \otimes \varphi_{j_2} \otimes \cdots \otimes \varphi_{j_n}. \tag{4.31}$$

Performing a similar procedure, in which a Hadamard gate is applied to the jth qubit and is followed by two-qubit controlled gates for which the qubits satisfy $k > j$, gives us the following overall transformation:

$$\phi_j \mapsto H_n V_{n-1,n} H_{n-1} \cdots \left(\prod_{j=3}^{n} V_{2,j} \right) H_2 \left(\prod_{j=2}^{n} V_{1,j} \right) H_1 \phi_j$$

$$= \frac{1}{2^{n/2}} (\varphi_0 + e^{2i\pi 0.j_1 \ldots j_n}\varphi_1) \otimes \cdots \otimes (\varphi_0 + e^{2i\pi 0.j_n}\varphi_1). \tag{4.32}$$

We have used the notation $H_j = V_H \otimes I_{\bar{j}}$ to represent the application of a Hadamard gate to the jth qubit; $I_{\bar{j}}$ stands for the identity operator acting on all qubits but the jth. Altogether this sequence consists of $n + (n-1) + \cdots + 1 = \frac{1}{2}n(n-1)$ unitary gates. Comparing (4.30) and (4.32) we see that up to the order of qubits the states coincide. By applying $\frac{1}{2}n$ SWAP gates we can transform one state into the other; hence, we can provide a decomposition of the quantum Fourier gate into two-qubit gates $V_{j,k}$, SWAP gates and single-qubit Hadamard gates. Since both the $V_{j,k}$ and SWAP gates can be implemented by some fixed number of controlled NOT gates (see Exercise 4.33 above) we may conclude that the number of gates required increases quadratically in the input size measured in bits. Let us note, however, that classically the scaling is exponential. This quantum speed-up in Fourier transformation is a key ingredient of the Shor algorithm for prime factorization [129]. $\triangle$

Divisibility of channels

Next, we focus on channels which are not unitary. Is it possible to decompose a general channel $\mathcal{E}$ into a concatenation of some other channels $\mathcal{E}_1$ and $\mathcal{E}_2$? Trivially, we can always write $\mathcal{E} = \mathcal{E}' \circ \sigma_U$, where $\mathcal{E}' = \mathcal{E} \circ \sigma_{U^*}$ and σ_U is a unitary channel. Excluding such uninteresting compositions we can define the concept of *channel divisibility*, originally introduced and studied by Wolf and Cirac [145].

Definition 4.35 A channel $\mathcal{E}$ is *indivisible* if $\mathcal{E} = \mathcal{E}_1 \circ \mathcal{E}_2$ implies that either $\mathcal{E}_1$ or $\mathcal{E}_2$ is unitary. Otherwise, $\mathcal{E}$ is *divisible*.

It is easy to find examples of divisible channels. For instance, let $\mathcal{A}_\zeta$ be a contraction into a state ζ (see Example 4.10). Then $\mathcal{A}_\zeta \circ \mathcal{E} = \mathcal{A}_\zeta$ for all channels $\mathcal{E}$; hence, $\mathcal{A}_\zeta$ is divisible. A much more interesting fact is that there exist indivisible nonunitary channels, as we illustrate next.

Example 4.36 (*Indivisible channel*)
Channels are specific linear transformations on a vector space of operators; thus, for finite-dimensional quantum systems they can be represented by matrices (we will give more detail on this type of representation in Section 4.4, but it is not needed in this example). In analyzing the divisibility of channels determinants are a useful tool, since $\det(\mathcal{E}_1\mathcal{E}_2) = \det \mathcal{E}_1 \det \mathcal{E}_2$ for all channels $\mathcal{E}_1, \mathcal{E}_2$. The determinant of a matrix equals the product of its eigenvalues (counted with their multiplicities). Since a channel $\mathcal{E}$ maps positive operators into positive operators, it also maps selfadjoint operators into selfadjoint operators. Therefore we obtain

$$\mathcal{E}(X^*) = \tfrac{1}{2}(\mathcal{E}(X^* + X) - i\mathcal{E}(i(X^* - X)))$$
$$= \tfrac{1}{2}(\mathcal{E}(X^* + X)^* - i\mathcal{E}(i(X^* - X))^*)$$
$$= \tfrac{1}{2}(\mathcal{E}(X^*)^* + \mathcal{E}(X)^* - \mathcal{E}(X^*)^* + \mathcal{E}(X)^*) = \mathcal{E}(X)^*,$$

and the eigenvalue equation $\mathcal{E}(X) = \lambda X$ implies that $\mathcal{E}(X^*) = \mathcal{E}(X)^* = \bar{\lambda}X^*$. We conclude that either the eigenvalues of $\mathcal{E}$ are real or they come in complex conjugate pairs $\lambda, \bar{\lambda}$. Consequently, the determinant is real and thus, for channels $|\lambda| \leq 1$, we have $-1 \leq \det\mathcal{E} \leq 1$; $\det\mathcal{E} = 1$ only for unitary channels. This was proved in [145], and we will show it for the case of qubit channels in Section 4.6.

Suppose that $\mathcal{E}_{\min}$ is a channel such that $\det\mathcal{E}_{\min} < 0$ and that no other channel has a smaller determinant. If $\mathcal{E}_{\min} = \mathcal{E}_1 \circ \mathcal{E}_2$ then $\det\mathcal{E}_{\min} = \det\mathcal{E}_1 \det\mathcal{E}_2$. However, since $\det\mathcal{E}_{\min}$ is the minimal (and a negative) value and $|\det\mathcal{E}| \leq 1$ for all channels, it follows that either $\det\mathcal{E}_1 = 1$ and $\det\mathcal{E}_2 = \det\mathcal{E}_{\min}$, or $\det\mathcal{E}_2 = 1$ and $\det\mathcal{E}_1 = \det\mathcal{E}_{\min}$. Since $\det\mathcal{E} = 1$ holds only for unitary channels, the channel $\mathcal{E}_{\min}$ is indivisible. It will be argued in Section 4.6 that for $d = 2$ the channel

$$T \mapsto \tfrac{2}{3}\operatorname{tr}[T]I - \tfrac{1}{3}T$$

minimizes the determinant. This channel is also known to be unitarily equivalent to the best approximation of the (hypothetical) universal quantum NOT gate, discussed in Example 2.56.   △

4.3.3 *Disturbance and noise*

In subsection 3.4.3 we discussed distance measures (trace distance and fidelity) quantifying the distance apart of two quantum states. In this subsection we employ these measures to describe related aspects (called disturbance and noise) of the action of quantum channels and define the distances apart of channels themselves.

A channel influences the states of a quantum system. This can be understood as a *disturbance* of the system. The larger the difference between the channel's input state and output state, the larger the induced disturbance. Obviously, the identity channel is the only nondisturbing channel, because all the output states coincide with the input states. Which channel is the most disturbing? As one might expect, the answer depends on the figure of merit and there is thus no unique answer. But we can safely say that whatever (reasonable) measure is used, mutual orthogonality of the input and output states, $\varrho_{in}\varrho_{out} = O$, represents the worst case of state disturbance. The quantum NOT gate would be a good candidate for the most disturbing transformation (see Example 2.56), but we have seen that it is actually not completely positive and therefore not a channel.

Besides transforming states individually, the channels also change their relative arrangement, i.e. the separation between a pair of states is different before and after application of the channel. The following proposition shows that this aspect of a channel's influence has a tendency that justifies the concept of *noise* when speaking about quantum channels.

Proposition 4.37 All quantum channels are *contractive*, i.e. if $\mathcal{E}$ is a channel and ϱ_1, ϱ_2 are states then the trace distance D satisfies

$$D(\mathcal{E}(\varrho_1), \mathcal{E}(\varrho_2)) \leq D(\varrho_1, \varrho_2). \tag{4.33}$$

Proof The difference of two positive operators is a selfadjoint operator. Consequently, it follows from the spectral decomposition theorem (see subsection 1.3.3) that for two states ϱ_1, ϱ_2, we have $\varrho_1 - \varrho_2 = T_1 - T_2$, where T_1, T_2 are positive operators with mutually orthogonal supports; hence $\|T_1 - T_2\|_{tr} = tr[T_1] + tr[T_2]$. The operator $\varrho_1 - \varrho_2$ is traceless, therefore $tr[T_1] = tr[T_2]$. According to subsection 3.4.3 we can choose a projection C such that $\frac{1}{2}\|\mathcal{E}(\varrho_1) - \mathcal{E}(\varrho_2)\|_{tr} = tr[C\mathcal{E}(\varrho_1 - \varrho_2)]$. Using all these facts we find that

$$D(\varrho_1, \varrho_2) = \frac{1}{2}\|\varrho_1 - \varrho_2\|_{tr} = \frac{1}{2}\|T_1 - T_2\|_{tr} = \frac{1}{2}(tr[T_1] + tr[T_2]) = tr[T_1]$$
$$= tr[\mathcal{E}(T_1)] \geq tr[C\mathcal{E}(T_1)] \geq tr[C\mathcal{E}(T_1)] - tr[C\mathcal{E}(T_2)]$$
$$= tr[C\mathcal{E}(T_1 - T_2)] = tr[C\mathcal{E}(\varrho_1 - \varrho_2)] = D(\mathcal{E}(\varrho_1), \mathcal{E}(\varrho_2)).$$

Let us note that we have used only the fact that channels are positive and trace preserving, i.e. complete positivity is not needed. □

A similar fact can be proved for the fidelity, namely that

$$F(\mathcal{E}(\varrho_1), \mathcal{E}(\varrho_2)) \geq F(\varrho_1, \varrho_2). \qquad (4.34)$$

The proof of (4.34) can be found in, for instance, [104], p. 414.

We conclude that quantum channels cannot increase our ability to distinguish quantum states. This feature captures the intuitive meaning of noise. As in the case of disturbance, we can ask which channels are the least noisy and which are the noisiest. Clearly, the identity channel does not introduce any noise; however, it is not the only channel with such a property. On the one hand, any (reasonable) measure of difference is invariant under unitary channels: although unitary channels can be quite disturbing, they do not change the mutual relations between states. Their reversibility is a sign of their being noiseless. On the other hand, the noisiest channels are the single-point contractions introduced in Example 4.10, because they completely remove any initial differences between the states.

Exercise 4.38 Show that $D(\mathcal{E}(\varrho_1), \mathcal{E}(\varrho_2)) = D(\varrho_1, \varrho_2)$ if $\mathcal{E}$ is a unitary channel. [Hint: it may be helpful first to prove that $U\sqrt{T}U^* = \sqrt{UTU^*}$ for every positive operator $T \in \mathcal{L}(\mathcal{H})$ and unitary operator $U \in \mathcal{U}(\mathcal{H})$.]

How should we properly quantify the amount of disturbance and noise induced by a given channel? Is there any relation between these two concepts? The disturbance is of relevance when we are interested in the transfer of quantum states, whereas the noise is relevant when the difference between states is the feature to be transmitted, i.e. when the quantum systems are used to encode classical information.

The identity channel perfectly transfers both quantum and the classical information, and we certainly regard it as noiseless and disturbance-free. Unitary channels are noiseless but not disturbance-free. We conclude that noiseless channels can be disturbing and, depending on their figure of merit, they can even belong to the class of the most disturbing channels. For instance, if the disturbance is defined as the supremum of the difference between the input and the output states then any unitary channel transforming some pure state P_ψ into P_ϕ such that $\phi \perp \psi$ is an example of a most disturbing channel.

One possible route to quantify the disturbance of a channel is to take it as the difference between the channel and the identity channel. For these purposes we can use any measure of the difference between two channels. It is of special interest to know how well we can distinguish a given pair of channels $\mathcal{E}_1, \mathcal{E}_2$. We say they are *perfectly distinguishable* if there exists a test state $\omega \in \mathcal{S}(\mathcal{H} \otimes \mathcal{K})$, for some $\mathcal{K}$, such that the output states $\omega'_j = (\mathcal{E}_j \otimes \mathcal{I})(\omega)$ are perfectly distinguishable, i.e. $\omega'_1 \omega'_2 = O$. Reducing the certainty of conclusions, the success probability serves as the measure of the difference between the channels. As a result we obtain the formula

$$d(\mathcal{E}_1, \mathcal{E}_2) = \tfrac{1}{2} \sup_{\mathcal{K}} \sup_{\omega} \operatorname{tr}|(\mathcal{E}_1 \otimes \mathcal{I} - \mathcal{E}_2 \otimes \mathcal{I})(\omega)|, \qquad (4.35)$$

defining the so-called *completely bounded distance*. It achieves its maximum when the output states $\omega_1' = (\mathcal{E}_1 \otimes \mathcal{I})(\omega)$ and $\omega_2' = (\mathcal{E}_2 \otimes \mathcal{I})(\omega)$ are orthogonal one-dimensional projections, in which case $\|\omega_1' - \omega_2'\|_{\operatorname{tr}} = 2$ (see Lemma 2.62). In the following example we evaluate when this happens for unitary channels.

Example 4.39 (*Perfect discrimination of unitary channels*)
Consider a pair of unitary channels σ_U and σ_V on a finite-dimensional quantum system. Under what conditions can they be perfectly distinguished? Let us first note that if there exists a mixed test state $\omega \in \mathcal{S}(\mathcal{H} \otimes \mathcal{K})$ such that $\omega_U' = (\sigma_U \otimes \mathcal{I})(\omega)$, $\omega_V' = (\sigma_U \otimes \mathcal{I})(\omega)$ and $\omega_U' \omega_V' = 0$ then any pure state P_ψ from the support of ω could also act as a test state. Thus, we need to see when two vectors $\psi_U = (U \otimes I)\psi$ and $\psi_V = (V \otimes I)\psi$ are orthogonal for some vector $\psi \in \mathcal{H} \otimes \mathcal{K}$. A direct calculation gives

$$0 = \langle \psi_U | \psi_V \rangle = \langle \psi | (U^* V \otimes I) \psi \rangle = \operatorname{tr}[U^* V \varrho], \qquad (4.36)$$

where $\varrho = \operatorname{tr}_\mathcal{K}[P_\psi]$ is the reduced state of the system on which the channel acts. Let us denote by φ_j the eigenvectors of the unitary operator $U^* V$ associated with the eigenvalues $e^{i a_j}$ belonging to the unit circle in the complex plane. Then

$$0 = \operatorname{tr}[U^* V \varrho] = \sum_j e^{i a_j} \langle \varphi_j | \varrho \varphi_j \rangle. \qquad (4.37)$$

Since the numbers $q_j := \langle \varphi_j | \varrho \varphi_j \rangle$ are nonnegative and sum to 1, they define a probability distribution. The condition $0 = \sum_j e^{i a_j} q_j$ holds whenever 0 is contained in the convex hull of the eigenvalues of $U^* V$. Intuitively speaking, this happens if the eigenvalues are sufficiently spread over the unit circle. This result was derived by Acin [1]. A rather surprising consequence is that, for any pair of unitary channels σ_U and σ_V, there exists an integer n such that the channels $\sigma_U^{\otimes n}$ and $\sigma_V^{\otimes n}$ acting on $\mathcal{H}^{\otimes n}$ are perfectly distinguishable. Namely, the eigenvalues of the tensor product $(U^* V)^{\otimes n}$ are all the factors in the product $e^{i a_{j_1}} e^{i a_{j_2}} \cdots e^{i a_{j_n}} = e^{i(a_{j_1} + a_{j_2} + \cdots + a_{j_n})}$. For large enough n the factors give points in all four quadrants of the unit circle. Hence, the convex hull of the eigenvalues of $(U^* V)^{\otimes n}$ contains the origin. △

4.3.4 Conjugate channels

According to the unitary model (described by the dilation $\langle \mathcal{H}_E, U, \xi_E \rangle$), any quantum channel induces some action on the environment. As a result the final state of the environment depends on the actual input of the channel. This mapping (from initial states of the channel to final states of the environment) is obtained by taking the partial trace over the system instead of over the environment. This leads to the following definition.

Definition 4.40 A channel $\mathcal{E}' : T(\mathcal{H}) \rightarrow T(\mathcal{H}_E)$ is *conjugate* to a channel $\mathcal{E} : T(\mathcal{H}) \rightarrow T(\mathcal{H})$ if there exists a dilation $\langle \mathcal{H}_E, U, \xi \rangle$ of $\mathcal{E}$ such that

$$\mathcal{E}' = \text{tr}_S \circ \sigma_U \circ \mathcal{P}_\xi. \tag{4.38}$$

Here tr_S stands for the partial trace of the system and $\mathcal{P}_\xi$ denotes the addition of a factorized ancilla.

The fact that $\mathcal{E}'$ is a channel follows directly from its definition, because the mappings $\text{tr}_S, \sigma_U, \mathcal{P}_\xi$ are channels. Since the dilation is not unique it follows that a given channel $\mathcal{E}$ can have many conjugate channels.

Let $\mathcal{H}$ be a finite-dimensional Hilbert space and $\{\varphi_j\}$ an orthonormal basis for $\mathcal{H}$. We consider a dilation $\langle \mathcal{H}_E, U, \xi_E \rangle$ with $U = \sum_{j,k} A_{jk} \otimes |\varphi_k\rangle\langle\varphi_j|$, $\xi_E = |\varphi_1\rangle\langle\varphi_1|$, and set $R_j = A_{j1}$. The unitarity of U is guaranteed if we have that $\sum_j A^*_{jk} A_{jk'} = \delta_{kk'} I$ and $\sum_k A^*_{jk} A_{j'k} = \delta_{jj'} I$. Then we obtain for the channel $\mathcal{E}$ and its conjugate channel $\mathcal{E}'$ the following expressions:

$$\mathcal{E}(T) = \sum_j R_j T R^*_j; \tag{4.39}$$

$$\mathcal{E}'(T) = \sum_{jk} \text{tr}[R_j T R^*_k] |\varphi_j\rangle\langle\varphi_k|. \tag{4.40}$$

This shows that the abundance of the conjugate channels is related to the freedom in the operator-sum form of $\mathcal{E}$.

To make another observation on conjugate channels, suppose that

$$\mathcal{E}(T) = \sum_j A_j T A^*_j = \sum_r B_r T B^*_r.$$

For each $0 < \lambda < 1$, we can write a trivial convex combination

$$\mathcal{E}(T) = \lambda \sum_j A_j T A^*_j + (1 - \lambda) \sum_r B_r T B^*_r,$$

and formula (4.40) gives

$$\mathcal{E}'(T) = \lambda \sum_{j,k} \text{tr}[A_j T A^*_k] |\varphi_j\rangle\langle\varphi_k| + (1 - \lambda) \sum_{r,s} \text{tr}[B_r T B^*_s] |\psi_r\rangle\langle\psi_s|.$$

Thus $\mathcal{E}' = \lambda \mathcal{E}'_1 + (1 - \lambda)\mathcal{E}'_2$, but in general $\mathcal{E}'_1 \neq \mathcal{E}'_2$. This shows that the set of conjugate channels $\mathcal{E}'$ is convex. Notice, however, that a mixture of two conjugate channels makes sense only if the dilations under consideration have the same environment.

Example 4.41 *(Channels conjugate to a unitary channel)*
Let σ_U be a unitary channel. By (4.40), for a conjugate channel σ_U' we obtain

$$\sigma_U'(T) = \operatorname{tr}[T]\,|\varphi\rangle\langle\varphi|,$$

where the unit vector φ can be chosen arbitrarily. Thus, any contraction $\mathcal{A}_\varphi$ into a pure state $|\varphi\rangle\langle\varphi|$ is a conjugate channel for σ_U. The convexity of the set of conjugate channels (for a fixed environment) implies that contraction to any mixed state is also a channel conjugate to σ_U. Moreover, as a unitary channel has a unique operator-sum form (see Example 4.23), there are no other conjugate channels. $\triangle$

An interesting and important consequence of this example is that if a system evolves in a unitary way, no trace of the initial state ϱ is left in the final state of the environment. In other words, for unitary channels the environment is dynamically independent of the system.

If the environment is of the same dimension as the system then one can ask whether the relation $\mathcal{E}'$ conjugate to $\mathcal{E}$ is symmetric. In particular, is σ_U conjugate to a single-point contraction $\mathcal{A}_\varphi$? We assume that σ_U acts on a d-dimensional system and that $\varphi \in \mathcal{H}_d$. Suppose that $A_j = |\varphi\rangle\langle U^*\varphi_j|$, where the vectors $\{\varphi_j\}$ form an orthonormal basis of $\mathcal{H}_d$. Using the formula in (4.40) we find that

$$\mathcal{E}'(T) = \sum_{j,k} \operatorname{tr}[A_j T A_k^*]\,|\varphi_j\rangle\langle\varphi_k| = \sum_{j,k} \langle\varphi_j|UTU^*\varphi_k\rangle\,|\varphi_j\rangle\langle\varphi_k| = UTU^*.$$

In the general case, the symmetry of the conjugacy relation follows from the fact that the roles of the system and the environment can be exchanged after the dilation.

Exercise 4.42 Verify that Kraus operators $A_j = |\varphi\rangle\langle U^*\varphi_j|$ define the channel $\mathcal{A}_\varphi$. Find the associated dilation. [Hint: Set $\xi_E = |\varphi\rangle\langle\varphi|$ and use the derivation of (4.40).]

4.4 Parametrizations of quantum channels

Thus far we have shown that channels can be represented by their dilations or by a Kraus operator-sum form. Both these representations are ambiguous; however, there are situations in which the uniqueness of a representation is useful and important. In this section we will introduce several unique representations of quantum channels. We will assume all Hilbert spaces to be finite dimensional.

4.4.1 Matrix representation

Quantum channels are linear mappings on the vector space $T(\mathcal{H})$, and therefore they can be represented as matrices if operators are understood as vectors (recall subsection 2.1.3). In particular, let us fix an orthogonal operator basis $E_0, \ldots, E_{d^2-1}$ for a d-dimensional quantum system. A general operator $T = \sum_j t_j E_j$ can be expressed as a vector $\vec{t}$ with coefficients

$$t_j = \frac{1}{\operatorname{tr}\left[E_j^* E_j\right]} \operatorname{tr}\left[E_j^* T\right].$$

That is,

$$\mathcal{E}(T) = \sum_j \frac{1}{\operatorname{tr}\left[E_j^* E_j\right]} \operatorname{tr}\left[E_j^* \mathcal{E}(T)\right] E_j$$

$$= \sum_{j,k} \frac{1}{\operatorname{tr}\left[E_j^* E_j\right] \operatorname{tr}\left[E_k^* E_k\right]} \operatorname{tr}\left[E_j^* \mathcal{E}(E_k)\right] \operatorname{tr}\left[E_k^* T\right] E_j$$

$$= \sum_{j,k} \mathcal{E}_{jk} t_k E_j,$$

where the $\mathcal{E}_{jk} = \operatorname{tr}\left[E_j^* \mathcal{E}(E_k)\right] / \operatorname{tr}\left[E_j^* E_j\right]$ are the entries of the $d^2 \times d^2$ complex matrix describing the action of the channel $\mathcal{E}$ on operators represented by vectors $\vec{t}$ in the operator basis $\{E_j\}$.

For the matrix representation described the composition of two channels corresponds to matrix multiplication. A disadvantage of this representation is that the constraint of complete positivity is not translated into some nice feature of the matrix $[\mathcal{E}_{jk}]$, and so other, equivalent, forms must be exploited in order to verify this constraint.

Bloch representation

Choosing a basis of selfadjoint operators containing the identity $E_0 = I$ and satisfying $\operatorname{tr}\left[E_j^* E_j\right] = \operatorname{tr}[I] = d$ for all j reveals the Bloch representation of quantum states (see subsection 2.1.3), i.e. $\varrho = \frac{1}{d}(I + \vec{r} \cdot \vec{E})$ and $\vec{r} = \operatorname{tr}\left[\varrho \vec{E}\right] \in \mathbb{R}^{d^2-1}$ is the associated Bloch vector, setting $\vec{E} = (E_1, \ldots, E_{d^2-1})$. Let us stress a subtle feature: the vector $\vec{t}$ defined before is given by $(\frac{1}{d}, \frac{1}{d}\vec{r})$. Trace preservation implies that in this basis $\mathcal{E}_{0k} = \frac{1}{d}\operatorname{tr}[I\mathcal{E}(E_j)] = \delta_{0k}$, hence

$$t_0 \mapsto t_0' = \mathcal{E}_{00} t_0 = t_0; \tag{4.41}$$

$$t_j \mapsto t_j' = \sum_{k=1}^{d^2-1} (\mathcal{E}_{jk} t_k + \mathcal{E}_{k0} t_0). \tag{4.42}$$

Taking into account the particular form of $\vec{r}$ we obtain

$$t_0 \mapsto t_0' = \frac{1}{d}, \qquad \frac{1}{d}\vec{r} \mapsto \frac{1}{d}\vec{r}' = \frac{1}{d}(T\vec{r}+\vec{t}), \qquad (4.43)$$

where we define a $(d^2 - 1) \times (d^2 - 1)$ matrix T with entries $T_{jk} = \mathcal{E}_{jk}$ and a $(d^2 - 1)$-dimensional vector $\vec{t}$ with values $\tau_j = \mathcal{E}_{j0}$. As a result we find that, in the language of Bloch vectors, channels are represented by affine transformations $\vec{r} \mapsto \vec{r}' = T\vec{r}+\vec{t}$.

Example 4.43 (*Unitary channels in the Bloch representation*)
For unitary channels σ_U the vector $\vec{t}$ vanishes, because the components $\tau_j = \frac{1}{d}\operatorname{tr}[E_j U I U^*] = 0$, and $[T_U]_{jk} = \frac{1}{d}\operatorname{tr}[E_j U E_k U^*]$. We have

$$\sum_k [T_U]_{jk} [T_U^T]_{kl} = \sum_k [T_U]_{jk} [T_U]_{lk} = \frac{1}{d^2} \sum_k \operatorname{tr}[U^* E_j U E_k] \operatorname{tr}[E_k U^* E_l U]$$

$$= \frac{1}{d} \operatorname{tr}[U^* E_j U U^* E_l U] = \frac{1}{d} \operatorname{tr}[E_j E_l] = \delta_{jl} \qquad (4.44)$$

and similarly $T_U^T T_U = I$. In this derivation we have used the identity $X = \sum_k \operatorname{tr}[E_k^* X] E_k$ for $X = U^* E_l U$. Moreover, by definition $\det T_U$ is a product of all the eigenvalues of T_U. Let us denote by e^{ix_j} the eigenvalues of the unitary operator U; then $U\varphi_j = e^{ix_j}\varphi_j$ determines the eigenvectors $\varphi_j \in \mathcal{H}$. It is straightforward to verify that the eigenvalues of T_U are $\mu_{jk} = e^{i(x_j - x_k)}$ and its eigenvectors are the Bloch vectors corresponding to the operators $|\varphi_j\rangle\langle\varphi_k|$. Consequently, $\det T_U = \prod_{j,k} \mu_{jk} = 1$.

It follows that, in the Bloch parametrization, unitary channels are represented by special orthogonal matrices. However, the converse is not true unless $d = 2$. The easiest way to see this is to compare the number of real parameters describing a general unitary channel with the number of real parameters determining the special orthogonal matrices on a $(D = d^2 - 1)$-dimensional real vector space. The first-mentioned number is $d^2 - 1$ and the second is $\frac{1}{2}D(D - 1)$. They match only when $d = 2$. $\triangle$

4.4.2 The χ-matrix representation

In contrast with the Bloch representation, considered above, the operator-sum form is manifestly completely positive and that is a very important advantage. However, it is not unique; there is a certain freedom in the particular choice of Kraus operators. For simplicity let us consider an orthonormal operator basis $E_0, \ldots, E_{d^2-1}$ and express each Kraus operator A_n in this operator basis, as $A_n = \sum_r a_{nr} E_r$ with $a_{nr} = \operatorname{tr}[E_r^* A_n]$. Then the identity

$$\mathcal{E}(\varrho) = \sum_n A_n \varrho A_n^* = \sum_{rs} \sum_n a_{nr} a_{ns}^* E_r \varrho E_s^* = \sum_{rs} \chi_{rs} E_r \varrho E_s^* \qquad (4.45)$$

defines the χ-*matrix representation* of the channel $\mathcal{E}$. Let us note that the composition of quantum channels does not correspond to the multiplication of χ-matrices. The χ-matrix representation reduces the ambiguity of the operator-sum representation, as seen in the next exercise.

Exercise 4.44 Show that different Kraus decompositions of the same channel $\mathcal{E}$ define the same χ-matrix. [Hint: Recall the freedom in the choice of Kraus operators.]

Let us now derive the relation between the two representations of quantum channels that we have considered. A direct calculation shows that

$$\mathcal{E}_{jk} = \text{tr}\big[E_j^* \mathcal{E}[E_k]\big] = \sum \chi_{rs} \text{tr}\big[E_j^* E_r E_k E_s^*\big]. \qquad (4.46)$$

Defining $M_{jk,rs} = \text{tr}\Big[E_j^* E_r E_k E_s^*\Big]$, we obtain the following relation between the two representations of channel $\mathcal{E}$:

$$\mathcal{E}_{jk} = \sum_{r,s} M_{jk,rs} \chi_{rs}. \qquad (4.47)$$

The following proposition shows how the χ-matrix representation translates the complete positivity constraint.

Proposition 4.45 If $\mathcal{E}$ is a quantum channel then the corresponding χ-matrix is positive and $\text{tr}[\chi] = d$.

Proof For a quantum channel $\mathcal{E}(\varrho) = \sum_n A_n \varrho A_n^*$ in a d-dimensional quantum system, the corresponding χ-matrix acts on a d^2-dimensional complex vector space. For åll complex vectors $\vec{x} = (x_1, \ldots, x_{d^2})$,

$$\sum_{r,s} x_r^* \chi_{rs} x_s = \sum_n \sum_r (x_r^* a_{nr}) \sum_s (a_{ns}^* x_s) = \sum_n y_n y_n^* = \sum_n |y_n|^2 \geq 0,$$

hence the matrix is positive. Using the notation $\langle A|B\rangle_{\text{HS}} = \text{tr}[A^* B]$, a direct calculation gives

$$\sum_r \chi_{rr} = \sum_{r,n} \langle E_r|A_n\rangle_{\text{HS}} \langle A_n|E_r\rangle_{\text{H-S}} = \sum_n \langle A_n|E_r\rangle_{\text{HS}}$$

$$= \sum_n \text{tr}\big[A_n^* A_n\big] = \text{tr}[I] = d.$$

$\square$

4.4.3 Choi–Jamiolkowski isomorphism

In this subsection we will introduce yet another representation of channels, which turns out to be very close to the χ-matrix representation. We denote by $\mathcal{M}_d$ the matrix algebra of $d \times d$ complex matrices. Let us recall that for a d-dimensional quantum system the operators can be identified with $d \times d$ complex matrices; hence $\mathcal{M}_d = \mathcal{L}(\mathcal{H}_d)$.

Theorem 4.46 *(Choi's theorem)*
Let $\mathcal{E} : \mathcal{M}_d \to \mathcal{M}_{d'}$ be a positive linear mapping. Then the following statements are equivalent:

(i) $\mathcal{E}$ is completely positive;
(ii) $\mathcal{E}$ is d-positive, i.e. $\mathcal{E} \otimes \mathcal{I}_d$ is a positive map;
(iii) the matrix

$$
\Phi_{\mathcal{E}} = \begin{pmatrix} \mathcal{E}(|\varphi_1\rangle\langle\varphi_1|) & \cdots & \mathcal{E}(|\varphi_1\rangle\langle\varphi_d|) \\ \vdots & \ddots & \vdots \\ \mathcal{E}(|\varphi_d\rangle\langle\varphi_1|) & \cdots & \mathcal{E}(|\varphi_d\rangle\langle\varphi_d|) \end{pmatrix}
\tag{4.48}
$$

is positive, where φ_j is any orthonormal basis of the d-dimensional Hilbert space and $\mathcal{E}(|\varphi_j\rangle\langle\varphi_k|)$ is an element of $\mathcal{M}_{d'}$. The matrix $\Phi_{\mathcal{E}}$ is called the *Choi matrix* of $\mathcal{E}$.

Proof   The implication (i)$\Rightarrow$(ii) follows directly from the definitions.
   To show that (ii)$\Rightarrow$(iii), consider a positive matrix

$$
M = \sum_{jk} |\varphi_j \otimes \varphi_j\rangle\langle\varphi_k \otimes \varphi_k| \in \mathcal{M}_d \otimes \mathcal{M}_d.
$$

By the positivity of $\mathcal{E} \otimes \mathcal{I}_d$ the matrix $M' = (\mathcal{E} \otimes \mathcal{I}_d)(M)$ is positive. Since $M' = \mathcal{E}(|\varphi_j\rangle\langle\varphi_k|) \otimes |\varphi_j\rangle\langle\varphi_k| = \Phi_{\mathcal{E}}$, the positivity of $\Phi_{\mathcal{E}}$ follows. Hence the implication (ii)$\Rightarrow$(iii) holds.
   It remains to prove that (iii)$\Rightarrow$(i). Suppose that $\psi_l \in \mathbb{C}^{d'} \otimes \mathbb{C}^d$ are subnormalized eigenvectors of the positive operator $\Phi_{\mathcal{E}}$. They are linearly independent and form an orthogonal set in $\mathbb{C}^{d'} \otimes \mathbb{C}^d$, i.e. $l = 1, \ldots, n \leq dd'$. The tensor product can be regarded as a direct sum $\mathbb{C}^{d'} \oplus \cdots \oplus \mathbb{C}^{d'} = \mathbb{C}^{d'} \otimes \mathbb{C}^d$. Let $P_j : \mathbb{C}^{d'} \otimes \mathbb{C}^d \to \mathbb{C}^{d'}_j$ be a projection onto the jth 'copy' of $\mathbb{C}^{d'}$. Then

$$
\mathcal{E}\left[|\varphi_j\rangle\langle\varphi_k|\right] = P_j \Phi_{\mathcal{E}} P_k = \sum_l P_j |\psi_l\rangle\langle\psi_l| P_k = \sum_l |P_j \psi_l\rangle\langle P_k \psi_l|.
$$

Define $n \leq dd'$ operators $V_l : \mathbb{C}^d \to \mathbb{C}^{d'}$ by their action $V_l \varphi_j = P_j \psi_l$. With the help of these operators, we have

$$\mathcal{E}\left[|\varphi_j\rangle\langle\varphi_k|\right] = \sum_l |P_j\psi_l\rangle\langle P_k\psi_l| = \sum_l |V_l\varphi_j\rangle\langle V_l\varphi_k| = \sum_l V_l|\varphi_j\rangle\langle\varphi_k|V_l^*.$$

We thus have $\mathcal{E}(T) = \sum_l V_l T V_l^*$ for all $T \in \mathcal{M}_d$. Therefore, the mapping $\mathcal{E}$ can be written in the operator-sum form and it is thus completely positive.   □

Choi's theorem provides a relatively simple test of whether a given linear map $\mathcal{E} : \mathcal{L}(\mathcal{H}_d) \to \mathcal{L}(\mathcal{H}_{d'})$ between finite-dimensional systems is completely positive: it is sufficient to verify that the associated Choi matrix is positive. This test also has a physical interpretation. We have seen that $\Phi_{\mathcal{E}} = (\mathcal{E} \otimes \mathcal{I})[M]$, where $M : \mathcal{H}_d \otimes \mathcal{H}_d \to \mathcal{H}_d \otimes \mathcal{H}_d$ is a specific positive operator. Therefore, the operator

$$P_+ \equiv (\text{tr}[M])^{-1}M = \frac{1}{d}\sum_{j,k=1}^{d} |\varphi_j \otimes \varphi_k\rangle\langle\varphi_j \otimes \varphi_k| \tag{4.49}$$

is a state. Consequently, complete positivity can be tested by application of $\mathcal{E} \otimes \mathcal{I}_d$ on the single state P_+.

Exercise 4.47 Verify that P_+ is a pure state. [Hint: Calculate its purity.]

In the following theorem we shall see that state P_+ allows us to relate the channels on a d-dimensional system (having outputs in a d'-dimensional system) to states on a $(d' \times d)$-dimensional system, i.e. a system composed of a d'-dimensional system and a d-dimensional system.

Theorem 4.48 *(Choi–Jamiolkowski isomorphism)*
Suppose that $\mathcal{E} : \mathcal{L}(\mathcal{H}_d) \to \mathcal{L}(\mathcal{H}_{d'})$ is a linear map. Fix an orthonormal basis $\{\varphi_j\}$ for $\mathcal{H}_d$ and define P_+ as in (4.49). Then the mapping

$$\mathcal{J} : \mathcal{E} \mapsto \Omega_{\mathcal{E}} = (\mathcal{E} \otimes \mathcal{I})[P_+] \tag{4.50}$$

defines a so called *Choi–Jamiolkowski isomorphism* between the vector spaces of linear maps on a d-dimensional system and linear operators on a $(d' \times d)$-dimensional Hilbert space $\mathcal{L}(\mathcal{H}_{d'} \otimes \mathcal{H}_d)$. The inverse mapping is given by the relation

$$\mathcal{J}^{-1} : \Omega \mapsto \mathcal{E}_{\Omega} : \mathcal{E}_{\Omega}[X] = d\,\text{tr}_2[(I \otimes X^T)\Omega], \tag{4.51}$$

where tr_2 stands for the partial trace over the second Hilbert space. In particular, Choi's theorem (Theorem 4.46) implies that completely positive maps are transformed into positive operators on $\mathcal{H}_{d'} \otimes \mathcal{H}_d$ and that channels are isomorphic to a subset of the state space $\mathcal{S}(\mathcal{H}_{d'} \otimes \mathcal{H}_d)$ with elements satisfying $d\,\text{tr}_1[\Omega_{\mathcal{E}}] = I_d$.

Proof By definition the mapping $\mathcal{J}$ is linear, i.e. $\mathcal{J}(\mathcal{E}_1 + \lambda\mathcal{E}_2) = \mathcal{J}(\mathcal{E}_1) + \lambda\mathcal{J}(\mathcal{E}_2)$. Similarly, $\mathcal{J}^{-1}$ is linear also and hence the linear structures of the set of linear maps and the set of linear operators $\mathcal{L}(\mathcal{H}_d \otimes \mathcal{H}_d)$ are preserved. It is sufficient to prove that $\mathcal{J}^{-1}$, as defined above, is indeed the mapping inverse to $\mathcal{J}$, that is,

$$\mathcal{E}_{\Omega_\mathcal{E}}[X] = d \operatorname{tr}_2[(I \otimes X^T)\Omega_\mathcal{E}] = d \operatorname{tr}_2\left[(I \otimes X^T)(\mathcal{E} \otimes \mathcal{I})[P_+]\right]$$
$$= \sum_{j,k} \operatorname{tr}_2\left[(I \otimes X^T)\left(\mathcal{E}[|\varphi_j\rangle\langle\varphi_k|] \otimes |\varphi_j\rangle\langle\varphi_k|\right)\right]$$
$$= \sum_{j,k} \mathcal{E}[|\varphi_j\rangle\langle\varphi_k|] \operatorname{tr}\left[X^T |\varphi_j\rangle\langle\varphi_k|\right] = \sum_{j,k} \mathcal{E}[|\varphi_j\rangle\langle\varphi_k|] \langle\varphi_k| X^T \varphi_j\rangle$$
$$= \mathcal{E}\left[\sum_{j,k} \langle\varphi_j| X\varphi_k\rangle |\varphi_j\rangle\langle\varphi_k|\right] = \mathcal{E}[X]$$

for all $X \in \mathcal{L}(\mathcal{H}_d)$, i.e. $\mathcal{E}_{\Omega_\mathcal{E}} = \mathcal{E}$ for all linear maps $\mathcal{E}$. In a similar way we can prove that $\Omega_{\mathcal{E}_\Omega} = \Omega$ for all $\Omega \in \mathcal{L}(\mathcal{H}_d \otimes \mathcal{H}_d)$, i.e. $\mathcal{J} \circ \mathcal{J}^{-1} = \mathcal{J}^{-1} \circ \mathcal{J} = \operatorname{id}$. If the linear map $\mathcal{E}$ is trace preserving then

$$\operatorname{tr}_1[(\mathcal{E} \otimes \mathcal{I})(P_+)] = \frac{1}{d} \sum_{j,k} \operatorname{tr}\left[\mathcal{E}(|\varphi_j\rangle\langle\varphi_k|)\right] |\varphi_j\rangle\langle\varphi_k|$$
$$= \frac{1}{d} \sum_{j,k} \delta_{jk} |\varphi_j\rangle\langle\varphi_k| = \frac{1}{d} I,$$

thus, $\operatorname{tr}[\Omega_\mathcal{E}] = 1$. This proves that the Choi–Jamiolkowski isomorphism relates channels to a specific subset of states. $\square$

Exercise 4.49 Fix an orthonormal basis of $\mathcal{H}$. The operators $\{|\varphi_j\rangle\langle\varphi_k|\}_{j,k}$ form an orthonormal basis of $T(\mathcal{H})$. Show that with such a choice of operator basis the χ-matrix $\chi_\mathcal{E}$ and the Choi–Jamiolkowski matrix $\Omega_\mathcal{E}$ are proportional to each other.

The following proposition is a particular version of the Choi-Jamiolkowski isomorphism.

Proposition 4.50 Suppose that $\{\varphi_j\}$ is an orthonormal basis of $\mathcal{H}_d$ and write $\psi_+ = \frac{1}{\sqrt{d}} \sum_j \varphi_j \otimes \varphi_j \in \mathcal{H}_d \otimes \mathcal{H}_d$. Then the mapping

$$A \mapsto \psi_A = \sqrt{d}(A \otimes I)\psi_+ \tag{4.52}$$

defines an isomorphism between the Hilbert spaces $\mathcal{L}(\mathcal{H}_d \otimes \mathcal{H}_{d'})$ (endowed with the Hilbert–Schmidt inner product) and $\mathcal{H}_{d'} \otimes \mathcal{H}_d$.

Proof Let $\{\phi_k\}$ be the orthonormal basis of $\mathcal{H}_{d'}$. Then $A = \sum_{k,j} a_{jk} = |\phi_j\rangle\langle\varphi_k|$ is mapped into a vector $\psi_A = \sqrt{d}(A \otimes I)\psi_+ = \sum_k A(\varphi_k) \otimes \varphi_k =$

$\sum_{j=1}^{d} \sum_{k=1}^{d'} a_{jk}\phi_j \otimes \varphi_k$. It is straightforward to verify that this transformation is induced by the following identification of the basis elements of the linear spaces $\mathcal{L}(\mathcal{H}_d, \mathcal{H}_{d'})$ and $\mathcal{H}_{d'} \otimes \mathcal{H}_d$:

$$|\phi_j\rangle\langle\varphi_k| \leftrightarrow \phi_j \otimes \varphi_k. \qquad (4.53)$$

Thus, the mapping $A \mapsto \psi_A$ preserves the linear structure of the spaces. Moreover, the identity

$$\langle\psi_A|\psi_B\rangle = d \langle(A \otimes I)\psi_+|(B \otimes I)\psi_+\rangle = d \langle\psi_+|(A^*B \otimes I)\psi_+\rangle$$
$$= \sum_j \langle\varphi_j|A^*B\varphi_j\rangle = \mathrm{tr}[A^*B]$$

shows that the inner product is preserved as well; hence $A \mapsto \psi_A$ defines an isomorphism between Hilbert spaces.   □

Exercise 4.51 Use Proposition 4.50 to derive the Choi–Jamiolkowski isomorphism in Theorems 4.46 and 4.48. [Hint: Observe that the mapping $X \to AXB^*$ translates into the operator $|\psi_A\rangle\langle\psi_B|$.]

4.5 Special classes of channels

In this section we discuss some special classes of channels that are common in applications. Our main motivation is to demonstrate that it is possible to say much more about the properties of a given channel if it is known to belong to a certain subclass.

4.5.1 Strictly contractive channels

By Proposition 4.37 we know that any channel is contractive, i.e. the trace distance between two states cannot increase under the action of a channel. In this section we investigate the properties of strictly contractive channels. Our discussion follows essentially [120].

Definition 4.52 A channel $\mathcal{E}$ is *strictly contractive* if, for all states $\varrho_1, \varrho_2 \in \mathcal{S}(\mathcal{H})$, the inequality

$$\|\mathcal{E}(\varrho_1 - \varrho_2)\|_{\mathrm{tr}} \leq k \, \|\varrho_1 - \varrho_2\|_{\mathrm{tr}} \qquad (4.54)$$

is valid for some fixed number $0 \leq k < 1$.

Obvious examples of strictly contractive channels are the complete contractions discussed in Example 4.10. For a complete contraction $\mathcal{E}_F$, we get $\mathcal{E}_F(\varrho_1 - \varrho_2) = 0$ for all states ϱ_1, ϱ_2. Hence, they satisfy (4.54) with $k = 0$.

Example 4.53 (*Depolarizing channel*)

Let $\mathcal{H}$ be a finite-dimensional Hilbert space. An example of a class of strictly contractive channels is provided by the one-parameter family of *depolarizing channels*. For each $0 \leq p \leq 1$, we define

$$\mathcal{D}_p(\varrho) = p\frac{1}{d}I + (1-p)\varrho.$$

Hence, $\mathcal{D}_p$ is a convex combination of a contraction to a total mixture and the identity channel. For two states ϱ_1 and ϱ_2, we obtain

$$\|\mathcal{D}_p(\varrho_1 - \varrho_2)\|_{\mathrm{tr}} = (1-p)\|\varrho_1 - \varrho_2\|_{\mathrm{tr}}.$$

Thus, the contractivity factor k in (4.54) equals $1 - p$. This shows that the channel $\mathcal{D}_p$ is strictly contractive when $p \neq 0$. $\triangle$

Having one strictly contractive channel, it is easy to generate more. Namely, we have the following observation.

Proposition 4.54 Consider a pair of quantum channels $\mathcal{E}_1, \mathcal{E}_2$ and assume that $\mathcal{E}_1$ is strictly contractive. The compositions $\mathcal{E}_1 \circ \mathcal{E}_2$ and $\mathcal{E}_2 \circ \mathcal{E}_1$ and the convex combinations $t\mathcal{E}_1 + (1-t)\mathcal{E}_2$, with $0 < t \leq 1$, are strictly contractive.

Proof Let k be the contractivity factor as in (4.54). The strict contractiveness of $\mathcal{E}_1 \circ \mathcal{E}_2$ comes from the inequality

$$\|\mathcal{E}_1 \circ \mathcal{E}_2(\varrho_1 - \varrho_2)\|_{\mathrm{tr}} \leq \|\mathcal{E}_1(\varrho_1 - \varrho_2)\|_{\mathrm{tr}} \leq k\|\varrho_1 - \varrho_2\|_{\mathrm{tr}},$$

and similarly for $\mathcal{E}_2 \circ \mathcal{E}_1$. For a convex combination we obtain

$$\|(t\mathcal{E}_1 + (1-t)\mathcal{E}_2)(\varrho_1 - \varrho_2)\|_{\mathrm{tr}} \leq t\|\mathcal{E}_1(\varrho_1 - \varrho_2)\|_{\mathrm{tr}} + (1-t)\|\mathcal{E}_2(\varrho_1 - \varrho_2)\|_{\mathrm{tr}}$$
$$\leq (kt + 1 - t)\|\varrho_1 - \varrho_2\|_{\mathrm{tr}},$$

and this implies that $t\mathcal{E}_1 + (1-t)\mathcal{E}_2$ is a strictly contractive channel for $0 < t \leq 1$. $\square$

It is perhaps surprising that a channel $\mathcal{E} \otimes \mathcal{I}$ is not strictly contractive even if $\mathcal{E}$ is strictly contractive. This conclusion follows from the following observation.

Proposition 4.55 Let $\mathcal{E}_1$ and $\mathcal{E}_2$ be two channels. The tensor product $\mathcal{E}_1 \otimes \mathcal{E}_2$ is strictly contractive only if both $\mathcal{E}_1$ and $\mathcal{E}_2$ are strictly contractive.

Proof Suppose that $\mathcal{E}_2$ is not strictly contractive. Then there would be states ξ_1 and ξ_2, $\xi_1 \neq \xi_2$, such that

$$\|\mathcal{E}_2(\xi_1 - \xi_2)\|_{\mathrm{tr}} = \|\xi_1 - \xi_2\|_{\mathrm{tr}}.$$

Fix a state η and write $\varrho_1 = \eta \otimes \xi_1$ and $\varrho_2 = \eta \otimes \xi_2$. Then we obtain

$$\|\mathcal{E}_1 \otimes \mathcal{E}_2(\varrho_1 - \varrho_2)\|_{\mathrm{tr}} = \|\mathcal{E}_1(\eta) \otimes \mathcal{E}_2(\xi_1 - \xi_2)\|_{\mathrm{tr}} = \|\mathcal{E}_2(\xi_1 - \xi_2)\|_{\mathrm{tr}}$$
$$= \|\xi_1 - \xi_2\|_{\mathrm{tr}} = \|\varrho_1 - \varrho_2\|_{\mathrm{tr}},$$

which shows that the channel $\mathcal{E}_1 \otimes \mathcal{E}_2$ is not strictly contractive. $\qquad\square$

For a channel $\mathcal{E}$, each state $\varrho_0 \in \mathcal{S}(\mathcal{H})$ satisfying $\mathcal{E}(\varrho_0) = \varrho_0$ is called a *fixed point* of $\mathcal{E}$. An important property of strictly contractive channels is that they have a unique fixed point. The following theorem is a direct consequence of the Banach fixed point theorem.

Theorem 4.56 Let $\mathcal{E}$ be a strictly contractive channel. There is a *unique* state $\varrho_0 \in \mathcal{S}(\mathcal{H})$ such that $\mathcal{E}(\varrho_0) = \varrho_0$.

Exercise 4.57 Find the fixed point of the depolarizing channel D_p introduced in Example 4.53.

The following example demonstrates that there are channels with more than one fixed point.

Example 4.58 (*Fixed points of unitary channels*)
Let U be a unitary operator having $n \geq 2$ different eigenvalues. The eigenvectors ψ_j of U determine pure states $P_j = |\psi_j\rangle\langle\psi_j|$. Since the eigenvalues of U have modulus 1 (see subsection 1.2.4), we have $\sigma_U(P_j) = P_j$. Therefore every pure state P_j is a fixed point of σ_U, and also every mixed state of the form $\varrho = \sum_j \lambda_j P_j$ is a fixed point of σ_U. $\qquad\triangle$

4.5.2 Random unitary channels

Definition 4.59 A channel $\mathcal{E}$ is a *random unitary channel* if it can be expressed as a convex mixture of unitary channels, that is,

$$\mathcal{E}(\varrho) = \sum_j p_j U_j \varrho U_j^*, \tag{4.55}$$

where $0 \leq p_j \leq 1$ and $\sum_j p_j = 1$.

From this definition we see directly that the set of random unitary channels is convex. It is also closed under composition, i.e. whenever $\mathcal{E}_1$ and $\mathcal{E}_2$ are random unitary channels, $\mathcal{E}_1 \circ \mathcal{E}_2$ is a random unitary channel also. For any random unitary channel there exists a canonical dilation. Namely, if U is the controlled-U unitary operator $U = \sum_j U_j \otimes |\varphi_j\rangle\langle\varphi_j|$ for some collection of unitary operators $\{U_j\}$ then

$$\mathcal{E}(\varrho) = \sum_j \langle\varphi_j|\xi\varphi_j\rangle U_j \varrho U_j^*. \tag{4.56}$$

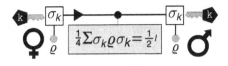

Figure 4.2 Private quantum channel. A shared cryptographic key k is used to encrypt the state ϱ by applying a unitary channel σ_k to the system and using the same channel to decrypt the quantum state.

Thus, for any state ξ the induced channel is random unitary. It is interesting that the induced channel $\mathcal{E}$ depends only on the diagonal entries $\langle\varphi_j|\xi\varphi_j\rangle$, which are, moreover, unaffected by U. Indeed, since $\xi' = \mathrm{tr}_S[U(\varrho \otimes \xi)U^*] = \sum_{j,k}\mathrm{tr}[U_j\varrho U_k^*]|\varphi_j\rangle\langle\varphi_k|$ is the final state of the environment, it follows that we have $\langle\varphi_j|\xi'\varphi_j\rangle = \langle\varphi_j|\xi\varphi_j\rangle$, for any ϱ and ξ.

Example 4.60 *(Private quantum channels)*
In subsection 5.2.4 we will discuss the one-time pad protocol enabling a secure communication of bit values (containing classical information). A *private quantum channel* is an analogous quantum cryptographic protocol aiming to transmit securely quantum states (quantum information) rather than classical states [5]. As for the one-time pad protocol, security is based on the existence of a shared classical key represented by identical sequences of random bits in the possession of both communicating parties. The protocol works as follows. Alice wants to transfer the states $\varrho_1, \dots, \varrho_n$ to Bob. A classical key is used to set the encoding channels $\mathcal{E}_1, \dots, \mathcal{E}_n$ used by Alice and the decoding channels $\mathcal{D}_1, \dots, \mathcal{D}_n$ used by Bob. In each run Bob receives a state $\mathcal{E}_j(\varrho_j)$ and, applying the appropriate decoding channel $\mathcal{D}_j$, he reveals the original state $\varrho_j = \mathcal{D}_j \circ \mathcal{E}_j(\varrho_j)$. See Figure 4.2.

We assume that ϱ_j is an arbitrary state. Therefore we must have $\mathcal{D}_j = \mathcal{E}_j^{-1}$, and this implies that $\mathcal{E}_j$ must be unitary (see Proposition 4.31). Thus $\mathcal{E}_j = \sigma_{U_j}$ and $\mathcal{D}_j = \sigma_{U_j^*}$ for some unitary operators U_j. The question is how many bits are needed to protect a single quantum bit, i.e. how many communication runs n are encoded in a sequence of bits of size N?

Let $\{U_1, \dots, U_m\}$ be a set of unitary operators defining the encoding and decoding channels. A sequence of N random bits forming the key defines a random sequence $U_{j_1}, \dots, U_{j_n}$ determining the sequence of encoding and decoding channels in n runs, i.e. $U_{j_l} \in \{U_1, \dots, U_m\}$. For anyone except Bob Alice is randomly applying unitary channels, hence she is implementing a random unitary channel constituting the private quantum channel

$$\mathcal{E}_{\mathrm{PQC}}(\varrho) = \sum_j p_j U_j \varrho U_j^*, \qquad (4.57)$$

where p_j is the probability of operator U_j in the random sequence $U_{j_1}, \ldots, U_{j_n}$ determined by the classical key. Security is achieved if the output state $\mathcal{E}_{\text{PQC}}(\varrho)$ is independent of ϱ, i.e. $\mathcal{E}_{\text{PQC}}(\varrho) = \sum_j p_j U_j \varrho U_j^* = \varrho_0$ for some fixed state ϱ_0.

Example 1.64 guarantees the existence of a random unitary channel with the required properties. In particular, a private quantum channel with $m = d^2$ mutually orthogonal unitary operators $\{U_j\}$ and

$$p_j = \frac{1}{d^2}$$

satisfies all requirements. As a result we obtain that if the encoding and decoding channels are associated with mutually orthogonal unitary operators then they constitute a private quantum channel with $\varrho_0 = \frac{1}{d}I$ and a key of length $N = n \log m = \log d^2 = 2n \log d$. In conclusion, the secure transmission of a d-dimensional quantum state requires $2n \log d$ classical bits of the cryptographic key. $\triangle$

Suppose that $\mathcal{H}$ is a finite-dimensional Hilbert space; then the identity operator I is a trace class operator and we can compute $\mathcal{E}(I)$ for any channel $\mathcal{E}$. If $\mathcal{E}$ is a random unitary channel then it follows from (4.55) that $\mathcal{E}(I) = I$, i.e. $\mathcal{E}$ is *unital*. The unitality of $\mathcal{E}$ is therefore a simple necessary condition for a channel to be random unitary. (Recall that every channel is unital in the Heisenberg picture but that here we require unitality in the Schrödinger picture.) The physical meaning of this condition is that the complete mixture $\frac{1}{d}I$ is a fixed point of any unital channel.

If $d = 2$ then the unital channels are exactly the random unitary channels, but in higher dimensions the latter constitute a proper subset of the former [92]. We refer to [8], [98] for further details on the connection between these two classes of channels.

4.5.3 Phase-damping channels

The concept of decoherence was originally introduced as a process standing behind the disappearance of interference patterns. In a sense, owing to decoherence an initial superposition changes into a mixture. The orthogonal pure states originally forming the superposition form a so-called *decoherence basis*. Nowadays, in the literature decoherence is sometimes understood as an arbitrary nonunitary dynamics. Therefore, we shall refer to the original concept as a family of *phase-damping channels*.

Definition 4.61 A channel $\mathcal{E}$ describes *phase damping* if its power series $\mathcal{E}, \mathcal{E}^2, \mathcal{E}^3, \ldots$, converges to the channel

$$\text{diag}_b : \varrho \mapsto \text{diag}_b(\varrho) = \sum_j \langle \varphi_j | \varrho | \varphi_j \rangle |\varphi_j\rangle \langle \varphi_j|, \qquad (4.58)$$

where $b = \{\varphi_1, \ldots, \varphi_d\}$ denotes the decoherence basis.

If the decoherence basis b is fixed and $\mathcal{E}_1$, $\mathcal{E}_2$ are phase-damping channels with respect to this basis then their convex combination $\lambda\mathcal{E}_1 + (1 - \lambda)\mathcal{E}_2$ and their composition $\mathcal{E}_1 \circ \mathcal{E}_2$ are also phase-damping channels with respect to the same basis b. Moreover, $\mathcal{E}_1$ and $\mathcal{E}_2$ commute, i.e. $\mathcal{E}_1 \circ \mathcal{E}_2 = \mathcal{E}_2 \circ \mathcal{E}_1$. If the decoherence basis of $\mathcal{E}_1$ is different from the decoherence basis of $\mathcal{E}_2$ then their convex combination and composition are no longer phase-damping channels.

Proposition 4.62 Consider a channel $\mathcal{E}(\varrho) = \sum_j A_j \varrho A_j^*$ expressed in the operator-sum form. It is a phase-damping channel if and only if the Kraus operators A_j commute with the projections $e_1, \ldots, e_d$, $e_j = |\varphi_j\rangle\langle\varphi_j|$. It follows that for a phase-damping channel its Kraus operators A_j mutually commute, i.e. $[A_j, A_k] = 0$ for all j, k.

Proof Since $\mathcal{E}(e_j) = e_j$, it follows that $\mathcal{E}(T) = T$ for all operators T belonging to the subalgebra generated by the projections $e_1, \ldots, e_d$. This subalgebra $\mathcal{A}_b$ is a maximal commutative subalgebra of the algebra of bounded operators $\mathcal{L}(\mathcal{H})$. Consequently, the identity

$$(\mathcal{E}(T) - T)(\mathcal{E}(T) - T)^* = 0$$

holds for all $T \in \mathcal{A}_b$. Using the relations $\mathcal{E}(T)\mathcal{E}(T)^* = TT^* = \mathcal{E}(TT^*)$ and $\sum_j A_j^* A_j = I$, we get the identity

$$0 = \sum_j \left(A_j TT^* A_j^* - A_j T A_j^* T^* - T A_j T^* A_j^* - T A_j^* A_j T^*\right)$$

$$= \sum_j [T, A_j][T, A_j]^*.$$

On the one hand, this identity holds if and only if $[T, A_j] = 0$ for all $T \in \mathcal{A}_b$ and all j. Moreover, because the subalgebra $\mathcal{A}_b$ is the maximal commutative subalgebra, it follows that the Kraus operators A_j must mutually commute, too.

On the other hand, the identities $[A_j, e_n] = 0$ imply that $[A_j, T] = 0$ for all $T \in \mathcal{A}_b$ and, consequently, the above arguments can be reversed to prove that $\mathcal{E}(T) = \sum_j A_j T A_j^* = T$ for all $T \in \mathcal{A}_b$, including the projectors $e_1, \ldots, e_d$. $\square$

Proposition 4.63 (*Dilations of phase-damping channels.*) If $\mathcal{E}$ is a phase damping channel then $\mathcal{E}[\varrho] = \mathrm{tr}_{\mathrm{env}}\left[U(\varrho \otimes \xi)U^*\right]$ where $U = \sum_j |\varphi_j\rangle\langle\varphi_j| \otimes U_j$ is a controlled-U unitary transformation with the system playing the role of the control system. The vectors φ_j form the decoherence basis.

Proof The preservation of the decoherence basis elements φ_j, i.e. the identity $U(\varphi_j \otimes \psi) = \varphi_j \otimes \psi'$, implies that we need $\psi' = U_j\psi$ in order to preserve the scalar product. Using this condition for all j, the operator U is defined on the whole Hilbert space and takes the form of a so-called controlled-U transformation,

$$U = \sum_j |\varphi_j\rangle\langle\varphi_j| \otimes U_j. \tag{4.59}$$

The decohering system plays the role of the control system and the environment is the target system. ☐

Exercise 4.64 Find an example of U for which both the system and the environment undergo a phase-damping process. Is it possible to find a U such that the decoherence bases for both systems coincide? [Hint: Analyze the controlled-NOT gate and observe the importance of the commutativity of $\{U_j\}$.]

4.6 Example: qubit channels

In this section we will demonstrate that even for the simplest possible quantum system, namely one involving qubits, a complete parametrization of the channels takes some effort. Our presentation is based on [124].

Recall from subsection 2.1.3 that the state space of qubits can be nicely represented as a Bloch sphere. Hence, qubit channels can be illustrated as mappings on the Bloch sphere. Adopting this point of view, a qubit channel is presented by a 4×4 matrix corresponding to an affine mapping. In particular, let us choose the (subnormalized) operator basis $\sigma_0 \equiv I, \sigma_x, \sigma_y, \sigma_z$. Then the matrix elements $\mathcal{E}_{jk} = \frac{1}{2}\text{tr}[\sigma_j \mathcal{E}[\sigma_k]]$ form a (real) matrix

$$\mathcal{E} = \begin{pmatrix} 1 & 0 & 0 & 0 \\ \tau_x & T_{xx} & T_{xy} & T_{xz} \\ \tau_y & T_{yx} & T_{yy} & T_{yz} \\ \tau_z & T_{zx} & T_{zy} & T_{zz} \end{pmatrix}, \tag{4.60}$$

and the channel acts as $\vec{r} \mapsto \vec{r}' = T\vec{r} + \vec{\tau}$.

Suppose that $A_k = \sum_l a_{kl}\sigma_l$ are the Kraus operators of $\mathcal{E}$. Then

$$T_{jj} = \sum_k \text{tr}[\sigma_j A_k \sigma_j A_k^*] = \sum_{k,l,l'} a_{kl}a_{kl'}^* \text{tr}[\sigma_j \sigma_l \sigma_j \sigma_{l'}] = \sum_l j_l q_l,$$

where $j_l = \pm$ according to $\sigma_j \sigma_l \sigma_j = \pm \sigma_l$, and $q_l = \sum_k a_{kl}a_{kl}^* > 0$. In particular, since $T_{00} = q_0 + q_1 + q_2 + q_3 = 1$ it follows that $T_{jj} = 2(q_0 + q_j) - 1$ for $j = 1, 2, 3$. We can derive the inequalities

$$T_{11} \pm T_{22} \le T_{00} \pm T_{33}, \qquad -T_{11} \pm T_{22} \le T_{00} \mp T_{33}, \tag{4.61}$$

which fix the values of T_{11}, T_{22} and T_{33} to form a tetrahedron with vertices $(1, 1, 1)$, $(1, -1, -1)$, $(-1, 1, -1)$ and $(-1, -1, 1)$. These conditions are necessary but not sufficient for a channel to be completely positive.

Reducing the number of parameters

In order to formulate the sufficient conditions we will reduce the family of channels that we need to investigate. The singular value decomposition theorem (see Theorem 1.66) implies that $T = Q_1 D_{sv} Q_2$, where the Q_j are matrices of orthogonal rotations ($Q_j^T = Q_j^{-1}$) and $D_{sv} = \text{diag}\{\mu_1, \mu_2, \mu_3\}$ is the diagonal matrix of the singular values of T. The question is whether this decomposition can be interpreted as concatenation of three channels. For a general orthogonal matrix, $\det Q = \pm 1$. However, in Example 4.43 we showed that unitary channels are associated with special orthogonal matrices (rotations) R_U, for which $\det R_U = 1$. Fortunately, for the case of qubits the correspondence is one-to-one, i.e. each special orthogonal matrix is associated with some unitary channel σ_U. The only remaining problem is the negative determinant of Q_j. But every three-dimensional orthogonal matrix Q is either a rotation ($\det Q = 1$) or can be written as a product of some rotation R and an inversion $-I$, i.e. $Q_j = \pm R_j$, where $\det R_j = 1$. Thus, $T = Q_1 D_{sv} Q_2 = R_1 (\pm D_{sv}) R_2 = (R_1 R_2) R_2^T (\pm D_{sv}) R_2 = RS$, where $S = R_2^T (\pm D_{sv}) R_2$ is semidefinite (either positive or negative) and $R = R_1 R_2$ is a rotation.

The rotations S_k associated with the Pauli unitary channels $\sigma_k[\varrho] = \sigma_k \varrho \sigma_k$ are diagonal matrices with $[S_k]_{jj} = -1$ if $j \neq k$ and $[S_k]_{kk} = 1$. On composing any rotation S_k with $D_{sv} = \text{diag}\{\mu_1, \mu_2, \mu_3\}$, two values μ_j will change their sign; hence the semidefiniteness of D_{sv} is lost in $S_k D_{sv}$. By definition, if D_{sv} determines a quantum channel then $S_k D_{sv}$ must describe a quantum channel, too. Hence, the semidefiniteness of S can be replaced by its self-adjointness. Since every selfadjoint real matrix can be diagonalized by some rotation R_V it follows that $T = RS = R R_V^T D R_V$, where $D = \{\lambda_1, \lambda_2, \lambda_3\}$ is a diagonal matrix composed of the eigenvalues of S. Let us stress that the λ_j are not singular values of T. In fact they are not necessarily positive. However, $|\lambda_j| = \mu_j$ are the singular values of T.

As a result we obtain $T = R_U D R_V$ for suitable rotations R_U, R_V and $D = \text{diag}\{\lambda_1, \lambda_2, \lambda_3\}$, where $|\lambda_1|$, $|\lambda_2|$ and $|\lambda_3|$ are the singular values of T. In such a case the action of the channel $\mathcal{E}$ can be written as

$$\mathcal{E}(\varrho) = (\sigma_U \circ \mathcal{D} \circ \sigma_V)(\varrho) = U\mathcal{D}(V\varrho V^*)U^*$$

with $\mathcal{D} : \vec{r} \mapsto \vec{r}' = D\vec{r} + \vec{t}$; thus

$$\vec{r} \mapsto \vec{r}' = R_U D R_V \vec{r} + R_U \vec{t}. \tag{4.62}$$

In other words, every qubit channel $\mathcal{E}$ is unitarily equivalent to one of the channels $\mathcal{D}$ forming a six-parameter family of channels. Although this assignment is not unique (the signs of any pair of diagonal elements can be changed by applying one of the S_k), most of the interesting properties of $\mathcal{E}$ are reflected in the properties of $\mathcal{D}_{\mathcal{E}}$. The following properties follow directly from definitions given earlier.

- A channel $\mathcal{E}$ is completely positive if and only if $\mathcal{D}_{\mathcal{E}}$ is completely positive.
- A channel $\mathcal{E}$ is strictly contractive if and only if $\mathcal{D}_{\mathcal{E}}$ is strictly contractive.
- A channel $\mathcal{E}$ is unital if and only if $\mathcal{D}_{\mathcal{E}}$ is unital.
- $\det \mathcal{E} = \det \mathcal{D}_{\mathcal{E}} = \lambda_1 \lambda_2 \lambda_3$. Note that the product $\lambda_1 \lambda_2 \lambda_3$ is fixed for all $\mathcal{D}_{\mathcal{E}}$ assigned to the same $\mathcal{E}$.

Complete positivity constraints for $\mathcal{D}_{\mathcal{E}}$

The positivity constraint of $\mathcal{D}$ (and $\mathcal{E}$) requires that $|\lambda_j| \leq 1$, because otherwise the length of the Bloch vector will increase and the negative region (lying outside the Bloch sphere) will be reached. In fact, the image of the pure states (the Bloch-sphere boundary) under $\mathcal{D}$ defines an ellipsoid,

$$\left(\frac{r_1' - t_1}{\lambda_1}\right)^2 + \left(\frac{r_2' - t_2}{\lambda_2}\right)^2 + \left(\frac{r_3' - t_3}{\lambda_3}\right)^2 = 1. \tag{4.63}$$

The interpretation is that each quantum channel transforms the Bloch sphere into an ellipsoid (inside the original Bloch sphere), but the converse is not true: not all ellipsoids included in the Bloch sphere correspond to some quantum channel. In this case the inequalities from (4.61), written in the compact form

$$|T_{11} \pm T_{22}| \leq |1 \pm T_{33}|, \tag{4.64}$$

reduce to $|\lambda_1 \pm \lambda_2| \leq |1 \pm \lambda_3|$. Without giving all the details we will state the necessary and sufficient conditions [124] for the complete positivity of $\mathcal{D}$.

Proposition 4.65 A channel $\mathcal{D}$ is completely positive if and only if the following inequalities hold:

$$(\lambda_1 + \lambda_2)^2 \leq (1 + \lambda_3)^2 - t_3^2, \tag{4.65}$$

$$(\lambda_1 - \lambda_2)^2 \leq (1 - \lambda_3)^2 - t_3^2, \tag{4.66}$$

$$\left[1 - |\vec{\lambda}|^2 - |\vec{t}|^2\right]^2 \geq 4[\lambda_1^2(t_1^2 + t_2^2) + \lambda_2^2(t_2^2 + t_3^2)$$

$$+ \lambda_3^2(t_3^2 + t_1^2) - 2\lambda_1 \lambda_2 \lambda_3] \tag{4.67}$$

Proof The result follows from the conditions on the positivity of the associated Choi matrix:

$$\Phi_{\mathcal{D}} = \frac{1}{2} \begin{pmatrix} 1 + t_3 + \lambda_3 & t_1 - it_2 & 0 & \lambda_1 + \lambda_2 \\ t_1 + it_2 & 1 - t_3 - \lambda_3 & \lambda_1 - \lambda_2 & 0 \\ 0 & \lambda_1 - \lambda_2 & 1 + t_3 - \lambda_3 & t_1 - it_2 \\ \lambda_1 + \lambda_2 & 0 & t_1 + it_2 & 1 - t_3 + \lambda_3 \end{pmatrix}.$$

$\square$

Determinant

We showed in Example 4.43 that $\det \sigma_U = 1$. Using $|\lambda_j| \leq 1$ and $\det \mathcal{E} = \det \mathcal{D}_{\mathcal{E}}$, it follows that $|\det \mathcal{E}| \leq 1$. Clearly, $\det \mathcal{E} = 1$ requires that $|\lambda_j| = 1$ for all j. Let us recall that the λ_j are real. If $|\lambda_j| = 1$ then $t_j = 0$, because otherwise we can find a pure state (whose only nonvanishing entry is $r_j = \pm 1$) that is transformed into a negative operator (lying outside the Bloch sphere). In summary, $\det \mathcal{D}_{\mathcal{E}} = \lambda_1 \lambda_2 \lambda_3 = 1$ implies that either $\lambda_j = 1$ for all $j = 1, 2, 3$ or $\lambda_j = -1$ for all j but one, for which $\lambda_j = 1$. These are exactly the unitary channels induced by Pauli operators. In conclusion, if for a qubit channel $\det \mathcal{E} = 1$ then $\mathcal{E}$ is unitary.

Further, we want to identify the channels minimizing the determinant. Clearly, the value $\det \mathcal{E} = -1$ requires that either $\lambda_j = -1$ for all $j = 1, 2, 3$ or $\lambda_j = 1$ for all j but one, for which $\lambda_j = -1$. Unfortunately, neither solution satisfies the necessary condition $|\lambda_1 \pm \lambda_2| \leq |1 \pm \lambda_3|$. Let us note that $\vec{\lambda}$ must be contained inside the tetrahedron when $\vec{t} \neq \vec{0}$ also. Since the determinant does not depend on $\vec{t}$, it is sufficient to investigate only whether the parameter space of $\vec{\lambda}$ is the tetrahedron. After a little algebra one can find that the minimum of $\det \mathcal{D}$ (and also of $\det \mathcal{E}$) is achieved for $\lambda_1 = \lambda_2 = \lambda_3 = -\frac{1}{3}$, when $\det \mathcal{E} = -\frac{1}{27}$. Let us note that in this case we have necessarily $\vec{t} = \vec{0}$.

Unital channels

Let us now recall that a channel $\mathcal{E}$ is called unital if $\mathcal{E}(I) = I$. In the Bloch representation, unital channels are those for which $\vec{\tau} = \vec{t} = \vec{0}$. This reduces the number of parameters to three.

Example 4.66 *(Pauli channels)*
A Pauli channel is defined as a convex combination of Pauli operators, i.e.

$$\mathcal{E}_{\text{Pauli}}(\varrho) = \sum_j q_j \sigma_j \varrho \sigma_j, \tag{4.68}$$

where $0 \leq q_j \leq 1$ and $\sum_j q_j = 1$. It is clear that these channels are unital. Since each Pauli unitary channel is itself associated with a diagonal matrix D, it follows that an arbitrary Pauli channel is also diagonal. In particular, $D_{\text{Pauli}} = \text{diag}\{\lambda_1, \lambda_2, \lambda_3\}$ such that

$$\lambda_j = 2(q_0 + q_j) - 1.$$

Owing to the normalization of the probability distribution $\{q_j\}$ it follows that for unital channels the conditions $|\lambda_1 \pm \lambda_2| \leq |1 \pm \lambda_3|$ are both necessary and sufficient for the complete positivity of $\mathcal{E}$. Thus, the set of all unital channels is unitarily equivalent to the set of Pauli channels.   △

It is an interesting fact that all Pauli channels commute with each other although the Pauli operators themselves do not commute. For certain values of $\vec{\lambda}$ or probabilities $\{q_j\}$ we can distinguish several families of Pauli channels:

- *Depolarizing channels*: $q_1 = q_2 = q_3 = q$ and $q_0 = 1 - 3q$, i.e. $\lambda_j = 1 - 4q$. Under the action of a depolarizing channel the Bloch sphere is symmetrically shrunk to a ball of radius $1 - 4q$.
- *Phase-damping channels*: $q_1 = q_2 = 0$ and $q_0 = q, q_3 = 1 - q$, i.e. $\lambda_1 = \lambda_2 = 2q - 1$ and $\lambda_3 = 1$. Under the action of a phase-damping channel the zth component of the Bloch vector is preserved; thus the eigenvectors of σ_z determine the decoherence basis.

Amplitude-damping channels

A prototype of nonunital channels is the family of *amplitude-damping channels* used to model the evolution of a system with two energy levels (a qubit) into its ground state $P_\varphi = \frac{1}{2}(I + \sigma_z)$. The amplitude-damping channel was introduced in Example 4.28 as an example of an extremal channel. In the basis of Pauli operators it takes the matrix form

$$\mathcal{E}_{AD} = \begin{pmatrix} 1 & 0 & 0 & 0 \\ 0 & \sqrt{1-\theta} & 0 & 0 \\ 0 & 0 & \sqrt{1-\theta} & 0 \\ \theta & 0 & 0 & 1-\theta \end{pmatrix}.$$

Clearly, the only fixed point of an amplitude-damping channel is associated with the Bloch vector $\vec{r} = (0, 0, 1)$ corresponding to the ground state P_φ. Consequently, on applying this channel again and again the Bloch ball is shrunk into this fixed point.

Exercise 4.67 (*How to find the operator-sum representation*)
Here we describe how to find an operator-sum representation for channels expressed in some matrix form. The idea is based on the following observation.

Since $(A \otimes I)P_+(A^* \otimes I) = |\psi_A\rangle\langle\psi_A|$ is a one-dimensional projection, the Choi matrix $\Omega_{\mathcal{E}}$ of a channel $\mathcal{E}$ can be expressed as

$$\Omega_{\mathcal{E}} = (\mathcal{E} \otimes \mathcal{I})(P_+) = \sum_j (A_j \otimes I)P_+(A_j^* \otimes I)$$

$$= \sum_j |\psi_j\rangle\langle\psi_j|.$$

Let us stress that the vectors ψ_j are not necessarily normalized, because $\|\psi_j\| = \mathrm{tr}\left[A_j^* A_j\right]$. Thus, by expressing the Choi matrix as a sum of operators $|\psi_j\rangle\langle\psi_j|$ we may determine the associated operators A_j forming the operator-sum decomposition of the channel. In particular, the canonical decomposition of $\Omega_{\mathcal{E}}$ into eigenvalues and eigenvectors gives a collection of mutually orthogonal (in the Hilbert–Schmidt sense) Kraus operators of $\mathcal{E}$. Perform this procedure for amplitude-damping channels, starting from the above matrix form. [Hint: This is not difficult, but it is quite a long calculation. To find the Choi matrix use the Bloch vector representation of the maximally mixed state $P_+ = \frac{1}{4}(I \otimes I + \sigma_x \otimes \sigma_x - \sigma_y \otimes \sigma_y + \sigma_z \otimes \sigma_z)$. Finally compare your result with Example 4.28.]

5

Measurement models and instruments

Until now we have treated measurement apparatuses as devices taking quantum systems as their inputs and producing measurement outcomes as outputs. At this level of description, measurement apparatuses are fully described by observables, as discussed in Chapter 3. The measurement outcome distribution is determined by the particular input state and the observable describing the measurement device.

In this chapter we explore a mathematical framework suitable for the description of measurement apparatuses for which, in addition to the measurement outcomes, the measured systems are available after the measurement is performed. In such cases we may try to obtain more information on the system by measuring some other observable, or, perhaps by filtering out some preferred output states we may use our measurement apparatus as a preparator. Clearly, in this kind of situation we need a description of measurement apparatuses that is more detailed than that of observables. Therefore, two more refined concepts, *measurement models* and *instruments*, will be investigated in this chapter.

There is no customary name for a measurement model. As used here, it was first formalized in a general way by Ozawa [106], [107] who referred to the *measurement process*. A comprehensive general reference on quantum measurements is the monograph by Busch *et al.* [35], where measurement models are called *premeasurements*. Some other authors use the term *indirect measurements*.

5.1 Three levels of description of measurements

It is convenient to think of three levels of description when considering measurements. At the first level, we concentrate only on measurement outcome probabilities and the description is thus given by observables. At the second level we want to include also post-measurement states in the description. This will lead to the notion of an *instrument*. The third level adds still more detail to the description, such as specifications of the measurement interaction and probe observable, and we call this kind of description a measurement model.

In this section we start from the most detailed description, i.e. measurement models (subsection 5.1.1), then continue to the coarser levels (subsection 5.1.2) and finally show that all these different descriptions can fit together perfectly (subsection 5.1.3).

5.1.1 Measurement models

Let us first think in an informal way about a typical measurement procedure. A measurement procedure starts by coupling the system of interest to a measurement apparatus, or probe. After some time the system and the probe are decoupled, and a measurement is carried out on the probe only. Owing to the coupling stage the system and the probe have become correlated and the measurement outcome distribution gives us information about the initial state of the system. The whole procedure can be interpreted as a measurement on the system corresponding to some observable A. The following definition formalizes this outline.

Definition 5.1 Let A be an observable on a system associated with a Hilbert space $\mathcal{H}$ and having an outcome space $(\Omega, \mathcal{F})$. This is the observable that we are intending to measure. Suppose that:

- $\mathcal{K}$ is the Hilbert space associated with the *probe system*;
- ξ is an *initial state* of the probe system;
- $\mathcal{V}$ is a channel from $\mathcal{T}(\mathcal{H} \otimes \mathcal{K})$ to $\mathcal{T}(\mathcal{H} \otimes \mathcal{K})$, which describes the *measurement interaction* between the system and the probe;
- F is an observable implemented on the probe system and assumed to have the same outcome space $(\Omega, \mathcal{F})$ as A, known as the *pointer observable*.

The quadruple $\mathcal{M} = \langle \mathcal{K}, \xi, \mathcal{V}, \mathsf{F} \rangle$ is a *measurement model* of A (see Figure 5.1) if it satisfies the *probability reproducibility condition*

$$\mathrm{tr}\big[\varrho \mathsf{A}(X)\big] = \mathrm{tr}\big[\mathcal{V}(\varrho \otimes \xi)(I \otimes \mathsf{F}(X))\big] \tag{5.1}$$

for all $X \in \mathcal{B}(\Omega)$ and $\varrho \in \mathcal{S}(\mathcal{H})$.

The probability reproducibility condition simply means that a measurement of the observable F of the probe system (after its interaction with the system of interest) leads to the same probabilities as if a measurement of A were performed directly on the system. It is important to note that we can also interpret the probability reproducibility condition in the opposite order: any quadruple $\langle \mathcal{K}, \xi, \mathcal{V}, \mathsf{F} \rangle$ defines a unique observable A by equation (5.1). This is actually what happens

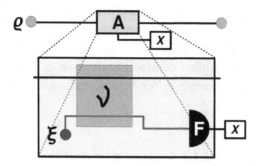

Figure 5.1 A measurement model; see Definition 5.1.

when we start to describe some measurement setting; we do not fix A at the beginning but calculate it from the given measurement model.

Exercise 5.2 Verify that the probability reproducibility condition (5.1) holds for all states $\varrho \in \mathcal{S}(\mathcal{H})$ if and only if it holds for all pure states. [Hint: Every mixed state can be written as a convex combination of pure states.]

As one can see from (5.1), the measurement interaction channel $\mathcal{V}$ takes part in the procedure only through its action on the subspace $\mathcal{T}(\mathcal{H}) \otimes \xi$ of $\mathcal{T}(\mathcal{H} \otimes \mathcal{K})$, since the initial state ξ of the probe system is fixed. For this reason, in order to define a measurement model it is enough to specify the measurement interaction only on a subset of states. Of course, we need to guarantee that the overall measurement interaction is a channel since otherwise its physical realizability would be questionable.

A common (but restricted) way to define a measurement interaction is the following. Fix an orthonormal basis $\{\varphi_j\}_{j=1}^d$ in $\mathcal{H}$ and let $\psi_0 \in \mathcal{K}$ be the initial vector state of the probe system. If $\{\psi_j\} \subset \mathcal{K}$ is a set of d vector states then the mapping

$$\varphi_j \otimes \psi_0 \mapsto \psi_j \otimes \psi_j \tag{5.2}$$

has an extension to a unitary operator on $\mathcal{H} \otimes \mathcal{K}$. Namely, any orthonormal set can be extended to form a complete orthonormal basis of the whole Hilbert space. If we extend both $\{\varphi_j \otimes \psi_0\}$ and $\{\varphi_j \otimes \psi_j\}$ to orthonormal bases of $\mathcal{H} \otimes \mathcal{K}$ then the mapping in (5.2) extends to a unitary operator (see Proposition 1.49). This type of construction is used in Example 5.3 below.

Example 5.3 *(Stern–Gerlach measurement on a spin-$\frac{1}{2}$ particle.)*
Let us take another look at the Stern–Gerlach apparatus discussed for the first time in Example 3.3. A spin-$\frac{1}{2}$ particle is described by the tensor product Hilbert space $\mathbb{C}^2 \otimes L^2(\mathbb{R}^3)$ and can be understood as a bipartite composite system consisting of the spin and spatial degrees of freedom. During the Stern–Gerlach experiment an

external magnetic field couples the spin degrees of freedom to the spatial degrees of freedom. This means that, depending on the spin state, the particle passing through the Stern–Gerlach apparatus is deflected either up or down.

We denote by φ_+ and φ_- the vector states in $\mathbb{C}^2$ associated with the situations when the particles are deflected only up or down, respectively. These states are assumed to be orthogonal and hence form an orthonormal basis of $\mathbb{C}^2$. Let $\psi_0 \in L^2(\mathbb{R}^3)$ be the initial vector state of the spatial part. The channel $\mathcal{V}$ describing the measurement coupling acts as follows:

$$\varphi_+ \otimes \psi_0 \qquad \mapsto \qquad \varphi_+ \otimes \psi_+,$$
$$\varphi_- \otimes \psi_0 \qquad \mapsto \qquad \varphi_- \otimes \psi_-,$$

where $\psi_\pm$ are fixed vector states in $L^2(\mathbb{R}^3)$. The initial spin state is assumed to be pure and we can therefore write it as $\varphi = a\varphi_+ + b\varphi_-$ for some $a, b \in \mathbb{C}$. The initial compound state is thus $\varphi \otimes \psi_0$. After the measurement coupling the compound state becomes

$$\varphi \otimes \psi_0 \qquad \mapsto \qquad a\varphi_+ \otimes \psi_+ + b\varphi_- \otimes \psi_-. \qquad (5.3)$$

A detector measures the presence of the particle in a certain region, hence it corresponds to the pointer observable measuring the position of the particle. In Stern–Gerlach measurements a screen is used as a detector of particles in the plane orthogonal to the incoming beam of particles. We are interested in whether the particle is observed in the upper or lower half-plane of the screen. Thus in the ideal case the pointer observable consists of the two projections $P_\pm$, and the probabilities $p(\pm)$ for observing a particle in the upper (+) or lower (−) half-plane are

$$p(\pm) = |a|^2 \langle \psi_+ | P_\pm | \psi_+ \rangle + |b|^2 \langle \psi_- | P_\pm | \psi_- \rangle, \qquad (5.4)$$

where

$$\langle \psi | P_+ | \psi \rangle = \int_{\text{upper h.p.}} |\psi(\vec{r})|^2 \, d\vec{r}, \qquad \langle \psi | P_- | \psi \rangle = \int_{\text{lower h.p.}} |\psi(\vec{r})|^2 \, d\vec{r}$$

for every $\psi \in L^2(\mathbb{R}^3)$. Defining the effects $E_\pm$ on $\mathbb{C}^2$ as

$$E_\pm = \langle \psi_+ | P_\pm | \psi_+ \rangle | \varphi_+ \rangle \langle \varphi_+ | + \langle \psi_- | P_\pm | \psi_- \rangle | \varphi_- \rangle \langle \varphi_- |, \qquad (5.5)$$

we obtain the following expression for the probabilities $p(\pm)$ in terms of only the initial spin state φ:

$$p(\pm) = \langle \varphi | E_\pm | \varphi \rangle. \qquad (5.6)$$

The resulting observable A is thus given by

$$\mathsf{A}(+) = E_+, \qquad \mathsf{A}(-) = E_-.$$

The properties of A depend on the vectors $\psi_\pm$. In particular, A is a spin component observable $S^{\vec{n}}$ (see Example 3.3) if and only if $\langle \psi_\pm | P_\pm | \psi_\pm \rangle = 1$ and $\langle \psi_\mp | P_\pm | \psi_\mp \rangle = 0$.

This description is already much more informative than the earlier one, in Example 3.3. Naturally, most details, such as the connection of the unitary interaction to the magnetic field and the concrete form of the pointer observable, are still hidden. For a more realistic description of the Stern–Gerlach measurement we refer to Chapter VII in [34]. $\triangle$

5.1.2 Instruments

If a system still exists after a measurement, one may try to perform some other measurement to gain more information about the original state of the system. Alternatively, we can use a measurement procedure as a (conditional) state preparator. In both cases, we need to know not only the measurement outcome probabilities but also the influence of the first measurement on the system and, consequently, on later measurements. We do not need to know all the details of the measurement model used, and often such a detailed description is not available or is too complicated to be helpful. The concept of an *instrument* neatly captures the relevant description of the measurement process for the above-mentioned situations. Instruments were introduced by Davies and Lewis [52] and the book [51] is still to be recommended. The connection between measurement models and instruments was clarified by Ozawa [107].

Let us assume that, after measuring the observable A by means of a measurement model $\mathcal{M} = \langle \mathcal{K}, \xi, \mathcal{V}, \mathsf{F} \rangle$, we perform a measurement of another observable B on the system. Altogether we find a joint probability distribution of the values of A and B, and we denote by $p_\varrho(\mathsf{A} \in X \,\&\, \mathsf{B} \in Y)$ the probability that the A-measurement gives an outcome from the set X and B-measurement gives an outcome from the set Y. The subindex refers to the initial state ϱ of the system. We do not write a measurement model for B but consider it as a direct measurement on the system. Hence the joint probability is given by

$$p_\varrho(\mathsf{A} \in X \,\&\, \mathsf{B} \in Y) = \mathrm{tr}\big[\mathcal{V}(\varrho \otimes \xi)(\mathsf{B}(Y) \otimes \mathsf{F}(X))\big], \tag{5.7}$$

valid for all X and Y.

Let us fix the set X for the present. We note that

$$\mathrm{tr}\big[\mathcal{V}(\varrho \otimes \xi)(\mathsf{B}(Y) \otimes \mathsf{F}(X))\big] = \mathrm{tr}\big[\mathsf{B}(Y)\mathrm{tr}_\mathcal{K}[\mathcal{V}(\varrho \otimes \xi)(I \otimes \mathsf{F}(X))]\big],$$

and it turns out to be useful to define

$$\mathcal{I}_X^\mathcal{M}(\varrho) := \mathrm{tr}_\mathcal{K}\big[\mathcal{V}(\varrho \otimes \xi)(I \otimes \mathsf{F}(X))\big]. \tag{5.8}$$

The operator $\mathcal{I}_X^{\mathcal{M}}(\varrho)$ is uniquely determined by the requirement that

$$p_\varrho(\mathsf{A} \in X \ \& \ \mathsf{B} \in Y) = \text{tr}\big[\mathcal{I}_X^{\mathcal{M}}(\varrho)\mathsf{B}(Y)\big] \tag{5.9}$$

for all observables B and all outcome sets Y. The mapping $\varrho \mapsto \mathcal{I}_X^{\mathcal{M}}(\varrho)$ has a unique linear extension to the vector space of trace class operators $\mathcal{T}(\mathcal{H})$, and we will therefore consider $\mathcal{I}_X^{\mathcal{M}}$ as a linear mapping from $\mathcal{T}(\mathcal{H})$ to $\mathcal{T}(\mathcal{H})$.

The following properties are straightforward to verify from (5.8):

(ins1) for each X, the mapping $\mathcal{I}_X^{\mathcal{M}}$ is an operation;
(ins2) $\text{tr}\big[\mathcal{I}_\Omega^{\mathcal{M}}(\varrho)\big] = 1$ and $\mathcal{I}_\emptyset^{\mathcal{M}}(\varrho) = O$ for each $\varrho \in \mathcal{S}(\mathcal{H})$;
(ins3) if $\varrho \in \mathcal{S}(\mathcal{H})$ and $\{X_j\}$ is a sequence of mutually disjoint sets then

$$\text{tr}\big[\mathcal{I}_{\cup_j X_j}^{\mathcal{M}}(\varrho)\big] = \sum_j \text{tr}\big[\mathcal{I}_{X_j}^{\mathcal{M}}(\varrho)\big].$$

Since

$$p_\varrho(\mathsf{A} \in X) = p_\varrho(\mathsf{A} \in X \ \& \ \mathsf{B} \in \Omega') = \text{tr}\big[\mathcal{I}_X^{\mathcal{M}}(\varrho)\mathsf{B}(\Omega')\big],$$

where Ω' stands for the sample space of observable B, we conclude that

$$p_\varrho(\mathsf{A} \in X) = \text{tr}\big[\mathcal{I}_X^{\mathcal{M}}(\varrho)\big].$$

For a fixed set X the relation $p_\varrho(\mathsf{A} \in X \ \& \ \mathsf{B} \in Y) = \text{tr}\big[\mathcal{I}_X^{\mathcal{M}}(\varrho)\mathsf{B}(Y)\big]$ implies that $\mathcal{I}_X^{\mathcal{M}}(\varrho)$ determines the (unnormalized) conditional output state of the measured system, on which the measurement of B is performed. In other words, the mapping $X \mapsto \mathcal{I}_X^{\mathcal{M}}$ contains information on the observable A and, in addition, on the conditional state transformation associated with the observation of the outcome from X. Abstracting its properties leads us to the following definition.

Definition 5.4 A mapping $\mathcal{I}$ from an outcome space $(\Omega, \mathcal{F})$ to the set of operations on $\mathcal{T}(\mathcal{H})$ is called an *instrument* if it satisfies the conditions (ins1)–(ins3).

If the set Ω of measurement outcomes is finite or countably infinite, we use the shorthand notation $\mathcal{I}_x \equiv \mathcal{I}_{\{x\}}$. As in the case of discrete observables, this type of instrument is completely determined by the operations $\mathcal{I}_x$, $x \in \Omega$. If we take a Kraus decomposition for each operation $\mathcal{I}_x$ then the collection of all these Kraus operators gives a Kraus decomposition for the channel $\mathcal{I}_\Omega$. If, however, we start from a channel $\mathcal{E}$, fix one of its Kraus decompositions $\mathcal{E}(\varrho) = \sum_x K_x \varrho K_x^*$ and define $\mathcal{I}_x(\varrho) = K_x \varrho K_x^*$ then we obtain a discrete instrument $\mathcal{I}$ such that $\mathcal{I}_\Omega = \mathcal{E}$. Although discrete instruments have this kind of intimate connection to the Kraus form of channels, one should remember that an instrument need not be discrete while the Kraus decomposition is always countable. Hence, in the general case we have to deal with the defining conditions (ins1)–(ins3).

We have seen previously that each measurement model $\mathcal{M}$ defines a unique instrument $\mathcal{I}^{\mathcal{M}}$, and we say that $\mathcal{I}^{\mathcal{M}}$ is the *instrument induced by* $\mathcal{M}$. The following fundamental theorem due to Ozawa [107] tells us that the converse also holds.

Theorem 5.5 (*Ozawa's theorem*)
For every instrument $\mathcal{I}$, there exists a measurement model $\mathcal{M} = \langle \mathcal{K}, \xi, \mathcal{V}, \mathsf{F} \rangle$ such that $\mathcal{I} = \mathcal{I}^{\mathcal{M}}$. In addition, it is possible to choose $\mathcal{M}$ such that ξ is a pure state, $\mathcal{V}$ is a unitary channel and F is a sharp observable.

Proof   We prove the theorem only in the simplest case, when the instrument $\mathcal{I}$ has a finite number of outcomes and the Hilbert space $\mathcal{H}$ is finite dimensional.

Each $\mathcal{I}_x$ has a Kraus decomposition $\mathcal{I}_x(\varrho) = \sum_j A_j \varrho A_j^*$ containing at most $(\dim \mathcal{H})^2$ Kraus operators. Summing all operations $\mathcal{I}_x$ gives a Kraus decomposition of the channel $\mathcal{I}_\Omega$, and this decomposition contains a finite number, say M, of operators.

We saw in subsection 4.2.2 that the channel $\mathcal{I}_\Omega$ has a dilation $\langle \mathcal{H}_E, U, \eta \rangle$ such that $\mathcal{I}_\Omega(\varrho) = \mathrm{tr}_E[U(\varrho \otimes |\eta\rangle\langle\eta|)U^*]$. There is some minimal dimension giving a minimal dilation for $\mathcal{I}_\Omega$, but we can always increase the size of $\mathcal{H}_E$ by tensoring it with another Hilbert space. We can hence assume that $\dim \mathcal{H}_E = M$ and fix an orthonormal basis $\{\varphi_j\}_{j=1}^M$ for $\mathcal{H}_E$. The measurement model consists of the probe system $\mathcal{H}_E$, the pointer state $|\eta\rangle\langle\eta|$, the unitary channel σ_U and a pointer observable, to be determined. For every $j = 1, \ldots, M$, the condition $\langle \psi \otimes \varphi_j | U\phi \otimes \eta \rangle = \langle \psi | B_j \phi \rangle$, required to hold for all $\psi, \phi \in \mathcal{H}$, determines an operator B_j on $\mathcal{H}$. It is straightforward to verify that $\mathcal{I}_\Omega(\varrho) = \sum_{j=1}^M B_j \varrho B_j^*$. The two sets $\{A_j\}$ and $\{B_j\}$ are Kraus decompositions of the same channel $\mathcal{I}_\Omega$ and, by Proposition 4.25, they are connected by a $M \times M$ unitary matrix $[u_{jk}]$. We define a new orthonormal basis for $\mathcal{H}_E$ by setting $\varphi_j' = \sum_k u_{jk}\varphi_k$, and this determines a sharp observable $\mathsf{F}'(j) = |\varphi_j'\rangle\langle\varphi_j'|$. We then have $\mathrm{tr}_{\mathcal{K}}\left[\mathcal{V}(\varrho \otimes |\eta\rangle\langle\eta|)(I \otimes \mathsf{F}'(j))\right] = A_j \varrho A_j^*$ for every state $\varrho \in \mathcal{S}(\mathcal{H})$. The pointer observable F completing the measurement model $\mathcal{M}$ is obtained by grouping the outcomes of F' into suitable collections. $\square$

5.1.3 Compatibility of the three descriptions

Let $\mathcal{I}$ be an instrument defined on an outcome space $(\Omega, \mathcal{F})$. For each $X \in \mathcal{F}$, we denote by $\mathcal{I}_X^*$ the dual (i.e. the Heisenberg picture) of the operation $\mathcal{I}_X$. The instrument $\mathcal{I}$ defines a unique observable $\mathsf{A}^{\mathcal{I}}$ by the formula

$$\mathsf{A}^{\mathcal{I}}(X) = \mathcal{I}_X^*(I), \tag{5.10}$$

required to hold for every $X \in \mathcal{F}$. Another way to write this formula is

$$\mathrm{tr}\left[\varrho \mathsf{A}^{\mathcal{I}}(X)\right] = \mathrm{tr}\left[\mathcal{I}_X(\varrho)\right], \tag{5.11}$$

which is required to hold for all $X \in \mathcal{F}$ and $\varrho \in \mathcal{S}(\mathcal{H})$. This connection between $\mathcal{I}$ and $\mathsf{A}^{\mathcal{I}}$ means that they describe the same measurement outcome probabilities.

We can also start the other way round and have a fixed observable A. Any instrument $\mathcal{I}$ is called A-*compatible* if $\mathsf{A} = \mathsf{A}^{\mathcal{I}}$. Each particular A-compatible instrument $\mathcal{I}$ describes a certain way of measuring A, leading to a certain kind of state transformation. Example 5.6 below demonstrates that every observable A has (infinitely many) A-compatible instruments.

Example 5.6 (*Trivial instrument*)
Let A be an observable and fix a state ξ. Then the formula

$$\mathcal{I}_X(\varrho) = \text{tr}\big[\varrho \mathsf{A}(X)\big]\xi \tag{5.12}$$

defines an A-compatible instrument. In fact, we have

$$\text{tr}\big[\mathcal{I}_X(\varrho)\big] = \text{tr}\big[\varrho \mathsf{A}(X)\big]\text{tr}[\xi] = \text{tr}\big[\varrho \mathsf{A}(X)\big],$$

showing that the compatibility condition (5.11) holds. Instruments which have the form (5.12) for some observable A and some state ξ are called *trivial instruments*. (The motivation for this name will be explained later, in Example 5.7.) Since the state ξ is arbitrary, we conclude that for each observable A there exist infinitely many A-compatible instruments. $\triangle$

To check that the different compatibility relationships are consistent, let $\mathcal{M}$ be a measurement model. It defines an instrument $\mathcal{I}^{\mathcal{M}}$ through formula (5.8). However, $\mathcal{I}^{\mathcal{M}}$ defines an observable $\mathsf{A}^{\mathcal{I}^{\mathcal{M}}}$ by the condition (5.10). Combining these two formulas we get

$$\text{tr}\Big[\varrho \mathsf{A}^{\mathcal{I}^{\mathcal{M}}}(X)\Big] = \text{tr}\big[\mathcal{I}_X^{\mathcal{M}}(\varrho)\big] = \text{tr}\big[\text{tr}_{\mathcal{K}}\big[\mathcal{V}(\varrho \otimes \xi)I \otimes \mathsf{F}(X)\big]\big]$$
$$= \text{tr}\big[\mathcal{V}(\varrho \otimes \xi)I \otimes \mathsf{F}(X)\big].$$

This is nothing other than the probability reproducibility condition (5.1) between $\mathcal{M}$ and the observable $\mathsf{A}^{\mathcal{I}^{\mathcal{M}}}$. Therefore, the way in which we have defined the compatibility relationships between observables, instruments and measurement models is indeed consistent.

It is important to note that the three classes of objects are not in one-to-one correspondence. Starting from an observable A, there is a collection $[\mathcal{I}]$ of A-compatible instruments. The set $[\mathcal{I}]$ is an equivalence class in the set of all instruments. Similarly, for each instrument $\mathcal{I}$, there is a collection $[\mathcal{M}]$ of measurement models that induce $\mathcal{I}$. This is again an equivalence class in the set of all measurement models. The physical explanation of these one-to-many relations is that measurements of an observable A can affect the system in many different ways, while different measurement settings can influence the system in a similar way.

We conclude that the three different perspectives can be adopted to describe a measurement apparatus. These perspectives lead to three different mathematical objects: measurement models, instruments and observables. These different layers of description are connected in the following way:

$$\mathcal{M} \to \mathcal{I}^{\mathcal{M}}, \quad \mathcal{I} \to \mathsf{A}^{\mathcal{I}}, \qquad \mathsf{A} \to [\mathcal{I}], \quad \mathcal{I} \to [\mathcal{M}],$$

meaning that whereas each measurement model uniquely defines an instrument and each instrument uniquely defines an observable, the opposite relations are not unique. Each observable defines an equivalence class of instruments and each instrument defines an equivalence class of measurement models.

5.2 Disturbance caused by a measurement

It is one of the basic lessons of quantum theory that a measurement on a quantum system in general causes an unavoidable disturbance in the sense that after the measurement the system is in a different state than that before the measurement. The concept of a *conditional output state* is useful for analyzing this phenomenon, and this will be discussed in subsection 5.2.1. In subsection 5.2.2 we give a simple and precise formulation of the idea that disturbance is unavoidable in any non-trivial quantum measurement. The other extreme case is presented in 5.2.3, where we show that an important class of observables can be measured only in a way such that all subsequent measurements become redundant. In subsection 5.2.4 we take a look on the famous BB84 quantum key distribution protocol, in order to demonstrate how one can sometimes benefit from the inescapable measurement disturbance.

5.2.1 Conditional output states

Let us go back to the setting of subsection 5.1.2, where we had two observables A and B. We will measure first observable A and after that observable B. Putting our measurements together, we find a joint probability distribution of the values of A and B, and for an initial state ϱ this is given by

$$p_\varrho(\mathsf{A} \in X \,\&\, \mathsf{B} \in Y) = \mathrm{tr}\big[\mathcal{I}_X(\varrho)\mathsf{B}(Y)\big],$$

where $\mathcal{I}$ is an A-compatible instrument depending on how the A-measurement is performed. The conditional probability $p_\varrho(\mathsf{B} \in Y | \mathsf{A} \in X)$ can be written in the form

$$p_\varrho(\mathsf{B} \in Y | \mathsf{A} \in X) = \frac{p_\varrho(\mathsf{A} \in X \ \& \ \mathsf{B} \in Y)}{p_\varrho(\mathsf{A} \in X)} = \frac{\mathrm{tr}\big[\mathcal{I}_X(\varrho)\mathsf{B}(Y)\big]}{\mathrm{tr}\big[\mathcal{I}_X(\varrho)\big]}$$

$$\equiv \mathrm{tr}\big[\tilde{\varrho}_X \mathsf{B}(Y)\big].$$

The state

$$\tilde{\varrho}_X = \frac{1}{\mathrm{tr}\big[\mathcal{I}_X(\varrho)\big]}\mathcal{I}_X(\varrho)$$

is called a *conditional output state*.

It is worth noting two things. First, the conditional output state $\tilde{\varrho}_X$ is defined only when $\mathrm{tr}\big[\varrho\mathsf{A}(X)\big] = \mathrm{tr}\big[\mathcal{I}_X(\varrho)\big] \neq 0$. The reason is simply that we cannot define the conditional probability $p_\varrho(\mathsf{B} \in Y | \mathsf{A} \in X)$ if $p_\varrho(\mathsf{A} \in X) = 0$. As a second point, notice that the mapping $\varrho \mapsto \tilde{\varrho}_X$ is not, in general, linear. However, it is linear (and hence a channel) when $X = \Omega$, since in this case $\tilde{\varrho}_\Omega = \mathcal{I}_\Omega(\varrho)$. We say that $\tilde{\varrho}_\Omega$ is the *unconditional output state*.

Example 5.7 (*Conditional output states for a trivial instrument*)
Let us recall Example 5.6 and consider again the trivial instrument $\mathcal{I}$ defined in (5.12). We then have

$$\tilde{\varrho}_X = \frac{1}{\mathrm{tr}\big[\mathcal{I}_X(\varrho)\big]}\mathcal{I}_X(\varrho) = \xi \tag{5.13}$$

for all X and ϱ such that $\mathrm{tr}\big[\mathcal{I}_X(\varrho)\big] \neq 0$. Hence, the conditional output state depends on neither the particular measurement outcome nor the input state.

The reason for the name 'trivial instrument' can now be explained. Suppose that we make an A-measurement and after that a B-measurement. If the A-measurement corresponds to a trivial instrument, then we get

$$p_\varrho(\mathsf{B} \in Y \mid \mathsf{A} \in X) = \mathrm{tr}\big[\tilde{\varrho}_X \mathsf{B}(Y)\big] = \mathrm{tr}\big[\xi \mathsf{B}(Y)\big] = p_\xi(\mathsf{B} \in Y).$$

This is just the same as measuring the trivial observable $Y \mapsto \mathrm{tr}\big[\xi\mathsf{B}(Y)\big]I$ in the state ϱ. In other words, the B-measurement does not contain any information on the input state ϱ. All measurements following the trivial instrument are 'trivialized'. $\triangle$

If the set Ω of measurement outcomes is finite or countably infinite, the conditional output state corresponding to an outcome x is

$$\tilde{\varrho}_x = \frac{1}{\mathrm{tr}\big[\mathcal{I}_x(\varrho)\big]}\mathcal{I}_x(\varrho), \tag{5.14}$$

where as before we set $\tilde{\varrho}_x \equiv \tilde{\varrho}_{\{x\}}$ and $\mathcal{I}_x \equiv \mathcal{I}_{\{x\}}$. The following example illustrates an instrument that can be employed as a conditional state preparator.

Example 5.8 (*Conditional state preparation*)

Let A be an observable with a countable outcome set Ω. For each outcome $x \in \Omega$, we fix a state ξ_x. The formula

$$\mathcal{I}_x(\varrho) = \text{tr}\big[\varrho A(x)\big] \xi_x \qquad (5.15)$$

defines an A-compatible instrument. The conditional output states are simply $\widetilde{\varrho}_x = \xi_x$. We call this type of instrument a conditional state preparator. Although the state preparation is probabilistic, the measurement outcome indicates which state is obtained. Clearly, trivial instruments are special cases of conditional state preparators for which the conditional output state is always the same.

The action of the instrument in (5.15) can be mimicked in a very simple way. After the measurement of A is performed and a measurement outcome x is obtained, the system is destroyed and another system of the same kind is prepared in the state ξ_x. This system is then adopted as the output system. $\triangle$

If A is a discrete observable then we can write

$$\widetilde{\varrho}_\Omega = \sum_{x \in \Omega} \text{tr}\big[\varrho A(x)\big] \widetilde{\varrho}_x, \qquad (5.16)$$

which expresses that, as expected, the unconditional output state $\widetilde{\varrho}_\Omega$ is the weighted sum of the conditional output states $\widetilde{\varrho}_x$. A formula of the type (5.16) can be developed also for general observables; however, this requires additional mathematical machinery, and we refer the reader to the original article of Ozawa [108] for further details.

5.2.2 No information without disturbance

Is it possible to perform a measurement of an observable A without causing any disturbance? The answer to this question is the content of this section. Following [25] we show that the linearity of operations implies that it is impossible to gain any information without disturbing the system.

Let us start from the requirement that the obtained measurement outcome of A does not make any difference to the measurement outcome distribution of a subsequently measured observable B. This would mean that

$$p_\varrho(B \in Y | A \in X) = p_\varrho(B \in Y) \qquad (5.17)$$

for all $X, Y \in \mathcal{F}$ and $\varrho \in \mathcal{S}(\mathcal{H})$. Since B can be chosen to be any observable, this is equivalent to requiring that $\widetilde{\varrho}_X = \varrho$ for all $X \in \mathcal{F}$ and $\varrho \in \mathcal{S}(\mathcal{H})$. Written in terms of the related A-compatible instrument $\mathcal{I}$, the requirement is

$$\mathcal{I}_X(\varrho) = c_X(\varrho)\varrho \qquad \forall X \in \mathcal{F}, \varrho \in \mathcal{S}(\mathcal{H}), \qquad (5.18)$$

where $c_X(\varrho)$ is a nonnegative number possibly depending on X and ϱ.

Owing to the linearity of $\mathcal{I}_X$ it follows that the number $c_X(\varrho)$ does not depend on ϱ. To see this, let ϱ_1 and ϱ_2 be two different states. Since $\mathcal{I}_X$ is linear, we obtain on the one hand

$$\mathcal{I}_X(\varrho_1 + \varrho_2) = \mathcal{I}_X(\varrho_1) + \mathcal{I}_X(\varrho_2) = c_X(\varrho_1)\,\varrho_1 + c_X(\varrho_2)\,\varrho_2,$$

and, on the other hand,

$$\mathcal{I}_X(\varrho_1 + \varrho_2) = c_X(\varrho_1 + \varrho_2)\,(\varrho_1 + \varrho_2) = c_X(\varrho_1 + \varrho_2)\,\varrho_1 + c_X(\varrho_1 + \varrho_2)\,\varrho_2.$$

Comparing these two equations we see that $c_X(\varrho_1) = c_X(\varrho_1 + \varrho_2) = c_X(\varrho_2)$. Hence, the number $c_X(\varrho)$ is the same for all states ϱ and we can set $c_X(\varrho) \equiv c_X$. Taking the trace on both sides of (5.18), we obtain

$$\mathrm{tr}\big[\varrho A(X)\big] = \mathrm{tr}\big[\mathcal{I}_X(\varrho)\big] = c_X. \tag{5.19}$$

As this is true for all states ϱ, we have in particular

$$c_X = \mathrm{tr}\big[|\varphi\rangle\langle\varphi|A(X)\big] = \langle\varphi|A(X)\varphi\rangle$$

for all unit vectors $\varphi \in \mathcal{H}$. Using Proposition 1.21, this implies that $A(X) = c_X I$. Thus, A is necessarily a trivial observable, which does not provide any information on the state of the system (see subsection 3.5.1).

Exercise 5.9 We saw in subsection 4.1.1 that the linearity of channels leads to the no-cloning theorem. The above 'no information without disturbance' theorem is linked conceptually with the no-cloning theorem: the first implies the second. Namely, assume that the cloning transformation $\varrho \mapsto \varrho \otimes \varrho$ is possible. Explain how one could thus perform any desired measurement with no disturbance of the initial state.

One could argue that the requirement (5.17) is too strong and that the conclusions drawn from it are therefore irrelevant. We could start from the milder requirement that performing a measurement of A does not affect any other measurement performed later on the system. This would mean that

$$p_\varrho(B \in Y | A \in \Omega) = p_\varrho(B \in Y) \tag{5.20}$$

for all $Y \in \mathcal{F}$ and $\varrho \in \mathcal{S}(\mathcal{H})$, hence we would require (5.17) only for the total set Ω and not for all X. This would be to say that there is an A-compatible instrument $\mathcal{I}$ satisfying

$$\mathcal{I}_\Omega(\varrho) = \varrho \tag{5.21}$$

for all $\varrho \in \mathcal{S}(\mathcal{H})$.

Let us consider a fixed outcome set $X \in \mathcal{F}$. For a pure state P we have $P = \mathcal{I}_\Omega(P) = \mathcal{I}_X(P) + \mathcal{I}_{\neg X}(P)$; hence by Proposition 1.63 we can conclude that $\mathcal{I}_X(P) = c_X(P)P$ for some nonnegative number $c_X(P)$. We need to show that this number does not depend on P.

Let us first consider two orthogonal pure states $|\psi\rangle\langle\psi|$ and $|\psi_\perp\rangle\langle\psi_\perp|$, where $\psi, \psi_\perp \in \mathcal{H}$ are orthogonal unit vectors. We write $\varphi = \frac{1}{\sqrt{2}}(\psi + \psi_\perp)$ and $\varphi_\perp = \frac{1}{\sqrt{2}}(\psi - \psi_\perp)$. It is straightforward to verify that

$$|\psi\rangle\langle\psi| + |\psi_\perp\rangle\langle\psi_\perp| = |\varphi\rangle\langle\varphi| + |\varphi_\perp\rangle\langle\varphi_\perp|. \tag{5.22}$$

Hence, by the linearity of $\mathcal{I}_X$ we obtain

$$\mathcal{I}_X(|\psi\rangle\langle\psi|) + \mathcal{I}_X(|\psi_\perp\rangle\langle\psi_\perp|) = \mathcal{I}_X(|\varphi\rangle\langle\varphi|) + \mathcal{I}_X(|\varphi_\perp\rangle\langle\varphi_\perp|)$$

and therefore

$$c_1|\psi\rangle\langle\psi| + c_2|\psi_\perp\rangle\langle\psi_\perp| = c_3|\varphi\rangle\langle\varphi| + c_4|\varphi_\perp\rangle\langle\varphi_\perp| \tag{5.23}$$

for some nonnegative numbers $c_1, c_2, c_3, c_4 \in \mathbb{R}$. Suppose that one of these, say c_4, is nonzero. Combining (5.22) and (5.23) leads to

$$(c_1 - c_4)|\psi\rangle\langle\psi| + (c_2 - c_4)|\psi_\perp\rangle\langle\psi_\perp| + (c_4 - c_3)|\varphi\rangle\langle\varphi| = 0. \tag{5.24}$$

Even if the three vectors ψ, $\psi_\perp$ and φ are linearly dependent, the corresponding three rank-1 operators are linearly independent. (Any linear combination of $|\psi\rangle\langle\psi|$ and $|\psi_\perp\rangle\langle\psi_\perp|$ commutes with $|\psi\rangle\langle\psi|$, but this is not true for $|\varphi\rangle\langle\varphi|$.) We thus get $c_1 = c_2 = c_3 = c_4$ and conclude that $c_X(P_1) = c_X(P_2)$ whenever P_1, P_2 are two orthogonal pure states.

To treat nonorthogonal pure states, we start from two unit vectors $\psi, \varphi \in \mathcal{H}$ that are neither parallel nor orthogonal. We can then find two vectors $\psi_\perp$ and $\varphi_\perp$ that are orthogonal to ψ, φ, respectively, and that are linear combinations of ψ and φ. Both $\{\psi, \psi_\perp\}$ and $\{\varphi, \varphi_\perp\}$ are orthogonal bases of the same two-dimensional subspace. This implies that

$$|\psi\rangle\langle\psi| + |\psi_\perp\rangle\langle\psi_\perp| = |\varphi\rangle\langle\varphi| + |\varphi_\perp\rangle\langle\varphi_\perp|, \tag{5.25}$$

and we continue the calculation in the same way as previously. We then see that $c_X(P_1) = c_X(P_2)$ whenever P_1, P_2 are two nonorthogonal pure states.

In summary, the number $c_X(P)$ does not depend on P, and this implies that $\mathcal{I}_X(P) = c_X P$ for all pure states P. By the linearity of $\mathcal{I}_X$ we then have $\mathcal{I}_X(\varrho) = c_X \varrho$ for all states ϱ. We are thus in the same situation as that corresponding to (5.19), and the rest follows.

We have seen that even the milder condition (5.20) can be satisfied only if A is a trivial observable. We conclude that in order to acquire at least some nontrivial information a measurement must produce some disturbance.

Exercise 5.10 Instead of (5.21), let us start from the more general assumption that

$$\mathcal{I}_\Omega(\varrho) = U\varrho U^* \tag{5.26}$$

for some unitary operator U. Prove that this is possible only if A is a trivial observable. [Hint: Apply the previous result to the instrument $\varrho \mapsto U^* \mathcal{I}_X(\varrho) U$.]

5.2.3 Disturbance in a rank-1 measurement

We recall from subsection 1.3.2 that an effect $E \in \mathcal{E}(\mathcal{H})$ is *rank*-1 if there is a one-dimensional projection P and a number $0 < e \le 1$ such that $E = eP$. Alternatively, we can write $E = |\psi\rangle\langle\psi|$ for some vector $\psi \in \mathcal{H}$ with $0 < \|\psi\| \le 1$. A discrete observable A is called a *rank*-1 *observable* if each effect $\mathsf{A}(x)$ is rank-1.

The class of rank-1 observables forms an important subset of all discrete observables. For instance, sharp observables associated with an orthonormal basis (Example 3.28) and symmetric informationally complete (SIC) observables (subsection 3.3.2) are all rank-1 observables. The following proposition shows that for rank-1 observables all the compatible instruments are conditional state preparators. We will follow the proof presented in [74].

Proposition 5.11 Let A be a rank-1 observable. If $\mathcal{I}$ is an A-compatible instrument then it is of the form

$$\mathcal{I}_x(\varrho) = \mathrm{tr}\big[\varrho \mathsf{A}(x)\big] \xi_x, \tag{5.27}$$

for some set of states $\{\xi_x\}$.

Proof Let $\mathcal{I}$ be an A-compatible instrument and fix an outcome x. We write $\mathsf{A}(x) = eP$, where P is a one-dimensional projection P and e is a number $0 < e \le 1$. We choose Kraus operators $\{K_j\}$ for the operation $\mathcal{I}_x$, so that

$$\mathcal{I}_x(\varrho) = \sum_j K_j \varrho K_j^*, \qquad \mathcal{I}_x^*(I) = \sum_j K_j^* K_j = \mathsf{A}(x). \tag{5.28}$$

The last equation implies that for each j the operator inequality $K_j^* K_j \le \mathsf{A}(x)$ holds. Since $\mathsf{A}(x)$ is rank-1, there is a number $0 < k_j \le 1$ such that $K_j^* K_j = k_j \mathsf{A}(x)$ (see Proposition 1.63). From $\sum_j K_j^* K_j = \mathsf{A}(x)$ it follows that $\sum_j k_j = 1$.

Let $K_j = V_j |K_j|$ be the polar decomposition of K_j (see Theorem 1.37). Here V_j is a partial isometry and

$$|K_j| = \sqrt{K_j^* K_j} = \sqrt{k_j}\sqrt{\mathsf{A}(x)} = \sqrt{ek_j}\, P.$$

For every state ϱ, we then obtain

$$K_j \varrho K_j^* = ek_j V_j P \varrho P V_j^* = ek_j\, \mathrm{tr}\big[\varrho P\big] V_j P V_j^* = \mathrm{tr}\big[\varrho \mathsf{A}(x)\big] k_j V_j P V_j^*$$

and hence

$$\mathcal{I}_x(\varrho) = \text{tr}[\varrho A(x)] \sum_j k_j V_j P V_j^*.$$

It remains to show that the operator $V_j P V_j^*$ is a state for each j, implying that the convex sum $\sum_j k_j V_j P V_j^* =: \xi_x$ is also a state. If $\dim \mathcal{H} < \infty$ then V_j is unitary and we see immediately that $V_j P V_j^*$ is a state. If $\dim \mathcal{H} = \infty$, we notice that the operator $V_j^* V_j$ is the projection on the closure of $\text{ran}(|K_j|)$; thus $V_j^* V_j = P$. Therefore

$$\text{tr}[V_j P V_j^*] = \text{tr}[V_j^* V_j P] = \text{tr}[P] = 1.$$

Since the operator $V_j P V_j^* = V_j P (V_j P)^*$ is positive and has trace 1, it is a state.

$\square$

Suppose we measure first a rank-1 observable A and then some other observable B. By Proposition 5.11 the instrument $\mathcal{I}$ describing the A-measurement is necessarily a conditional state preparator of the form (5.27) for some set of states $\{\xi_x\}$. The joint probability distribution is thus given by

$$p_\varrho(\text{A} \in X \ \& \ \text{B} \in Y) = \text{tr}[\mathcal{I}_X(\varrho) B(Y)] = \sum_{x \in X} \text{tr}[\varrho A(x)] \, \text{tr}[\xi_x B(Y)].$$

We see that the joint probability distribution can be calculated after the first measurement since the numbers $\text{tr}[\xi_x B(Y)]$ do not depend on the initial state ϱ. Hence, the B-measurement is redundant and does not give any new information on the initial state of the system.

5.2.4 Example: BB84 quantum key distribution

The fact that a measurement necessarily disturbs the system observed can be exploited in cryptographic protocols as a tool to identify the presence of an eavesdropper. Let us discuss the most profound example of quantum cryptography – the so-called BB84 quantum key distribution protocol originally proposed in 1984 by Bennett and Brassard [16].

The goal of a key distribution protocol is not to communicate private information but merely to establish a secret key that can be used afterwards for (classical) perfectly secure communication via the *one-time pad* protocol originally introduced by G. Vernam [137].

Example 5.12 (*One-time pad*)
In 1926 Vernam proposed the first (and so far the only) provably secure cryptographic protocol, known as the *one-time pad*, or *Vernam cipher*.

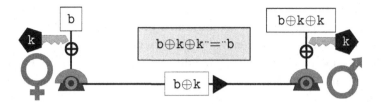

Figure 5.2 Schematic description of the one-time pad.

The one-time pad protocol needs a prerequisite called a *cryptographic key*, which is shared by the communication parties, Alice and Bob. This key is used to lock and unlock the confidential message (the so-called *plaintext*). In practice the key is represented by a random string of symbols (characters), for instance, by the sequence of bit values 00101001... The message itself is represented by another string of bits, 11001100... Binary addition $\oplus$, defined as $0 \oplus 0 = 1 \oplus 1 = 0$ and $0 \oplus 1 = 1 \oplus 0 = 1$, is used as the enciphering and deciphering operation. Let us denote by k the bits of the key and by b the bits of the message. The identity

$$b \oplus k \oplus k = b \qquad (5.29)$$

guarantees that the deciphered message equals the original plaintext. See Figure 5.2. For instance,

$$11001100 \oplus 00101001 = 11100101$$
$$11100101 \oplus 00101001 = 11001100$$

If the key is a random string then the transmitted ciphertext is a random string too. Therefore, no one intercepting the ciphertext has access to any meaningful information except the length of the message. And any message of this length is equally probable. This is why perfect security is achieved. Without going into further details let us note that in order to maintain security the key cannot be reused.

The shared key forms an ensemble of pairs of classical bits described by a joint probability $p_{00} = p_{11} = \frac{1}{2}$ and $p_{01} = p_{10} = 0$, i.e. a *classically maximally correlated state*. In quantum formalism this state can be expressed as $\omega_{\text{key}} = (|\varphi_0 \otimes \varphi_0\rangle\langle\varphi_0 \otimes \varphi_0| + \frac{1}{2}|\varphi_1 \otimes \varphi_1\rangle\langle\varphi_1 \otimes \varphi_1|)$. The modulo addition described above can be understood either as a conditioned application of one of the logical gates I and NOT, depending on the value of the bit b, or as a two-valued measurement comparing the bit values of the bits b and k and producing outcome 0 if b = k and outcome 1 if b $\neq$ k. The one-time pad implements the secure transfer of a state of the classical bit b from Alice to Bob exploiting an ideal (public) classical channel and a classical key represented by ω_{key}.

Any of these three classical resources (plaintext, channel, key) can be replaced by its quantum counterpart. We demonstrated in Example 4.60 that by replacing the classical information represented by a bit b with the quantum information represented by a qubit in a state ϱ, replacing the classical public channel by a quantum channel and doubling the length of the shared classical key we were able to transfer securely qubit states from Alice to Bob. In the following chapter we will describe two other quantum generalizations of the one-time pad. By replacing the classical keys by quantum (entangled) keys we can design protocols either for *superdense coding* (if the aim is to communicate classical information via a quantum channel and a shared quantum key, see Example 6.7) or for *quantum state teleportation* (if the aim is to communicate quantum information, but no quantum channel is available), as described in Example 6.62.   △

The success of the one-time pad strongly depends on the method of distribution of the key. The protocol is secure only if distribution is secure and hidden from others. So, the crucial question is how to communicate the key itself. It is not difficult to transfer a random string from Alice to Bob, but the problem is how to do it in a secure way, so that no one else has access to the distributed key. One natural option is that Alice and Bob meet each other, prior to distance communication of the private information, and agree on the key personally. This is a safe method but very inefficient and inconvenient. Surprisingly, quantum physics provides promising alternatives that overcome the practical problem of *at-a-distance* key distribution.

The BB84 key distribution protocol, which we are going to describe next, establishes in a secure way a cryptographic key (a sequence of random bits) between Alice and Bob. First we fix an orthonormal basis $\{\psi_{0,+}, \psi_{1,+}\}$ of a two-dimensional Hilbert space $\mathcal{H}$, and then define another orthonormal basis by

$$\psi_{0,\times} = \tfrac{1}{\sqrt{2}}(\psi_{0,+} + \psi_{1,+}), \quad \psi_{1,\times} = \tfrac{1}{\sqrt{2}}(\psi_{0,+} - \psi_{1,+}).$$

Here, the subindices $+, \times$ label the orthonormal bases and the bits 0, 1 label the vectors in the given basis. We note that these two orthonormal bases are mutually unbiased (see subsection 3.2.3).

The BB84 protocol consists of the following steps (see Figure 5.3).

1. In each run r, Alice and Bob independently and randomly choose orthonormal bases, either $+$ or $\times$. We denote by j_r Alice's choice and by k_r Bob's choice (hence $j_r, k_r \in \{+, \times\}$).
2. Alice generates a random bit $x_r \in \{0, 1\}$ and sends the vector state ψ_{x_r, j_r} to Bob.

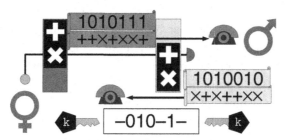

Figure 5.3 Schematic description of the BB84 quantum key distribution protocol.

3. Bob performs a sharp measurement in the basis k_r and therefore records an outcome $y_r \in \{0, 1\}$, with conditional probability

$$|\langle \psi_{y_r,k_r} | \psi_{x_r,j_r} \rangle|^2$$

depending on the vector state ψ_{x_r,j_r} received from Alice. We have

$$|\langle \psi_{y_r,k_r} | \psi_{x_r,j_r} \rangle|^2 = \delta_{j_r,k_r} \delta_{x_r,y_r} + \tfrac{1}{2}(1 - \delta_{j_r,k_r}). \tag{5.30}$$

This shows that if Alice and Bob chose the same orthonormal basis (i.e. $j_r = k_r$) then the bits x_r, y_r are identical. If, however, Alice and Bob are using different orthonormal bases (i.e. $j_r \neq k_r$) then the bits x_r, y_r are completely independent.

4. After the previous steps have been repeated sufficiently (e.g. 100 runs), Alice and Bob compare their choices of orthonormal bases by means of classical public communication (e.g. using Twitter). If they have chosen different bases in the rth run (i.e. $j_r \neq k_r$) then that particular run is ignored, meaning that the bits x_r and y_r are discarded. In this procedure the original bit sequences $\vec{x} = x_1 \cdots x_n$ and $\vec{y} = y_1 \cdots y_n$ are transformed into smaller bit sequences $\vec{x}'$ and $\vec{y}'$. The modified sequences $\vec{x}'$ and $\vec{y}'$ are identical and form the rough key. The goal of the key distribution protocol has been achieved; Alice and Bob share identical sequences of random bits.

So far we have presented a procedure in which Alice and Bob establish a sequence of perfectly correlated bits. This can also be done, and much more easily, without quantum tricks. However, the security of the quantum procedure is the important added value. A potential adversary Eve is always considered to have unlimited resources but we still assume that her possibilities are restricted by physical laws. Her aim is to learn the key in a way such that no one recognizes her presence. Let us note that if classical systems are measured then any introduced disturbance can be undone by the eavesdropper, at least in theory; thus her presence cannot be detected. In the case of BB84 the general strategy of the eavesdropper must be based on the coupling of her system to the transmitted particles. After that she can wait until Alice and Bob publicly announce the orthonormal bases that they have used in each run and measure her system in order to estimate the bit values. From Alice's and Bob's point of view Eve acts as a specific pointer observable, hence realizing a measurement model of some observable on systems transmitted from Alice to Bob.

However (and this is crucial!), there is no error-free measurement distinguishing among the four possible vector states $\psi_{0,+}$, $\psi_{1,+}$, $\psi_{0,\times}$, $\psi_{1,\times}$, because they are not mutually orthogonal. As we have discussed previously, some disturbance is necessarily introduced. Consequently, the transmitted and received states are different and there is a nonvanishing probability that the rough keys $\vec{x}'$, $\vec{y}'$ do not match perfectly. In order to find any differences, Alice and Bob release random bits from their rough keys and compare their bit values publicly. Fortunately, no information is transferred in the key distribution and thus this step does not reduce security. In the ideal and strict version of the protocol, any difference in the rough keys implies that the whole key must be discarded and the whole process restarted.

For the details of how much of the key must be released in order to detect the eavesdropper with a very high probability we refer to the overview paper by Gisin *et al.* [65]. Proofs of unconditional security of BB84 can be found in [97] and [130]. There have been several alternative proposals for quantum key distribution and also for quantum secret communication. The common idea of all these protocols is that security, in particular the detection of adversaries, is based on the unavoidable disturbance caused by nontrivial measurements.

Example 5.13 (*Simple attack on BB84*)
Let us consider the simplest version of an attack on the BB84 key distribution protocol. Let us assume that an adversary, Eve, is measuring each transmitted particle independently and trying to guess what the state is. One option for her is to switch randomly between the same bases as those used by Alice and Bob (we assume that Eve knows these bases). Let us denote by l_r the orthonormal basis chosen by Eve in the rth run and by z_r her outcome value. Eve tries to learn the correct bit values when Alice's and Bob's bit values coincide. If Eve gets an outcome z_r then she sends to Bob the vector state ψ_{z_r, l_r}.

The conditional probability that $j_r = k_r = l_r$ (all three choose the same basis) under the condition $j_r = k_r$ (Alice and Bob choose the same basis) is $\frac{1}{2}$. In these cases $x_r = y_r = z_r$, and we conclude that Eve can learn half the bits of the rough key. This could be already enough for Eve to guess the confidential message later sent by Alice to Bob using the one-time pad.

However, there are also cases when Alice and Bob have chosen the same orthonormal basis but Eve chooses the wrong one. The conditional probability that $j_r = k_r \neq l_r$ under the condition $j_r = k_r$ is $\frac{1}{2}$. Alice and Bob then expect to have identical bits but fail to do so in half the cases, since Bob has performed his measurement on the vector state ψ_{z_r,l_r} instead of the correct one, ψ_{j_r,x_r}. As a result, one quarter of the bits in the rough key are different. If Alice and Bob publish sufficiently many bits from their rough key for comparison, they will notice Eve's presence with a high probability.

$\triangle$

5.3 Lüders instruments

Lüders instruments are perhaps the most common instruments occurring in applications. They have a simple structure and serve as a good illustration of our earlier concepts. In subsection 5.3.1 we will discuss a particular type of measurement model introduced by John von Neumann in his influential book [138]. Induced instruments were studied in greater detail by Lüders [95]. In subsection 5.3.2 we define the Lüders instrument for any discrete observable, and in subsection 5.3.3 a particular feature of this instrument is explored.

5.3.1 Von Neumann's measurement model

Let us start with a discussion of a specific kind of measurement model for a discrete sharp observable associated with an orthonormal basis. Let $\{\varphi_j\}$ be an orthonormal basis for $\mathcal{H}$ and A the associated sharp observable, i.e., $\mathsf{A}(j) = |\varphi_j\rangle\langle\varphi_j|$ (recall Example 3.28. To construct a measurement model for A, we fix a Hilbert space $\mathcal{K}$ with the same dimension as $\mathcal{H}$. We fix an orthonormal basis $\{\phi_j\}$ for $\mathcal{K}$, and choose the pointer observable F to be the sharp observable associated with this basis, i.e., $\mathsf{F}(j) = |\phi_j\rangle\langle\phi_j|$. The probe is initially in a pure state $|\phi_1\rangle\langle\phi_1|$, and the measurement interaction is given by the unitary operator V on $\mathcal{H} \otimes \mathcal{K}$, defined as

$$V(\varphi_j \otimes \phi_1) = \varphi_j \otimes \phi_j \qquad \forall j. \tag{5.31}$$

This requirement does not specify V uniquely, but only the action on the vectors $\varphi_j \otimes \phi_1$ is relevant (see subsection 5.1.1).

If the system is initially in a vector state $\varphi = \sum c_i \varphi_i$, then after the measurement coupling the composition of the system and the probe will be in the state

$$|V(\varphi \otimes \phi_1)\rangle\langle V(\varphi \otimes \phi_1)| = \sum_{l,j} \bar{c}_l c_j \, |\varphi_l\rangle\langle\varphi_j| \otimes |\phi_l\rangle\langle\phi_j|. \tag{5.32}$$

On the one hand, after measuring the probe we observe outcome k with probability

$$p(k) = \text{tr}\big[|V(\varphi \otimes \phi_1)\rangle\langle V(\varphi \otimes \phi_1)|\, I \otimes \mathsf{F}(k)\big]$$
$$= \text{tr}\Big[\Big(\sum_{l,j} \bar{c}_l c_j |\varphi_l\rangle\langle\varphi_j| \otimes |\phi_l\rangle\langle\phi_j|\Big)(I \otimes |\phi_k\rangle\langle\phi_k|)\Big]$$
$$= \sum_j |c_j|^2\, \text{tr}\big[|\varphi_j\rangle\langle\varphi_k| \otimes |\phi_j\rangle\langle\phi_k|\big] = |c_k|^2.$$

On the other hand, we have

$$\text{tr}\big[|\varphi\rangle\langle\varphi|\mathsf{A}(k)\big] = |\langle\varphi|\varphi_k\rangle|^2 = |c_k|^2,$$

and the probability reproducibility condition is therefore satisfied. We conclude that $\mathcal{M} = \langle \mathcal{K}, \phi_1, V, \mathsf{F}\rangle$ is a measurement model of A.

Let us then calculate the instrument $\mathcal{I}^{\mathcal{M}}$ corresponding to $\mathcal{M}$. We obtain

$$\mathcal{I}_k^{\mathcal{M}}(|\varphi\rangle\langle\varphi|) = \text{tr}_{\mathcal{K}}\big[|V(\varphi \otimes \phi)\rangle\langle V(\varphi \otimes \phi)|(I \otimes \mathsf{F}(k))\big]$$
$$= \sum_j |c_j|^2\, \text{tr}_{\mathcal{K}}\big[|\varphi_j\rangle\langle\varphi_k| \otimes |\phi_j\rangle\langle\phi_k|\big]$$
$$= |c_k|^2|\varphi_k\rangle\langle\varphi_k| = \text{tr}\big[|\varphi\rangle\langle\varphi|\mathsf{A}(k)\big]\, \mathsf{A}(k).$$

The action on a general state ϱ can be thus written in the form

$$\mathcal{I}_k^{\mathcal{M}}(\varrho) = \text{tr}\big[\varrho\mathsf{A}(k)\big]\, \mathsf{A}(k). \tag{5.33}$$

Since $\mathsf{A}(k)$ are one-dimensional projections, formula (5.33) can be written in an equivalent form

$$\mathcal{I}_k^{\mathcal{M}}(\varrho) = \mathsf{A}(k)\varrho\mathsf{A}(k). \tag{5.34}$$

This instrument is the *Lüders instrument* of A.

Exercise 5.14 Verify that (5.33) and (5.34) are equivalent. [Hint: Dirac notation may be useful.]

If the system is initially in the vector state φ_k, then the outcome k will be observed with certainty. Moreover, $\mathsf{A}(k)\varphi_k = \varphi_k$ implies that $\mathsf{A}(k)|\varphi_k\rangle\langle\varphi_k|\mathsf{A}(k) = |\varphi_k\rangle\langle\varphi_k|$. We conclude that the Lüders instrument of A satisfies

$$\text{tr}\big[\varrho\mathsf{A}(k)\big] = 1 \quad\Longrightarrow\quad \mathsf{A}(k)\varrho\mathsf{A}(k) = \varrho \tag{5.35}$$

for all outcomes k and states ϱ. In other words, if the probability of an outcome is 1, the corresponding state is not disturbed by the associated Lüders intrument. This feature is called *ideality*, and one can show that it is a characteristic feature of Lüders intruments; if an instrument associated to a discrete sharp observable is ideal, then it is a Lüders instrument. We refer to [35] for further details and references.

5.3.2 Lüders instrument for a discrete observable

As a generalization of (5.34), we make the following definition.

Definition 5.15 The *Lüders instrument* $\mathcal{I}^L$ for a discrete observable A is defined as follows:

$$\mathcal{I}_x^L(\varrho) = \mathsf{A}(x)^{1/2}\varrho\mathsf{A}(x)^{1/2}. \tag{5.36}$$

The fact that $\mathcal{I}^L$ is an A-compatible instrument is easy to see. First, from Example 4.7 we conclude that each $\mathcal{I}_x^L$ is an operation. Since

$$\mathrm{tr}\big[\mathcal{I}_x^L(\varrho)\big] = \mathrm{tr}[\mathsf{A}(x)^{1/2}\varrho\mathsf{A}(x)^{1/2}] = \mathrm{tr}\big[\varrho\mathsf{A}(x)\big], \tag{5.37}$$

we see that $\sum_x \mathrm{tr}\big[\mathcal{I}_x^L(\varrho)\big] = \mathrm{tr}\big[\varrho\big]$ and thus that $\mathcal{I}^L$ is A-compatible.

Exercise 5.16 Why do we need the square roots in formula (5.36)? Let A be a discrete observable (not necessarily sharp) and define a mapping $x \mapsto \mathcal{I}_x(\varrho) = \mathsf{A}(x)\varrho\mathsf{A}(x)$. This would also be a generalization of (5.34). Find an explicit example showing that $\mathcal{I}$ is not, in general, A-compatible (it is not even an instrument).

Starting from the Lüders instrument $\mathcal{I}^L$, we can easily construct more A-compatible instruments. Namely, let us fix a channel $\mathcal{E}_x$ for every x. Then the formula

$$\mathcal{I}_x(\varrho) = (\mathcal{E}_x \circ \mathcal{I}_x^L)(\varrho) = \mathcal{E}_x\big(\mathsf{A}(x)^{1/2}\varrho\mathsf{A}(x)^{1/2}\big) \tag{5.38}$$

defines an A-compatible instrument. This type of instrument can be interpreted as a sequential procedure; first we implement a Lüders measurement of A and then, depending on the measurement outcome x, a channel $\mathcal{E}_x$ is applied.

Proposition 5.17 Let A be a discrete observable on a finite-dimensional Hilbert space $\mathcal{H}$. Every A-compatible instrument is of the form

$$\mathcal{I}_x = \mathcal{E}_x \circ \mathcal{I}_x^L, \tag{5.39}$$

where $\mathcal{I}^L$ is an A-compatible Lüders instrument and $\{\mathcal{E}_x\}$ is a set of channels.

Proof We will give a proof only for a simple special case and refer the reader to [72] for a general proof. Let $\mathcal{I}$ be an A-compatible instrument. We assume that an operation $\mathcal{I}_x$ is determined only by a single Kraus operator K_x; hence

$$\mathcal{I}_x(\varrho) = K_x\varrho K_x^*, \qquad K_x^*K_x = \mathsf{A}(x).$$

The latter equation implies that

$$|K_x| = (K_x^*K_x)^{1/2} = \mathsf{A}(x)^{1/2} \tag{5.40}$$

and therefore that the polar decomposition of K_x is $V_x A(x)^{1/2}$. where V_x is a unitary operator. We conclude that $\mathcal{I}_x = \sigma_{V_x} \circ \mathcal{I}_x^L$. In the more general case, an operation $\mathcal{I}_x$ is determined by several Kraus operators and the proof has to be modified. □

A related result was presented by Ozawa [109]. If A is a discrete *sharp* observable (on a finite- or infinite-dimensional Hilbert space) then every A-compatible instrument is of the form

$$\mathcal{I}_x = \mathcal{E} \circ \mathcal{I}_x^L, \tag{5.41}$$

where $\mathcal{I}^L$ is an A-compatible Lüders instrument and $\mathcal{E}$ is a channel. Thus in this case the instrument $\mathcal{I}$ can be interpreted as a sequential procedure of the Lüders measurement of A followed by a fixed channel $\mathcal{E}$.

5.3.3 Lüders' theorem

Let A and B be two discrete observables. Suppose that we make a Lüders measurement of A and, after that, perform a measurement of B. We can then ask whether the measurement of B is disturbed by the measurement of A. We consider B not to be disturbed if the measurement outcome probabilities of B-measurement do not depend on whether A has been measured first. Hence, if $\mathcal{I}^L$ is the A-compatible Lüders instrument, the nondisturbance condition reads

$$\text{tr}\big[\mathcal{I}_\Omega^L(\varrho)B(y)\big] = \text{tr}\big[\varrho B(y)\big]$$

and is required to hold for all outcomes y and states ϱ (compare with (5.20)). Concerning this problem the following theorem was first observed by Lüders [95].

Theorem 5.18 (*Lüders' theorem*)
Let A and B be discrete observables and suppose that A is sharp. A Lüders measurement of A does not disturb B if and only if A and B commute.

Proof Assume that A and B commute, i.e. $A(x)B(y) = B(y)A(x)$ for all x, y. We obtain

$$\text{tr}\big[\mathcal{I}_\Omega^L(\varrho)B(y)\big] = \sum_x \text{tr}\big[A(x)\varrho A(x)B(y)\big] = \sum_x \text{tr}\big[A(x)\varrho B(y)\big]$$

$$= \text{tr}\big[\varrho B(y)\big].$$

Therefore, B is not disturbed by the A-measurement.

Further let us start from the assumption that the measurement statistics of B are not altered by the measurement of A. To proceed in this direction, we use also the assumption that A is sharp. Then the nondisturbance assumption means that

$$\sum_x \text{tr}\big[A(x)\varrho A(x)B(y)\big] = \text{tr}\big[\varrho B(y)\big] \tag{5.42}$$

for every state ϱ, so that

$$\sum_x A(x)B(y)A(x) = B(y). \tag{5.43}$$

For each x', we obtain (by multiplying both sides of (5.43) by $A(x')$ on the left)

$$A(x')B(y) = \sum_x A(x')A(x)B(y)A(x) = A(x')B(y)A(x') \tag{5.44}$$

and similarly (by multiplying both sides of (5.43) by $A(x')$ on the right)

$$B(y)A(x') = \sum_x A(x)B(y)A(x)A(x') = A(x')B(y)A(x'). \tag{5.45}$$

Here we have used the fact that $A(x')A(x) = A(\{x\} \cap \{x'\}) = \delta_{x,x'}A(x')$; see Proposition 3.29. A comparison of (5.44) and (5.45) shows that A and B commute.
□

Is there a version of Lüders' theorem for a general pair of discrete observables? First, suppose that A and B commute, i.e. $A(x)B(y) = B(y)A(x)$ for all x, y. We recall from Theorem 1.33 that this implies that $A(x)^{1/2}B(y) = B(y)A(x)^{1/2}$ for all x, y. Hence, as in the proof of Theorem 5.18, we see that a Lüders measurement of A does not disturb B.

Now let us consider the converse: do two discrete observables commute if their Lüders measurements do not disturb them mutually? The answer to this question is negative and a (quite tricky) counterexample was given by Arias *et al.* [7]. This means that there are observables A and B which do not commute but for which, nevertheless, a Lüders measurement of A does not disturb B.

Even though there is no general Lüders theorem covering all pairs of discrete observables, the statement can be extended to certain classes of observables. All these kinds of result are commonly referred to as generalized Lüders' theorems. In particular, there is a version of Lüders' theorem for a pair of discrete observables in finite-dimensional Hilbert space. Namely, suppose that A and B are discrete observables on a finite-dimensional Hilbert space. Then a Lüders measurement of A does not disturb B if and only if A and B commute. This result was first presented by Busch and Singh [31].

5.3.4 Example: mean king's problem

Once upon a time . . . Alice (a quantum physicist) was sailing on an ocean. A big storm surprised her and she ended up on an island ruled by king Brutus. Brutus loved cats. One day he learned about a cat being alive and dead at once and he started to hate all quantum physicists. Nevertheless, he decided to give Alice a last chance. Here is his deadly challenge.

I shall take a spin-$\frac{1}{2}$ system and perform a spin component measurement in the x, y or z direction, thereby getting an outcome 1 (up) or -1 (down). Your task is to correctly guess the outcome I obtain. You are allowed to prepare the system before I measure it. In the morning, I shall make the choice, perform the measurement and give you back the measured system. For the whole day you will have full access to my royal laboratory. However, at sunset I shall close and lock the doors and there will be no possibility of taking anything out of the laboratory. I shall tell you my choice of measurement and you must immediately make your final guess of the outcome. You have only one chance.

What are the chances for Alice to survive this game? Before continuing the story, let us stress that some details must be specified more precisely. After Alice gets access to the royal laboratory she must first investigate the measurement apparatuses and conclude to which type of instrument they are related. The task is possible only if Alice learns this information. We assume that the king performs Lüders measurements and that Alice finds this out. Thus, the states corresponding to the eigenvectors $\varphi_{+,b}$ and $\varphi_{-,b}$ of σ_b are the conditional output states corresponding to the outcomes ± 1 when a measurement in the direction $b = x, y, z$ is performed.

The measurement outcome obtained by the king will be random in whichever way Alice prepares the system. Therefore, Alice's chances of survival must depend on the fact that she will learn the king's choice at the end. The question is whether this information comes too late. We will not analyze the problem in its whole generality but demonstrate that Alice can find the king's outcome with certainty. This problem was originally considered in [136]. Here we present a slightly different solution. Suppose Alice prepares two spin-$\frac{1}{2}$ systems in the singlet state

$$|\psi_-\rangle\langle\psi_-|, \qquad \psi_- = \tfrac{1}{\sqrt{2}}\left(\varphi_{+,b} \otimes \varphi_{-,b} - \varphi_{-,b} \otimes \varphi_{+,b}\right),$$

which has the same form in any orthonormal basis. Alice keeps the first system and gives the second system to the king. Suppose that the king performs a measurement of S^b. Alice gets the second system back and, depending on whether the measurement outcome is $+1$ or -1, the final state of the composite system is

$$|\Phi_{b,\pm}\rangle\langle\Phi_{b,\pm}|, \qquad \text{where } \Phi_{b,+} = \varphi_{-,b} \otimes \varphi_{+,b} \quad \text{and} \quad \Phi_{b,-} = \varphi_{+,b} \otimes \varphi_{-,b}.$$

Alice can count on the additional information she gets after performing her measurement. Let us define an orthonormal basis $\{\theta_1, \theta_2, \theta_3, \theta_4\}$ of $\mathbb{C}^2 \otimes \mathbb{C}^2$ in the following way:

$$\theta_1 = \tfrac{1}{\sqrt{2}}\varphi_{+,z} \otimes \varphi_{-,z} + \tfrac{1}{2}e^{-i\pi/4}\varphi_{+,z} \otimes \varphi_{+,z} - \tfrac{1}{2}e^{i\pi/4}\varphi_{-,z} \otimes \varphi_{-,z},$$

$$\theta_2 = \tfrac{1}{\sqrt{2}}\varphi_{+,z} \otimes \varphi_{-,z} - \tfrac{1}{2}e^{-i\pi/4}\varphi_{+,z} \otimes \varphi_{+,z} + \tfrac{1}{2}e^{i\pi/4}\varphi_{-,z} \otimes \varphi_{-,z},$$

$$\theta_3 = \tfrac{1}{\sqrt{2}} \varphi_{-,z} \otimes \varphi_{+,z} + \tfrac{1}{2} e^{i\pi/4} \varphi_{+,z} \otimes \varphi_{+,z} - \tfrac{1}{2} e^{-i\pi/4} \varphi_{-,z} \otimes \varphi_{-,z},$$

$$\theta_4 = \tfrac{1}{\sqrt{2}} \varphi_{-,z} \otimes \varphi_{+,z} - \tfrac{1}{2} e^{i\pi/4} \varphi_{+,z} \otimes \varphi_{+,z} + \tfrac{1}{2} e^{-i\pi/4} \varphi_{-,z} \otimes \varphi_{-,z}.$$

Alice's observable A is the sharp nondegenerate observable associated with this basis, i.e. $A(k) = |\theta_k\rangle\langle\theta_k|$ for $k = 1, 2, 3, 4$. The following table contains the conditional probability distributions (columns) measured by Alice for all possibilities of the king's measurement choice and outcome:

	$x, +1$	$x, -1$	$y, +1$	$y, -1$	$z, +1$	$z, -1$
1	$\frac{1}{2}$	0	$\frac{1}{2}$	0	0	$\frac{1}{2}$
2	0	$\frac{1}{2}$	0	$\frac{1}{2}$	0	$\frac{1}{2}$
3	0	$\frac{1}{2}$	$\frac{1}{2}$	0	$\frac{1}{2}$	0
4	$\frac{1}{2}$	0	0	$\frac{1}{2}$	$\frac{1}{2}$	0

The numbers in this table are the overlaps $|\langle\theta_k|\Phi_{b,\pm}\rangle|^2$. The last two columns are easy to calculate as the vectors θ_k are given in terms of $\varphi_{+,z}$ and $\varphi_{-,z}$. To calculate the other columns, one can use the expansions

$$\varphi_{+,x} = \tfrac{1}{\sqrt{2}}\left(\varphi_{+,z} + \varphi_{-,z}\right), \quad \varphi_{-,x} = \tfrac{1}{\sqrt{2}}\left(\varphi_{+,z} - \varphi_{-,z}\right),$$

$$\varphi_{+,y} = \tfrac{1}{\sqrt{2}}\left(\varphi_{+,z} + i\varphi_{-,z}\right), \quad \varphi_{-,y} = \tfrac{1}{\sqrt{2}}\left(\varphi_{+,z} - i\varphi_{-,z}\right).$$

It is straightforward to verify that if b is announced by the king then the outcomes $1, 2, 3, 4$ uniquely match the king's results ± 1. For instance, if the king decides to measure σ_x then only first two columns of the table are of interest for Alice, and outcomes 1 and 4 indicate with certainty that king's outcome was $+1$. Similarly, outcomes 2 and 3 imply that the king's the outcome was -1. In summary, Alice can survive with certainty the 'deadly challenge' of the king.

5.4 Repeatable measurements

Given an observable A, we can ask what kind of A-compatible instruments exist. It should be emphasized, perhaps, that there are no preferred A-compatible instruments. Different instruments simply describe different kinds of measurement of A, which influence the system in different ways. Of course, we may want to perform the measurement of A in a certain way and wish to influence the system in some way suitable for our purposes. For instance, one possibility is that we use our measurement to prepare the system. Then, a typical requirement is that this preparation procedure makes the measurement outcome in further measurements completely predictable; we always get the same result as in the first measurement. This leads to the notion of a repeatable measurement.

An overview on repeatable measurements was presented in [33]. An extensive discussion on repeatability and related concepts can be found in [35].

5.4.1 Repeatability

To formulate the above idea of a measurement device as a preparator, let A be a discrete observable. Suppose that we perform a measurement of A in a state ϱ and that the measurement is described by an A-compatible instrument $\mathcal{I}$. Let x be a measurement outcome which has a nonzero probability of occurring. The conditional output state corresponding to the measurement outcome x is

$$\tilde{\varrho}_x = \frac{1}{\text{tr}\big[\mathcal{I}_x(\varrho)\big]}\mathcal{I}_x(\varrho), \tag{5.46}$$

where we write $\mathcal{I}_x \equiv \mathcal{I}_{\{x\}}$.

Assume, then, that after recording some measurement outcome x we repeat the measurement using the same system. Naturally, the measurement outcome now depends on the properties of the instrument $\mathcal{I}$. We require that we obtain the same measurement outcome x as in the first measurement, with certainty. Hence, we set the repeatability condition

$$\text{tr}\big[\tilde{\varrho}_x A(x)\big] = 1. \tag{5.47}$$

This leads us to the following definition.

Definition 5.19 An instrument $\mathcal{I}$ is *repeatable* if $\text{tr}\big[\tilde{\varrho}_x A(x)\big] = 1$ holds whenever $\text{tr}\big[\varrho A(x)\big] \neq 0$.

Repeatability is thus a property of instruments and it does not depend on the other details of a measurement model. Sometimes we say that a measurement model $\mathcal{M}$ is repeatable, but this means that the induced instrument $\mathcal{I}^{\mathcal{M}}$ is repeatable. In the following examples we will demonstrate that there are both kinds of instrument, repeatable and nonrepeatable.

Example 5.20 (*The Lüders instrument of a sharp observable is repeatable*)
In (5.34) we defined the Lüders instrument related to a sharp observable A. This instrument is clearly repeatable. Namely, the conditional outcome states are $\tilde{\varrho}_x = |\psi_x\rangle\langle\psi_x|$ and hence $\text{tr}\big[\tilde{\varrho}_x A(x)\big] = \text{tr}\big[\tilde{\varrho}_x\big] = 1$.   △

Example 5.21 (*The trivial instrument is not repeatable*)
Let A be a discrete observable, ξ a state and $\mathcal{I}$ the trivial instrument determined by ξ (recall Example 5.6). For each state ϱ and measurement outcome x such that $\mathrm{tr}[\varrho A(x)] \neq 0$, we then have $\tilde{\varrho}_x = \xi$. Thus, if the trivial instrument $\mathcal{I}$ is repeatable then $\mathrm{tr}[\xi A(x)] = 1$ whenever $A(x) \neq O$. However, since

$$1 = \mathrm{tr}[\xi] = \sum_x \mathrm{tr}[\xi A(x)] = \sum_{x:A(x)\neq O} 1,$$

the repeatability condition can hold only if there is only a single nonzero effect in the range of A. In this case, the observable A is of the following form:

$$A(x) = I \text{ for some outcome } x, \qquad A(y) = O \text{ for all outcomes } y \neq x. \quad (5.48)$$

We conclude that trivial instruments are never repeatable unless the corresponding observable A is of the banal form (5.48). $\triangle$

The repeatability condition can be written in different but equivalent forms.

Proposition 5.22 For an instrument $\mathcal{I}$, the following conditions are equivalent:

(i) $\mathcal{I}$ is repeatable;
(ii) $\mathrm{tr}[\mathcal{I}_x(\mathcal{I}_x(\varrho))] = \mathrm{tr}[\mathcal{I}_x(\varrho)]$ for every outcome x and every state ϱ;
(iii) $\mathrm{tr}[\mathcal{I}_y(\mathcal{I}_x(\varrho))] = 0$ for all outcomes $x \neq y$ and every state ϱ;
(iv) $\mathcal{I}_y(\mathcal{I}_x(\varrho)) = 0$ for all outcomes $x \neq y$ and every state ϱ.

Proof Using equations (5.11) and (5.46) we see that the repeatability condition (5.47) for $\mathcal{I}$ is equivalent to

$$\mathrm{tr}[\mathcal{I}_x(\mathcal{I}_x(\varrho))] = \mathrm{tr}[\mathcal{I}_x(\varrho)], \qquad (5.49)$$

which must hold whenever $\mathrm{tr}[\mathcal{I}_x(\varrho)] \neq 0$. If $\mathrm{tr}[\mathcal{I}_x(\varrho)] = 0$ then both sides in (5.49) give 0. Therefore, (i) and (ii) are equivalent.

The equivalence of (ii) and (iii) follows immediately by noticing that

$$\sum_y \mathrm{tr}[\mathcal{I}_y(\mathcal{I}_x(\varrho))] = \mathrm{tr}[\mathcal{I}_\Omega(\mathcal{I}_x(\varrho))] = \mathrm{tr}[\mathcal{I}_x(\varrho)].$$

Here we have used the fact that $\mathcal{I}_\Omega$ is trace preserving.

It is clear that on the one hand (iv) implies (iii). On the other hand, since $\mathcal{I}_y(\mathcal{I}_x(\varrho))$ is a positive trace class operator we can have $\mathrm{tr}[\mathcal{I}_y(\mathcal{I}_x(\varrho))] = 0$ only if $\mathcal{I}_y(\mathcal{I}_x(\varrho)) = 0$. Thus, (iii) and (iv) are equivalent. $\square$

Which observables admit repeatable instruments? The following result was reported in [35].

Proposition 5.23 Suppose that A is a discrete observable. Then there exists an A-compatible repeatable instrument if and only if every nonzero effect $A(x)$ has eigenvalue 1.

Proof The repeatability condition $\mathrm{tr}\big[\widetilde{\varrho}_x A(x)\big] = 1$ implies that any nonzero effect $A(x)$ must have eigenvalue 1.

In the other direction, assume that each nonzero effect $A(x)$ has eigenvalue 1. For each outcome x satisfying $A(x) \neq O$, choose a unit vector $\psi_x \in \mathcal{H}$ such that $A(x)\psi_x = \psi_x$. Then define an instrument $\mathcal{I}$ as

$$\mathcal{I}_x(\varrho) = \mathrm{tr}\big[\varrho A(x)\big]\,|\psi_x\rangle\langle\psi_x|.$$

This instrument is an A-compatible conditional state preparator. Using condition (ii) in Proposition 5.22 it easy to see that $\mathcal{I}$ is repeatable:

$$\mathrm{tr}\big[\mathcal{I}_x(\mathcal{I}_x(\varrho))\big] = \mathrm{tr}\big[\varrho A(x)\big]\,\mathrm{tr}[\mathcal{I}_x(|\psi_x\rangle\langle\psi_x|)] = \mathrm{tr}\big[\varrho A(x)\big]\,\mathrm{tr}\big[|\psi_x\rangle\langle\psi_x|A(x)\big]$$
$$= \mathrm{tr}\big[\mathcal{I}_x(\varrho)\big]. \qquad \square$$

In conclusion, we have seen that repeatable measurements are possible but set some requirements for the observable in question. And, even if an observable admits repeatable measurements, it always admits nonrepeatable measurements as well.

5.4.2 Wigner–Araki–Yanase theorem

Eugene Wigner, the father of symmetry principles in quantum theory, was the first to discover that a conservation law puts constrains on the measurements that are possible [143]. Wigner's observation was later formulated as a theorem by Huzihiro Araki and Mutsuo Yanase [6]. This result is nowadays called the Wigner–Araki–Yanase (WAY) theorem. In the following we will present a simple formulation of the Wigner–Araki–Yanase theorem. A slightly more general version can be found in, for instance, [13] or [107]. Wigner's original argument is explained in [94].

We consider measurement models $\mathcal{M} = \langle \mathcal{K}, \xi, \mathcal{V}, \mathsf{F} \rangle$ with the following specifications.

(N1) The pointer observable F is a sharp nondegenerate observable associated with some orthonormal basis $\{\phi_j\}$ in $\mathcal{K}$.

(N2) The initial state ξ of the apparatus is a pure state $\xi = |\phi_0\rangle\langle\phi_0|$.

(N3) The measurement interaction $\mathcal{V}$ is given by a unitary operator V on $\mathcal{H} \otimes \mathcal{K}$.

We say that these types of measurement model are *normal*. The conditions (N1)–(N3) are quite typical, and normal measurement models are a representative class of all measurement models. For instance, the von Neumann measurement model presented in subsection 5.3.1 falls into this class.

The observable that we intend to measure is a sharp nondegenerate observable A with $A(j) = |\psi_j\rangle\langle\psi_j|$.

Proposition 5.24 If $\mathcal{M}$ is a normal measurement model for the observable A then there are vectors $\tilde{\psi}_j \in \mathcal{H}$ such that

$$V(\psi_j \otimes \phi_0) = \tilde{\psi}_j \otimes \phi_j. \tag{5.50}$$

Moreover, the vectors $\tilde{\psi}_j$ are orthogonal if $\mathcal{M}$ is repeatable.

Proof Since $\langle\psi_j|A(j)\psi_j\rangle = 1$, the probability reproducibility condition (5.1) gives

$$\langle V(\psi_j \otimes \phi_0)|I \otimes F(j)V(\psi_j \otimes \phi_0)\rangle = 1; \tag{5.51}$$

hence, we conclude that

$$I \otimes F(j)V(\psi_j \otimes \phi_0) = V(\psi_j \otimes \phi_0). \tag{5.52}$$

We write the vector $V(\psi_j \otimes \phi_0)$ in the tensor product basis $\{\psi_k \otimes \phi_l\}$, obtaining $V(\psi_j \otimes \phi_0) = \sum_{k,l} c_{k,l}^j \psi_k \otimes \phi_l$. Using (5.52) we obtain

$$V(\psi_j \otimes \phi_0) = I \otimes |\phi_j\rangle\langle\phi_j| \sum_{k,l} c_{k,l}^j \psi_k \otimes \phi_l = \sum_{k,l} c_{k,l}^j \langle\phi_j|\phi_l\rangle \psi_k \otimes \phi_j$$

$$= \left(\sum_k c_{k,j}^j \psi_k\right) \otimes \phi_j \equiv \tilde{\psi}_j \otimes \phi_j.$$

If the system is initially in the state $|\psi_j\rangle\langle\psi_j|$ then we get outcome j with probability 1, and the conditional output state is $|\tilde{\psi}_j\rangle\langle\tilde{\psi}_j|$. Assuming that $\mathcal{M}$ is repeatable, we then obtain

$$1 = \text{tr}[|\tilde{\psi}_j\rangle\langle\tilde{\psi}_j|A(j)] = \left|\langle\psi_j|\tilde{\psi}_j\rangle\right|^2.$$

This implies that $\tilde{\psi}_j = e^{i\theta_j}\psi_j$ for some $\theta_j \in \mathbb{R}$. In particular, the vectors $\tilde{\psi}_j$ are orthogonal. $\square$

The time evolution of the compound system during the measurement is given by the interaction channel $\varrho \mapsto \mathcal{V}(\varrho)$. A selfadjoint operator C acting on $\mathcal{H} \otimes \mathcal{K}$ is called a *conserved quantity* if

$$\text{tr}[\mathcal{V}(\varrho)C] = \text{tr}[\varrho C]$$

for all states ϱ. If C can be split into two selfadjoint operators, so that $C = C_1 \otimes I + I \otimes C_2$, then we say that C is an *additive* conserved quantity.

Theorem 5.25 (*Wigner–Araki–Yanase theorem*)
Let $\mathcal{M}$ be a normal measurement model for the observable A, and suppose that $C = C_1 \otimes I + I \otimes C_2$ is an additive conserved quantity. If $\mathcal{M}$ is repeatable then $[C_1, A(j)] = 0$ for every j.

Proof   For every $i \neq j$, on the one hand we have

$$\langle \psi_i \otimes \phi_0 | C(\psi_j \otimes \phi_0) \rangle = \langle \psi_i | C_1 \psi_j \rangle,$$

and, on the other hand,

$$\begin{aligned}
\langle \psi_i \otimes \phi_0 | C(\psi_j \otimes \phi_0) \rangle &= \langle V(\psi_i \otimes \phi_0) | CV(\psi_j \otimes \phi_0) \rangle \\
&= \langle \tilde{\psi}_i \otimes \phi_i \big| C(\tilde{\psi}_j \otimes \phi_j) \rangle \\
&= \langle \tilde{\psi}_i \big| C_1 \tilde{\psi}_j \rangle \langle \phi_i | \phi_j \rangle + \langle \tilde{\psi}_i \big| \tilde{\psi}_j \rangle \langle \phi_i | C_2 \phi_j \rangle \\
&= 0.
\end{aligned}$$

Therefore, $\langle \psi_i | C_1 \psi_j \rangle = \delta_{ij} \langle \psi_i | C_1 \psi_j \rangle$.
For every k, ℓ, we then have

$$\begin{aligned}
\langle \psi_k \big| [C_1, A(j)] \psi_\ell \rangle &= \langle \psi_k | C_1 \psi_j \rangle \langle \psi_j | \psi_\ell \rangle - \langle \psi_k | \psi_j \rangle \langle \psi_j | C_1 \psi_\ell \rangle \\
&= \delta_{kj} \delta_{j\ell} \langle \psi_k | C_1 \psi_j \rangle - \delta_{kj} \delta_{j\ell} \langle \psi_j | C_1 \psi_\ell \rangle = 0.
\end{aligned}$$

Thus $[C_1, A(j)] = 0$. $\square$

The lesson of the WAY theorem is that not all measurement models are physically allowable even if quantum theory contains their description. For instance, the conservation of angular momentum forbids a repeatable normal measurement model of a spin component.

5.4.3 Approximate repeatability

One could formulate the repeatability condition also in the general case, rather than just for discrete observables. In the general case the repeatability of an instrument $\mathcal{I}$ means that

$$\text{tr}\big[\mathcal{I}_X(\mathcal{I}_X(\varrho))\big] = \text{tr}\big[\mathcal{I}_X(\varrho)\big] \tag{5.53}$$

for every $X \in \mathcal{F}$ and $\varrho \in \mathcal{S}(\mathcal{H})$. However, according to a result of Ozawa [108], there is no A-compatible repeatable instrument unless A is discrete.

The fact that repeatable instruments do not exist for nondiscrete observables may seem controversial in view of the current quantum technology, where a single particle can be repeatedly localized with high precision. One can give the practical argument that in any real experiment there can be only a finite number of possible measurement outcomes. However, this issue can also be explained by modifying the repeatability condition slightly, which allows to gain some additional insight into the problem.

Overviews of relaxations of the repeatability condition can be found in [27] and [37]. To demonstrate this kind of approach, we will follow the seminal article of Davies and Lewis [52] and describe one possible way of relaxing the repeatability condition. For simplicity, we assume that the outcome space of observables and instruments is the Borel outcome space $(\mathbb{R}, \mathcal{B}(\mathbb{R}))$.

For every $\varepsilon > 0$ and $x \in \mathbb{R}$, we denote by $I_{x;\varepsilon}$ the closed interval centered in x and having length ϵ, i.e. $I_{x;\varepsilon} = \left[x - \frac{1}{2}\varepsilon, x + \frac{1}{2}\varepsilon \right]$. For every $X \subseteq \mathbb{R}$, we then write

$$X_\varepsilon = \bigcup_{x \in X} I_{x;\varepsilon}.$$

Thus, if the diameter of a set X (defined as the least upper bound of the distances between pairs of points in X) is $|X|$ then the diameter of X_ε is $|X| + \varepsilon$.

Definition 5.26 Let $\varepsilon > 0$. An instrument $\mathcal{I}$ is *ε-repeatable* if

$$\text{tr}\left[\mathcal{I}_{X_\varepsilon}(\mathcal{I}_X(\varrho))\right] = \text{tr}\left[\mathcal{I}_X(\varrho)\right] \tag{5.54}$$

for every $X \in \mathcal{B}(\mathbb{R})$ and every $\varrho \in \mathcal{S}(\mathcal{H})$.

The condition (5.54) has the following meaning. Assume that an outcome from a subset X is recorded. Then a repeated application of the same measurement gives an outcome from X_ε with probability 1. When ε is close to 0 then an ε-repeatable instrument is almost the same as a repeatable instrument. Formally, the definition of ε-repeatability makes sense also for $\varepsilon = 0$; in this case it would reduce to the usual definition of repeatability.

Exercise 5.27 Show that if an instrument $\mathcal{I}$ is ε-repeatable then it is ε'-repeatable for every $\varepsilon' \geq \varepsilon$.

To conclude the previous discussion and Exercise 5.27, for a given observable A we are interested in finding the smallest possible number ε such that there exists an A-compatible instrument $\mathcal{I}$ which is ε-repeatable.

Proposition 5.28 Let A be an observable and assume there is an $\varepsilon > 0$ such that, for every open interval X with diameter $|X| = \varepsilon$, the effect $A(X)$ has eigenvalue 1. Then there exists an ε-repeatable A-compatible instrument.

Proof For each $n \in \mathbb{Z}$, write $X_n = [n\varepsilon, (n+1)\varepsilon)$. Then (X_n) is a sequence of mutually disjoint sets and $\cup_n X_n = \mathbb{R}$. From the assumption it follows that, for each $n \in \mathbb{Z}$, we can choose a pure state ϱ_n such that $\text{tr}\left[\varrho_n A(X_n)\right] = 1$. The formula

$$\mathcal{I}_X(\varrho) := \sum_n \mathrm{tr}\big[\varrho A(X \cap X_n)\big]\, \varrho_n$$

defines an A-compatible instrument $\mathcal{I}$.

To prove that $\mathcal{I}$ is ε-repeatable, let $X \in \mathcal{B}(\mathbb{R})$ and $\varrho \in \mathcal{S}(\mathcal{H})$. We then obtain

$$\mathrm{tr}\big[\mathcal{I}_{X_\varepsilon}(\mathcal{I}_X(\varrho))\big] = \sum_n \sum_k \mathrm{tr}\big[\varrho_n A(X_\varepsilon \cap X_k)\big]\, \mathrm{tr}\big[\varrho A(X \cap X_n)\big]$$

$$= \sum_n \mathrm{tr}\big[\varrho_n A(X_\varepsilon)\big]\, \mathrm{tr}\big[\varrho A(X \cap X_n)\big].$$

If $X \cap X_n \neq \emptyset$ then $X_n \subseteq X_\varepsilon$. This implies that either $\mathrm{tr}\big[\varrho A(X \cap X_n)\big] = 0$ or $\mathrm{tr}\big[\varrho_n A(X_\varepsilon)\big] = 1$. Therefore

$$\mathrm{tr}\big[\mathcal{I}_{X_\varepsilon}(\mathcal{I}_X(\varrho))\big] = \sum_n \mathrm{tr}\big[\varrho A(X \cap X_n)\big] = \mathrm{tr}\big[\varrho A(X)\big].$$

$\square$

Example 5.29 (*Approximate repeatability of position measurement*)
Let us recall the canonical position observable Q introduced in Example 3.32. It is a sharp observable and thus each $Q(X) \neq O$ has eigenvalue 1. However, $Q(X) \neq O$ whenever X is a nonempty open interval. We conclude from Proposition 5.28 that there exists an ε-repeatable Q-compatible instrument for each $\varepsilon > 0$. There is thus no theoretical limit to the precision of repeatability of the position observable. $\triangle$

5.5 Programmable quantum processors

By changing the probe state ξ in a measurement model $\mathcal{M} = \langle \mathcal{K}, \xi, \mathcal{V}, \mathsf{F}\rangle$ (while keeping the other components fixed), we potentially implement different instruments, observables and channels on the system. In this sense the probe states can be seen as *quantum programs* and the triple $\langle \mathcal{K}, \mathcal{V}, \mathsf{F}\rangle$ as a *programmable quantum processor*. The paradigm of programmable quantum processors was introduced by Nielsen and Chuang [103]. Ten years later, a simple programmable quantum gate was realized in photonic experiments done at Olomouc [100]. To make things simpler, we will assume in this section that $\mathcal{H}$ is a finite-dimensional Hilbert space.

5.5.1 Programming of observables and channels

A programmable quantum processor can be used to realize instruments, observables and channels. Each program state ξ encodes all these objects; thus, a programmable processor $\langle \mathcal{K}, \mathcal{V}, \mathsf{F}\rangle$ induces mappings from the state space $\mathcal{S}(\mathcal{K})$ into the set of instruments, observables and channels, respectively, of the system $\mathcal{H}$.

All observables realized by a given quantum processor are given by the usual formula,

$$\text{tr}\big[\varrho A_\xi(X)\big] = \text{tr}\big[\mathcal{V}(\varrho \otimes \xi)(I \otimes \mathsf{F}(X))\big],\qquad (5.55)$$

where ξ runs over all probe states. All implemented observables A_ξ have the same outcome space as F, and we see also that if $\mathsf{F}(X) = O$ for some X then this implies that $A_\xi(X) = O$. In measure-theoretic terms this means that A_ξ is absolutely continuous with respect to F.

The realizable channels correspond to nonselective state changes. Hence, they are given by

$$\varrho \mapsto \mathcal{E}_\xi(\varrho) = \text{tr}_{\mathcal{K}}\big[\mathcal{V}(\varrho \otimes \xi)\big].\qquad (5.56)$$

The pointer observable plays no role in the realization of channels. Therefore, when our aim is to program only channels, we can identify programmable processors with channels $\mathcal{V}$ defined on $\mathcal{H} \otimes \mathcal{K}$.

Exercise 5.30 Suppose that a programmable processor $\langle \mathcal{K}, \mathcal{V}, \mathsf{F}\rangle$ can implement two channels $\mathcal{E}_1$ and $\mathcal{E}_2$. Show that it can also implement any mixture $\lambda\mathcal{E}_1 + (1-\lambda)\mathcal{E}_2$ of these channels. [Hint: Try to mix the probe states.]

Suppose that a collection of probe states ξ_j, now considered as programs, implements a collection of devices D_j (these are either instruments, observables or channels). We purify all states ξ_j to pure states P_j; we do this by adding a Hilbert space $\mathcal{K}'$ with the same dimension as $\mathcal{K}$ (see subsection 2.4.2). We define a new programmable processor $\langle \mathcal{K}\otimes\mathcal{K}', \mathcal{V}', \mathsf{F}'\rangle$ by setting $\mathcal{V}' = \mathcal{V}\otimes\mathcal{I}$ and $\mathsf{F}'(X) = \mathsf{F}(X)\otimes I$. Then the pure states P_j implement the devices D_j. We conclude that any collection of devices that can be programmed in a single processor can also be programmed using pure states.

5.5.2 Universal processor for channels

In this subsection we concentrate on the programmability of channels; hence a programmable quantum processor will be represented by the pair $\langle \mathcal{K}, \mathcal{V}\rangle$. We will assume further that $\mathcal{V}$ is a unitary channel σ_G, where G is a unitary operator defined on $\mathcal{H} \otimes \mathcal{K}$. This assumption is justified by the fact that by extending the probe system $\mathcal{K}$ we can dilate $\mathcal{V}$ to a unitary channel (recall Section 4.2).

Let us start with a simple observation related to simultaneous realization of a pair of quantum channels.

Proposition 5.31 Suppose that two channels $\mathcal{E}_1$ and $\mathcal{E}_2$ are realized on the same processor with vector states Ξ_1 and Ξ_2, respectively. There exist operator-sum representations $\mathcal{E}_1(X) = \sum_j A_j X A_j^*$ and $\mathcal{E}_2(X) = \sum_j B_j X B_j^*$ such that

$$\langle \Xi_1 | \Xi_2 \rangle\, I = \sum_j A_j^* B_j. \tag{5.57}$$

Proof Let us evaluate the action of a programmable processor $\langle \mathcal{K}, \sigma_G \rangle$ on a vector state $\psi \in \mathcal{H}$. The action of the unitary operator G on $\mathcal{H} \otimes \mathcal{K}$ can be written in the form $G = \sum_{j,k} E_{jk} \otimes |\varphi_j\rangle\langle\varphi_k|$, where the vectors φ_j form an orthonormal basis of $\mathcal{K}$ and the operators E_{jk} satisfy $\sum_j E_{jk}^* E_{jk'} = I\delta_{kk'}$.

Using a pure program state Ξ we get

$$G|\psi\rangle \otimes |\Xi\rangle = \sum_{j,k} E_{jk}|\psi\rangle \otimes \langle\varphi_k|\Xi\rangle\, |\varphi_j\rangle = \sum_j X_j|\psi\rangle \otimes |\varphi_j\rangle, \tag{5.58}$$

where $X_j = \sum_k E_{jk} \langle\varphi_k|\Xi\rangle$. Tracing out the program system we obtain the channel $\mathcal{E}(\varrho) = \sum_j X_j \varrho X_j^*$.

Starting with the vector states Ξ_1, Ξ_2 we get a particular operator-sum form for channels $\mathcal{E}_1, \mathcal{E}_2$ with operators A_j, B_j, respectively. Using the completeness of the orthonormal basis $\{\varphi_j\}$ and the property $\sum_j E_{jk}^* E_{jk'} = I\delta_{kk'}$ of the operators E_{jk}, we obtain

$$\sum_j A_j^* B_j = \sum_{j,k,k'} E_{jk}^* E_{jk'} \langle\Xi_1|\varphi_k\rangle \langle\varphi_k'|\Xi_2\rangle$$

$$= I\langle\Xi_1|\left(\sum_k |\varphi_k\rangle\langle\varphi_k|\right)|\Xi_2\rangle = \langle\Xi_1|\Xi_2\rangle\, I.$$

$\square$

It is natural to ask whether there exists a *universal programmable processor* that is able to encode all channels into states of the program register. That kind of device would certainly be very useful. This question was asked by Nielsen and Chuang [103] and they discovered the 'no-go' theorem formulated below.

Theorem 5.32 There is no programmable processor implementing all unitary channels.

Proof Applying Proposition 5.31 we see that, in order to implement two different unitary channels σ_U and σ_V, the corresponding program states must be orthogonal. Namely, we have

$$U^* V = cI \qquad \Leftrightarrow \qquad V = cU \text{ or } c = 0. \tag{5.59}$$

The first option, $V = cU$, means that U and V determine the same unitary channel. Hence, two different unitary channels must be encoded with orthogonal probe states. There is at most a countable number of orthogonal vectors in separable Hilbert spaces, but there are uncountably many different unitary channels. Therefore, it is impossible to realize all unitary channels on a single programmable processor. $\square$

Exercise 5.33 Show that a maximally mixed program state can encode only unital channels. [Hint: $\mathrm{tr}_{\mathcal{K}}[G(I \otimes I)G^*] = ?$.]

The following example illustrates that there are interesting one-parametric continuous sets of channels that can be implemented on a fixed quantum programmable processor.

Example 5.34 *(Programming phase-damping rates.)*
In subsection 4.5.3 we introduced the family of phase-damping channels. In particular, for a qubit system they take the form

$$\mathcal{E}_{\eta,\vec{n}}(\varrho) = \eta\varrho + (1 - \eta)(\vec{n} \cdot \vec{\sigma})\varrho(\vec{n} \cdot \vec{\sigma}), \tag{5.60}$$

where $0 \le \eta \le 1$ and $\|\vec{n}\| = 1$; hence $\vec{n} \cdot \vec{\sigma}$ is a unitary operator. Let us denote by φ_+, φ_- the eigenvectors of $\vec{n} \cdot \vec{\sigma}$ corresponding to the eigenvalues ± 1. A short calculation shows that $\langle \varphi_{\pm}|\varrho|\varphi_{\pm}\rangle = \langle \varphi_{\pm}|\mathcal{E}_{\eta}(\varrho)|\varphi_{\pm}\rangle$, and the parameter (called the *phase-damping rate*)

$$\gamma := \frac{|\langle \varphi_{\pm}|\mathcal{E}_{\eta}(\varrho)|\varphi_{\mp}\rangle|}{|\langle \varphi_{\pm}|\varrho|\varphi_{\mp}\rangle|} = \sqrt{1 - 4\eta(1 - \eta)} \tag{5.61}$$

determines the speed of the phase-damping process providing that the channel concatenates again and again. The smaller is the parameter γ, the faster is the phase-damping process.

Our goal is to investigate whether the phase-damping rate can be encoded into the state of the program register (the environment) if the phase-damping basis is held fixed. A direct calculation shows us that in this particular case (5.57) always holds, and

$$\langle \Xi_{\eta_1}|\Xi_{\eta_2}\rangle = \sqrt{\eta_1\eta_2} + \sqrt{(1 - \eta_1)(1 - \eta_2)}. \tag{5.62}$$

For $\eta = 0, 1$ the phase-damping channels are unitary ($U = \vec{n}\cdot\vec{\sigma}$ and I, respectively) and, of course, in this case the vector states Ξ_I, Ξ_U are orthogonal. Moreover, $\langle \Xi_I|\Xi_{\eta}\rangle = \sqrt{\eta}$ and $\langle \Xi_U|\Xi_{\eta}\rangle = \sqrt{1 - \eta}$. The program states

$$\Xi_{\eta} = \sqrt{\eta}\,\Xi_I + \sqrt{1 - \eta}\,\Xi_U \tag{5.63}$$

satisfy (5.62). Indeed, setting

$$G = I \otimes |\Xi_I\rangle\langle\Xi_I| + U \otimes |\Xi_U\rangle\langle\Xi_U|, \tag{5.64}$$

it is straightforward to verify that the program states Ξ_{η} encode the channels $\mathcal{E}_{\eta}$. As a result we obtain that the qubit phase-damping rates can be programmed using only a two-dimensional program space. $\triangle$

In general, which parameters of channels can be adjusted by changing the environment (the program) degrees of freedom and which require different interactions (a change in the programmable processor) is an interesting question.

Exercise 5.35 A single-point contraction is a quantum channel that maps the whole state space into a single quantum state, i.e. $\mathcal{A}_\xi : S(\mathcal{H}) \mapsto \xi$. Find a quantum processor implementing all single-point contractions on a d-dimensional quantum system. [Hint: There is a solution such that ξ encodes $\mathcal{A}_\xi$. Example 4.15 may be useful.]

5.5.3 Probabilistic programming

In the previous subsection we saw that no universal programmable quantum processor exists. Could a measurement of the output of the program register help to break this no-go theorem? By including a fixed pointer observable F in the description of the programmable processor and post-selecting the outputs associated with a particular outcome X we achieve the transformation

$$\varrho \mapsto \mathcal{T}_{\xi,X}(\varrho) = \frac{1}{\mathrm{tr}[\mathcal{I}_X^\xi(\varrho)]} \mathcal{I}_X^\xi(\varrho). \tag{5.65}$$

Although the realization of $\mathcal{T}_{\xi,X}$ is probabilistic, we know exactly when realization occurs and when it does not. As our goal is to implement quantum channels we are interested in outcomes, for which $\mathcal{T}_{\xi,X}$ is a valid quantum channel. Linearity requires that the probability $\mathrm{tr}[\mathcal{I}_X^\xi(\varrho)]$ is independent of ϱ; thus, the operation $\mathcal{I}_X^\xi$ is proportional to some trace-preserving map. Let us note that such a property is not required for all program states ξ. In fact, for a given programmable processor $\langle \mathcal{K}, \sigma_G, F \rangle$ we may define the set of *probabilistic programs* as the set of program states that implement channels in practice.

Example 5.36 (*Universal probabilistic programmable processor*)
Suppose that $\mathcal{K} = \mathcal{H} \otimes \mathcal{H}$ and set $G = V_{\mathrm{SWAP}} \otimes I$. That is, the program register consists of two d-dimensional systems, where $d = \dim \mathcal{H}$. Let F be a two-outcome observable associated with the effects P_+ and $I - P_+$, where P_+ is the projection onto a vector $\psi_+ = \frac{1}{\sqrt{d}} \sum_{j=1}^d \varphi_j \otimes \varphi_j$. Suppose that the program state is the Choi–Jamiolkowski state, i.e. $\Xi_{\mathcal{E}} = (\mathcal{E} \otimes \mathcal{I})(P_+) = \frac{1}{d} \sum_{j,k} \mathcal{E}(|\varphi_j\rangle\langle\varphi_k|) \otimes |\varphi_j\rangle\langle\varphi_k|$. Let us use the notation $\mathcal{H}_{\mathrm{data}} \otimes \mathcal{K}_{\mathrm{program}} = \mathcal{H}_1 \otimes \mathcal{H}_2 \otimes \mathcal{H}_3$. Starting with the data register in the state ϱ and observing the outcome associated with P_+, the data register undergoes the transformation $\varrho \mapsto \varrho'$ where

$$\varrho' = \frac{1}{p} \text{tr}_{23}[(I^1 \otimes P_+^{23})G(\varrho^1 \otimes \Xi_{\mathcal{E}}^{23})G^*] = \frac{1}{p} \text{tr}_{23}[(I^1 \otimes P_+^{23})(\varrho^2 \otimes \Xi_{\mathcal{E}}^{13})]$$

$$= \frac{1}{pd^2} \sum_{r,s,j,k} \mathcal{E}(|\varphi_j\rangle\langle\varphi_k|) \otimes \text{tr}[|\varphi_r\rangle\langle\varphi_s|\varrho] \otimes \text{tr}[|\varphi_r\rangle\langle\varphi_s|\varphi_j\rangle\langle\varphi_k|]$$

$$= \frac{1}{pd^2} \sum_{j,k} \mathcal{E}(|\varphi_j\rangle\langle\varphi_k|)\langle\varphi_j|\varrho|\varphi_k\rangle = \frac{1}{pd^2}\mathcal{E}(\varrho)$$

and $p = \text{tr}[(I \otimes P_+)G(\varrho \otimes \Xi_{\mathcal{E}})G^*] = 1/d^2$; thus

$$\varrho' = d^2 \text{tr}_{\mathcal{K}}[(I \otimes P_+)G(\varrho \otimes \Xi_{\mathcal{E}})G^*] = \mathcal{E}(\varrho). \qquad (5.66)$$

That is, the processor $\langle \mathcal{K}, \sigma_G \rangle$, F implements (with a fixed probability $1/d^2$) the encoding of an arbitrary channel into the associated Choi–Jamiolkowski state. In this sense the Choi–Jamiolkowski state can be used to store in a quantum fashion the information on a quantum channel.

If the outcome $I - P_+$ is recorded (it occurs with probability $1 - 1/d^2$) then

$$\varrho \mapsto \frac{d^2}{d^2-1}\left(\frac{1}{d}I - \frac{1}{d^2}\mathcal{E}(\varrho)\right) = \frac{1}{d^2-1}(dI - \mathcal{E}(\varrho)).$$

Unfortunately, this transformation cannot be corrected without some knowledge of $\mathcal{E}$. That is, there is no fixed universal channel mapping it into $\mathcal{E}$. Finally, we point out that the processor described is intimately related to the idea of quantum teleportation, which will be discussed in subsection 6.4.1. △

This example illustrates that universality in the programmability of quantum channels can be achieved in probabilistic settings; the processor described above is an example of a *universal probabilistic programmable processor*. It is an important feature that for probabilistic processors the implementation is *heralded*, meaning that we know whether it was successful or not.

The earlier probabilistic programming scheme can be used to program binary observables, as we discuss in the next example. It has still a flavor of universality; it programs all effects up to a factor.

Example 5.37 (*Implementation of binary observables*)
In Example 5.36 we showed that the Choi–Jamiolkowski state can be used as a program encoding a quantum channel. Suppose that the processor is the same as in Example 5.36. The general program states can be expressed as $\Xi_{\mathcal{F}} = (\mathcal{F} \otimes \mathcal{I})(P_+)$, where $\mathcal{F}$ is some trace-nonincreasing completely positive linear map (i.e. an operation). Using the same calculation as in Example 5.36 we find that

$$p_+(\varrho) = \text{tr}[(I \otimes P_+)G(\varrho \otimes \Xi_{\mathcal{F}})G^*] = \frac{1}{d^2}\text{tr}[\mathcal{F}(\varrho)], \qquad (5.67)$$

$$p_-(\varrho) = \mathrm{tr}\big[(I \otimes (I - P_+))G(\varrho \otimes \Xi_{\mathcal{F}})G^*\big] = 1 - p_+(\varrho). \qquad (5.68)$$

Assuming that $\mathcal{F}(\cdot) = \sum_n A_n \cdot A_n^*$ and setting $F = \sum_n A_n^* A_n$, we have $\mathrm{tr}\big[\mathcal{F}(\varrho)\big] = \mathrm{tr}\big[F\varrho\big]$; hence, the program state $\Xi_{\mathcal{F}}$ encodes an observable F', $I - F'$, where $F' = F/d^2$. In summary, the processor considered implements any binary observable X, $I - X$ satisfying $X \le I/d^2$. $\qquad \triangle$

6

Entanglement

In Section 2.4 we learned that a composite quantum system, composed of two subsystems A and B, is associated with a tensor product space $\mathcal{H}_A \otimes \mathcal{H}_B$. Composite quantum systems have already played an important role in many instances in this text, especially in open systems and measurement models. In this chapter we dive deeper into tensor product spaces, discovering some strange implications of the quantum theory of composite quantum systems.

As a result of its tensor product structure, quantum theory has embedded into it the phenomenon of *entanglement*, which is often seen as the foremost quantum feature. As early as in 1935 it puzzled Erwin Schrödinger, who introduced the German term *Verschränkung* [127] and Albert Einstein, who, together with Boris Podolski and Nathan Rosen, used the specific example of an entangled state to argue that quantum theory is incomplete [58]. In the 1960s John Bell demonstrated that the consequences of entanglement are incompatible with the Einstein–Podolsky–Rosen model of local realism [11], [12]. In recent years entanglement has been recognized as *the* resource for quantum information processing.

Nowadays entanglement theory is a highly developed subject, and we can present only the mathematical basics. For the reader with a deeper interest in entanglement we recommend the recent reviews by Horodecki *et al.* [80] and Plenio *et al.* [118], which also cover the quantum information aspects of entanglement. To simplify our discussion we will assume that the Hilbert spaces are finite dimensional.

6.1 Entangled bipartite systems

Entanglement is a feature of composite quantum systems. In the simplest case we have two subsystems, and the total system is then called bipartite. We will first consider only bipartite systems and then briefly discuss multipartite systems in subsection 6.4.2.

In the present section we introduce the mathematical separation of bipartite vectors, operators and channels into factorized and nonfactorized classes. Entangled vectors are easy to define. We have to be slightly more careful when we define entangled operators. Finally, the analogous classification for channels is a tricky topic.

6.1.1 Entangled vectors

In the following, $\mathcal{H}_A$ and $\mathcal{H}_B$ are two finite-dimensional Hilbert spaces. Their tensor product $\mathcal{H}_A \otimes \mathcal{H}_B$ was defined in subsection 1.3.5.

Definition 6.1 Let η be a vector in $\mathcal{H}_A \otimes \mathcal{H}_B$. The vector η is called

1. *factorized* if $\eta = \phi \otimes \psi$ for some vectors $\phi \in \mathcal{H}_A$, $\psi \in \mathcal{H}_B$,
2. *entangled* otherwise.

The following example gives a concrete illustration of entangled vectors. In particular, it demonstrates that entangled vectors exist, and it will serve as a typical example later.

Example 6.2 *(Existence of entangled vectors)*
Let $\varphi_1, \varphi_2 \in \mathcal{H}$ be two orthogonal unit vectors and let α, β be two nonzero complex numbers such that $|\alpha|^2 + |\beta|^2 = 1$. The unit vector

$$\eta = \alpha \varphi_1 \otimes \varphi_1 + \beta \varphi_2 \otimes \varphi_2 \in \mathcal{H} \otimes \mathcal{H} \tag{6.1}$$

is entangled. To see this, choose unit vectors $\varphi_3, \varphi_4, \ldots \in \mathcal{H}$ such that the set $\{\varphi_1, \varphi_2, \ldots\}$ is an orthonormal basis for $\mathcal{H}$. A general factorized vector in $\mathcal{H} \otimes \mathcal{H}$ then takes the form

$$\phi \otimes \psi = \left(\sum_j c_j \varphi_j \right) \otimes \left(\sum_k d_k \varphi_k \right),$$

where $c_1, c_2, \ldots$ and $d_1, d_2, \ldots$ are complex numbers. Comparing this expression with (6.1) we get the conditions

$$c_1 d_1 = \alpha, \qquad c_2 d_2 = \beta, \qquad c_1 d_2 = c_2 d_1 = 0.$$

Multiplying the first two and the last two equations together and noting that left-hand sides coincide, it follows that $\alpha\beta = 0$. But this contradicts the fact that both α and β were chosen to be nonzero numbers, and therefore η is entangled. △

Consider orthonormal bases $\{\varphi_j\}_{j=1}^{d_A}$ for $\mathcal{H}_A$ and $\{\psi_k\}_{k=1}^{d_B}$ for $\mathcal{H}_B$. A factorized vector has the form

$$\eta = \varphi \otimes \psi = \left(\sum_j c_j \varphi_j\right) \otimes \left(\sum_k d_k \psi_k\right) = \sum_{j,k} c_j d_k \varphi_j \otimes \psi_k$$

for some complex numbers c_j, d_k. If the vectors φ and ψ do not belong to the respective orthonormal bases then there are at least two coefficients c_j and similarly at least two coefficients d_k. Hence, at least for some values j, k with $j \neq k$, the product $c_j d_k$ is nonvanishing. On the basis of this observation, we conclude that all vectors of the form $\sum_{j=1}^{d} a_j \varphi_j \otimes \psi_j$, where $d = \min\{d_A, d_B\}$, are entangled provided that at least two coefficients a_j are nonzero. Notice that the entangled vector in Example 6.2 is just a special case of the above form. The following theorem says that any entangled vector can be written in this form if suitable orthonormal bases of $\mathcal{H}_A$ and $\mathcal{H}_B$ are chosen. This result, known as the *Schmidt decomposition*, is useful in many calculations related to entanglement.

Theorem 6.3 (*Schmidt decomposition*) For each vector $\psi \in \mathcal{H}_A \otimes \mathcal{H}_B$ there exist orthogonal bases $\{e_1, \ldots, e_{d_A}\}$ of $\mathcal{H}_A$ and $\{f_1, \ldots, f_{d_B}\}$ of $\mathcal{H}_B$ such that

$$\psi = \sum_{j=1}^{d} \sqrt{\lambda_j}\, e_j \otimes f_j, \tag{6.2}$$

where $d = \min\{d_A, d_B\}$ and $\lambda_1, \ldots, \lambda_d$ are decreasingly ordered nonnegative numbers forming the so-called *Schmidt vector* $\vec{\lambda}_\psi$.

The *Schmidt rank* $r_s(\psi)$ is defined as the number of nonvanishing elements in $\vec{\lambda}_\psi$, and the vector ψ is entangled if and only if $r_s(\psi) \geq 2$.

Proof The Schmidt decomposition (6.2) is a direct consequence of the singular-value decomposition given in subsection 1.3.3. In finite dimensions we can express singular-value decomposition as follows: any $n \times n$ matrix C can be written in the form $C = UDV$, where U, V are unitary matrices and D is a diagonal matrix with nonnegative entries.

Let us first consider the case $d_A = d_B \equiv d$. We fix orthonormal bases $\{\varphi_1, \ldots, \varphi_d\}$ and $\{\phi_1, \ldots, \phi_d\}$ of $\mathcal{H}_A$ and $\mathcal{H}_B$, respectively. A vector $\psi \in \mathcal{H}_A \otimes \mathcal{H}_B$ has a basis expansion $\psi = \sum_{j,k=1}^{d} c_{jk} \varphi_j \otimes \phi_k$. The complex numbers c_{jk} form a $d \times d$ matrix $C = [c_{jk}]$, which has a singular-value decomposition $C = UDV$. Here U, V are unitary matrices and D is a diagonal matrix with nonnegative entries $\sqrt{\lambda_l}$. We can assume that these numbers have been decreasingly ordered by permutations of the rows and columns of the unitary operators U and V as necessary. Writing the singular-value decomposition of C as

$$c_{jk} = \sum_{l,m=1}^{d} U_{jl}\sqrt{\lambda_l}\delta_{lm}V_{mk},$$

the vector ψ takes the form

$$\psi = \sum_{l=1}^{d} \sqrt{\lambda_l}\left(\sum_{j=1}^{d} U_{jl}\varphi_j\right) \otimes \left(\sum_{k=1}^{d} V_{lk}\phi_k\right).$$

The vectors $e_l \equiv \sum_{j=1}^{d} U_{jl}\varphi_j$ and $f_l \equiv \sum_{k=1}^{d} V_{lk}\phi_k$ form orthonormal bases of $\mathcal{H}_A$ and $\mathcal{H}_B$, since U and V are unitary operators.

If $d_A \neq d_B$ then we can consider the smaller Hilbert space as a subspace in a Hilbert space with same dimension as the larger Hilbert space. The rank of the matrix C is bounded by the dimension of the smaller Hilbert space. Since the rank of the diagonal matrix D is the same as that of C, it follows that D can have at most $d = \min\{d_A, d_B\}$ nonzero elements.

The fact that the vector ψ is entangled if and only if $r_s(\psi) \geq 2$ follows from our discussion prior to the statement of the theorem. $\square$

Given the Schmidt decomposition, we can immediately see some useful consequences. Suppose that the compound system consisting of subsystems A and B is in a pure state P_ψ. We write ψ in its Schmidt decomposition $\psi = \sum_j \sqrt{\lambda_j}e_j \otimes f_j$, and the reduced states of the subsystems are then given by

$$\varrho_A = \sum_j \lambda_j|e_j\rangle\langle e_j|, \qquad \varrho_B = \sum_j \lambda_j|f_j\rangle\langle f_j|.$$

This shows that the eigenvalues of ϱ_A and ϱ_B are the same except that the multiplicity of the eigenvalue 0 can differ. Thus, if the composite system is in a pure state, the reduced states of the subsystems have the same purity and entropy. Moreover, the Schmidt rank $r_s(\psi)$ of ψ equals the number of pure states in canonical convex decompositions of the reduced states ϱ_A and ϱ_B.

In the following we give two examples of entangled vectors, which are both related to specific choices of orthonormal bases for the tensor product space $\mathcal{H}\otimes\mathcal{H}$. In the first example we discuss the *exchange symmetry* of vectors on $\mathcal{H} \otimes \mathcal{H}$. In the second example we discuss the existence of an orthonormal basis consisting of entangled states only. The properties of this latter basis play a crucial role in the so-called *superdense coding protocol*, which doubles the rate of the (classical) one-time pad communication (Example 6.7).

Example 6.4 (*Symmetric and antisymmetric subspaces*)
Consider a bipartite system $\mathcal{H} \otimes \mathcal{H}$ and let $\varphi_1, \varphi_2 \dots$ form an orthonormal basis of $\mathcal{H}$. In Example 4.15 we defined the swap operator V_{SWAP} associated with the

exchange (relabelling) of the systems. The operator V_{SWAP} has two eigenvalues ± 1 (see Exercise 4.16). Depending on its behavior under the exchange transformation a vector $\psi \in \mathcal{H} \otimes \mathcal{H}$ belongs in one of three categories.

- It is an eigenvector of V_{SWAP} with eigenvalue 1; then we call it *symmetric*.
- It is an eigenvector of V_{SWAP} with eigenvalue -1; then we call it *antisymmetric*.
- It is not an eigenvector of V_{SWAP}.

Recall that eigenvectors corresponding to the same eigenvalue form a vector subspace (Exercise 1.25). The symmetric vectors form the *symmetric subspace* $\mathcal{H}_+$ of $\mathcal{H} \otimes \mathcal{H}$. Similarly, the antisymmetric vectors, form the *antisymmetric subspace* $\mathcal{H}_-$.

To find the dimensions of the subspaces $\mathcal{H}_+$ and $\mathcal{H}_-$, we construct a specific orthonormal basis for $\mathcal{H} \otimes \mathcal{H}$. First, we choose an orthonormal basis $\{\varphi_j\}_{j=1}^d$ for $\mathcal{H}$. For any pair of indices $j, k = 1, \ldots, d$ such that $j < k$, we define (unnormalized) vectors

$$\tilde{\psi}_{j\pm k} := \varphi_j \otimes \varphi_k \pm \varphi_k \otimes \varphi_j \in \mathcal{H} \otimes \mathcal{H}. \tag{6.3}$$

It follows from the discussion after Example 6.2 that all these vectors are entangled. For each $j = 1, \ldots, d$, we further define

$$\tilde{\psi}_j := \varphi_j \otimes \varphi_j \in \mathcal{H} \otimes \mathcal{H}. \tag{6.4}$$

It is straightforward to verify that all the vectors $\tilde{\psi}_{j\pm k}$, $\tilde{\psi}_j$ are mutually orthogonal. Therefore, we have d^2 nonzero mutually orthogonal vectors altogether, and d^2 is the dimension of $\mathcal{H} \otimes \mathcal{H}$. Thus, after normalization these vectors form an orthonormal basis of $\mathcal{H} \otimes \mathcal{H}$.

We now observe that $\tilde{\psi}_{j+k} \in \mathcal{H}_+$, $\tilde{\psi}_j \in \mathcal{H}_+$ and $\tilde{\psi}_{j-k} \in \mathcal{H}_-$. As a result, we conclude that the dimensions of the symmetric and antisymmetric subspaces equal $d_\pm \equiv \dim \mathcal{H}_\pm = \frac{1}{2}d(d \pm 1)$, respectively. In this particular orthonormal basis of $\mathcal{H} \otimes \mathcal{H}$, there are d factorized vectors and $d^2 - d$ entangled vectors. △

Example 6.5 (*Bell basis*)
Fix an orthonormal basis $\{\varphi_j\}$ for a d-dimensional Hilbert space $\mathcal{H}_d$ and consider an entangled vector $\psi_+ = \frac{1}{\sqrt{d}} \sum_{j=0}^{d-1} \varphi_j \otimes \varphi_j$ defined on the bipartite system $\mathcal{H}_d \otimes \mathcal{H}_d$. We define vectors $\psi_k = (I \otimes U_k)\psi_+$, where the U_k are unitary operators on $\mathcal{H}_d$. Owing to the Choi–Jamiolkowski isomorphism between the vectors in $\mathcal{H}_d \otimes \mathcal{H}_d$ and the operators on $\mathcal{H}_d$ (see subsection 4.4.3), we have the identity

$$\langle \psi_k | \psi_l \rangle = \frac{1}{d} \operatorname{tr}[U_k^* U_l]. \tag{6.5}$$

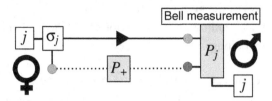

Figure 6.1 Superdense coding scheme transferring two bits of information per single use of qubit channel.

This formula implies that the vectors ψ_j are mutually orthogonal if and only if the corresponding unitary operators are orthogonal with respect to the Hilbert–Schmidt inner product. As we saw in Example 1.64, there exists an orthogonal operator basis composed of unitary operators only. Consequently, for suitable choices of unitary operators U_j the unit vectors $\psi_k = (I \otimes U_k)\psi_+$ are mutually orthogonal and form an orthonormal basis of $\mathcal{H}_d \otimes \mathcal{H}_d$. This kind of orthonormal basis is called a *Bell basis*. An observable associated with a Bell basis is called a *Bell observable*. △

Exercise 6.6 Consider a composite system consisting of a pair of qubits, i.e. $\mathcal{H} = \mathbb{C}^2 \otimes \mathbb{C}^2$. Verify by an explicit computation that the vectors $(\sigma_j \otimes I)\psi_+ \in \mathcal{H} \otimes \mathcal{H}$ are mutually orthogonal when the operators σ_j, $j = 0, 1, 2, 3$, are the Pauli operators defined in Example 2.21.

Example 6.7 (*Superdense coding*)
The success of the one-time pad protocol introduced in subsection 5.2.4 relies on the existence of a pre-shared cryptographic key represented by a maximally correlated state of two classical bits. In this example we will ask what would happen if the cryptographic key were quantum (see Figure 6.1). Could we increase the transmission rate and still preserve the security of the key?

Suppose that two spatially separated partners, Alice and Bob, share as the cryptographic key a pair of qubits and that the compound state is $P_+ = |\psi_+\rangle\langle\psi_+|$, where $\psi_+ = \frac{1}{\sqrt{2}}(\varphi \otimes \varphi + \varphi_\perp \otimes \varphi_\perp)$ and $\{\varphi, \varphi_\perp\}$ is an orthonormal basis for $\mathbb{C}^2$. Moreover, Alice and Bob are connected by an ideal qubit channel. Alice's encryption consists of the application of a particular channel to her qubit associated with the information she wants to submit. Bob, after receiving Alice's qubit, performs a two-qubit measurement in order to determine which channel (information) was chosen by Alice.

How much information can be securely transferred from Alice to Bob per single use of the given resources? It turns out that, following the so-called *superdense coding* protocol originally introduced by Bennett and Wiesner [16], they can communicate two bits of information per single use of the channel and the state P_+.

Moreover, the communication will be perfectly secure providing that the resources are certified. i.e. the maximally entangled state, the unitary channels applied and Bob's measurement are performing as claimed.

The success of superdense coding is based on the existence and properties of a Bell basis. The elements of a Bell basis are related by local unitary channels, i.e. $P_j = (\sigma_j \otimes I) P_+ (\sigma_j \otimes I) = |\psi_j\rangle\langle\psi_j|$, where $\sigma_0 = I$, $\sigma_1 = \sigma_x$, $\sigma_2 = \sigma_y$, $\sigma_3 = \sigma_z$ are the Pauli operators. We then have $\mathrm{tr}[P_j P_k] = \delta_{jk}$, and Alice can choose one of these four unitary channels to generate one of the pure states P_j. If Bob receives her system, he can identify perfectly which channel has been chosen by Alice by performing the corresponding Bell measurement. Since Alice has four possibilities she can communicate at most two bits of information to Bob.

The individual qubits of Alice and Bob are described by the total mixture, i.e., the reduced states are $\varrho_A = \varrho_B = \frac{1}{2}I$. The encrypting unitary channels do not change this feature. Consequently, the qubit communicated from Alice to Bob is independent of the information sent; thus it does not contain any information. In summary, the provided protocol of superdense coding not only doubles the transmission rate of the one-time pad with a classical key but also preserves the same degree of secrecy. $\triangle$

The following exercise is related to the existence of a superdense coding scheme for general quantum keys.

Exercise 6.8 Suppose Alice and Bob share a two-qubit quantum key described by a pure state P_ϕ. Find a maximal number of encrypting channels $\mathcal{E}_j$ such that the related states $\omega_j = (\mathcal{E}_j \otimes I)(P_\phi)$ are mutually perfectly distinguishable. [Hint: If $\omega P_\phi = O$ then also $P_\psi P_\phi = O$ for all $\psi \in \mathrm{supp}(\omega)$. Thus, we may assume that the ω_j are pure and the $\mathcal{E}_j$ are unitary channels. Use a Schmidt decomposition of ϕ and ψ to calculate how many orthogonal vector states can be produced.]

6.1.2 Entangled positive operators

Definition 6.9 Let $T \in \mathcal{L}_+(\mathcal{H}_A \otimes \mathcal{H}_B)$ be a positive operator. It is called

1. *factorized* if $T = T_A \otimes T_B$ for some $T_A \in \mathcal{L}_+(\mathcal{H}_A)$, $T_B \in \mathcal{L}_+(\mathcal{H}_B)$;
2. *separable* if T is a convex combination of factorized operators;
3. *entangled* if T is not separable.

Exercise 6.10 Show that a positive operator $T \in \mathcal{L}_+(\mathcal{H}_A \otimes \mathcal{H}_B)$ is separable if and only if $T = \sum_j T_j^A \otimes T_j^B$ for some positive operators $T_A \in \mathcal{L}_+(\mathcal{H}_A)$, $T_B \in \mathcal{L}_+(\mathcal{H}_B)$.

If we replace $\mathcal{L}_+(\mathcal{H})$ in the previous definition by $\mathcal{S}(\mathcal{H})$ (in the case of states) or by $\mathcal{E}(\mathcal{H})$ (in the case of effects) we get a classification of states or effects, respectively. Below we make the definition explicit in the case of states, which will be our main interest. The precise distinction between separable and entangled states is due to Werner [141]. The physical meaning of this division will become clearer in Section 6.2.

Definition 6.11 A state $\varrho \in \mathcal{S}(\mathcal{H}_A \otimes \mathcal{H}_B)$ is called

1. *factorized* if $\varrho = \varrho_A \otimes \varrho_B$; we denote by $\mathcal{S}^{\text{fac}}(\mathcal{H}_A \otimes \mathcal{H}_B)$ the set of factorized states;
2. *separable* if ϱ is a convex combination of factorized states; we denote by $\mathcal{S}^{\text{sep}}(\mathcal{H}_A \otimes \mathcal{H}_B)$ the set of all separable operators;
3. *entangled* if ϱ is not separable.

As we know, pure states are one-dimensional projections and hence associated with vectors. A pure state $|\eta\rangle\langle\eta| \in \mathcal{S}(\mathcal{H}_A \otimes \mathcal{H}_B)$, where $\eta \in \mathcal{H}_A \otimes \mathcal{H}_B$, is factorized if and only if η is a factorized vector. Since pure states are separable only if they are factorized, we can also draw the conclusion that the pure state $|\eta\rangle\langle\eta|$ is entangled if and only if the vector η is entangled. In this perspective, the next example may be surprising.

Example 6.12 (*Convex sum of entangled vectors*)
Let $\varphi_1, \varphi_2 \in \mathcal{H}$ be two orthogonal unit vectors and let α, β be two nonzero complex numbers such that $|\alpha|^2 + |\beta|^2 = 1$. We recall from Example 6.2 that the unit vectors

$$\eta_\pm = \alpha\varphi_1 \otimes \varphi_1 + \beta\varphi_2 \otimes \varphi_2 \in \mathcal{H} \otimes \mathcal{H}$$

are entangled. Consider the mixed state $T = \frac{1}{2}|\eta_+\rangle\langle\eta_+| + \frac{1}{2}|\eta_-\rangle\langle\eta_-|$. Although the vectors $\eta_\pm$ are entangled, the operator T is separable. Namely,

$$T = |\alpha|^2|\varphi_1 \otimes \varphi_1\rangle\langle\varphi_1 \otimes \varphi_1| + |\beta|^2|\varphi_2 \otimes \varphi_2\rangle\langle\varphi_2 \otimes \varphi_2| .$$

With this in mind, it may be found surprising by the reader that the states $T_\lambda = \lambda|\eta_+\rangle\langle\eta_+| + (1-\lambda)|\eta_-\rangle\langle\eta_-|$, $0 < \lambda < 1$, are entangled unless $\lambda = \frac{1}{2}$. This will be shown in subsection 6.3.4 (see Exercise 6.55), after we have learned some methods of identifying entanglement.   △

We learned in subsection 6.1.1 that to check whether a vector $\psi \in \mathcal{H} \otimes \mathcal{H}$ is entangled we need only write down its Schmidt decomposition. With this method we can therefore decide whether a pure state is entangled or separable. What about mixed states? Is there a similar test for them?

In the next example we demonstrate that there are separable states containing only entangled eigenvectors in their support. This indicates that canonical convex decompositions do not give any hint on the separability of mixed states.

Example 6.13 (*Separable state with entangled spectral decomposition*)
Let us consider the separable state

$$\omega = \tfrac{1}{2}(|\varphi_+ \otimes \varphi_+\rangle\langle\varphi_+ \otimes \varphi_+| + |\varphi \otimes \varphi\rangle\langle\varphi \otimes \varphi|) \tag{6.6}$$

defined on $\mathcal{H}_2 \otimes \mathcal{H}_2$, where $\varphi_+ = \tfrac{1}{\sqrt{2}}(\varphi + \varphi_\perp)$ and $\varphi, \varphi_\perp$ are orthogonal unit vectors. Let us assume that a factorized vector $\psi \otimes \psi'$ is a normalized eigenvector of ω associated with a nonzero eigenvalue μ, i.e. $\omega(\psi \otimes \psi') = \mu\psi \otimes \psi'$. We want to show that this leads to a contradiction.

First, direct derivation gives

$$\omega(\psi \otimes \psi') = \tfrac{1}{2}(x\varphi_+ \otimes \varphi_+ + y\varphi \otimes \varphi),$$

where $x = \langle\varphi_+|\psi\rangle\langle\varphi_+|\psi'\rangle$ and $y = \langle\varphi|\psi\rangle\langle\varphi|\psi'\rangle$. Our assumption is thus equivalent to the identity

$$x\varphi_+ \otimes \varphi_+ + y\varphi \otimes \varphi = 2\mu\psi \otimes \psi'.$$

Setting $\psi = c\varphi + d\varphi_\perp$ and $\psi' = c'\varphi + d'\varphi_\perp$ we obtain

$$2\mu cc' = y + \frac{x}{2}; \quad cd' = dc' = dd' = \frac{x}{4\mu}. \tag{6.7}$$

If $x \neq 0$ then necessarily $c, d, c', d' \neq 0$ and we get $c' = d' = x/(4d\mu)$ and also $d' = x/(4c\mu)$. Consequently, $y + \tfrac{1}{2}x = 2\mu cd' = \tfrac{1}{2}x$ implies that $y = 0$. However, by definition $y = cc'$, which is in contradiction with the assumption $x \neq 0$. Moreover, $x = 0$ requires that $d = d' = 0$; by normalization $c = c' = 1$, hence, $2\mu = y$ and $\psi = \psi' = \varphi$. However, this is in contradiction with the assumption $x = 0$, because $x = \langle\varphi_+|\varphi\rangle\langle\varphi_+|\varphi\rangle = \tfrac{1}{2}$. In summary, the conditions in (6.7) cannot be satisfied and we can conclude that the eigenvectors (associated with nonvanishing eigenvalues) of the operator ω are entangled. $\triangle$

By its definition the set of separable states is convex. The following proposition classifies its extremal elements.

Proposition 6.14 The extremal elements of $\mathcal{S}^{\mathrm{sep}}(\mathcal{H}_A \otimes \mathcal{H}_B)$ are one-dimensional factorized projections, i.e. operators of the form $P = |\varphi \otimes \psi\rangle\langle\varphi \otimes \psi|$ for some unit vectors $\varphi \in \mathcal{H}_A$ and $\psi \in \mathcal{H}_B$.

Proof The extremality of one-dimensional factorized projections follows from their extremality in the set of all states $\mathcal{S}(\mathcal{H}_A \otimes \mathcal{H}_B)$ (see Proposition 2.11). To prove that they are the only extremal elements, consider a separable state $\varrho = \sum_j p_j \xi_j \otimes \zeta_j$, where the ξ_j are states on $\mathcal{H}_A$ and the ζ_j are states on $\mathcal{H}_B$. Writing the states ξ_j, ζ_j in canonical convex decompositions we obtain

$$\varrho = \sum_{j,k,l} \pi_{jkl} |\varphi_{j,k}\rangle\langle\varphi_{j,k}| \otimes |\psi_{j,l}\rangle\langle\psi_{j,l}|, \tag{6.8}$$

where $0 \leq \pi_{jkl} \leq 1$ and $\varphi_{j,k} \in \mathcal{H}_A$, $\psi_{j,l} \in \mathcal{H}_B$ are unit vectors. Since $\mathrm{tr}[\varrho] = 1$, we have $\sum_{j,k,l} \pi_{jkl} = 1$ and (6.8) is therefore a convex decomposition. Hence, any separable state can be expressed as a convex decomposition of one-dimensional factorized projections. □

The proof of Proposition 6.14 shows not only that extremal separable states are one-dimensional factorized projections but also that any separable state can be expressed as a mixture of one-dimensional factorized projections. It follows that the set of separable states $\mathcal{S}^{\mathrm{sep}}(\mathcal{H}_A \otimes \mathcal{H}_B)$ is a closed subset of the set of all states $\mathcal{S}(\mathcal{H}_A \otimes \mathcal{H}_B)$ in the trace norm topology.

6.1.3 Nonlocal channels

Let us recall some notation from Chapter 4. We denote by $\mathcal{O}$ the set of all quantum operations. A set of channels is a subset $\mathcal{O}_\mathrm{c} \subset \mathcal{O}$ consisting of all trace-preserving operations.

Definition 6.15 A channel $\mathcal{E} : \mathcal{T}(\mathcal{H}_A \otimes \mathcal{H}_B) \to \mathcal{T}(\mathcal{H}_A \otimes \mathcal{H}_B)$ is:

1. *local* or *factorized* if $\mathcal{E} = \mathcal{E}_A \otimes \mathcal{E}_B$ and $\mathcal{E}_A$, $\mathcal{E}_B$ are channels defined on the subsystems A and B, respectively; we denote by $\mathcal{O}_\mathrm{c}^{\mathrm{fac}}$ the set of all local channels and by $co(\mathcal{O}_\mathrm{c}^{\mathrm{fac}})$ the convex hull of local channels;
2. *nonlocal* if $\mathcal{E}$ is not local;
3. *separable* if $\mathcal{E} = \sum_j \mathcal{F}_j^A \otimes \mathcal{F}_j^B$, where $\mathcal{F}_j^A$, $\mathcal{F}_j^B$ are operations on $\mathcal{H}_A$ and $\mathcal{H}_B$, respectively; we denote by $\mathcal{O}_\mathrm{c}^{\mathrm{sep}}$ the set of separable channels.

The physical meaning of 'local' in this context is clear: there are individual actions on the subsystems. The definition of a separable channel is, however, quite mathematical and lacks a direct physical interpretation. Although the definitions of separable states and separable channels look similar, there is a small yet important difference. While positive operators can be normalized to become density operators, an analogous 'normalization' procedure transforming operations to channels is not possible. In particular, a separable channel may not be expressible as a convex sum of factorized channels. This also means that the set of separable channels $\mathcal{O}_\mathrm{c}^{\mathrm{sep}}$ is not a convex hull of the set of factorized channels. Let us notice that a channel $\mathcal{E}$ is separable if and only if it can be expressed via factorized Kraus operators, i.e.

$$\mathcal{E}(\varrho) = \sum_j K_j \otimes L_j \varrho K_j^* \otimes L_j^* \quad \text{and} \quad \sum_j K_j^* K_j \otimes L_j^* L_j = I \otimes I.$$

This can be seen as follows: suppose that $\mathcal{E} = \sum_j \mathcal{F}_j^A \otimes \mathcal{F}_j^B$ where $\mathcal{F}_j^A(\varrho) = \sum_r K_{jr} \varrho K_{jr}^*$, and $\mathcal{F}_j^B(\varrho) = \sum_{r'} L_{jr'} \varrho L_{jr'}^*$. Then $\mathcal{E}(\varrho) = \sum_{j,r,r'} K_{jr} \otimes L_{jr'} \varrho K_{jr}^* \otimes L_{jr'}^*$ which is of the above mentioned form.

How do we implement a channel $\mathcal{E}$ which is a convex sum of factorized channels on spatially separated subsystems A and B? The implementation of such a channel $\mathcal{E} = \sum_j q_j \mathcal{E}_j^A \otimes \mathcal{E}_j^B$ is easy if the experimentalists (Alice and Bob) are allowed to exchange classical information in order to 'share' a random source which gives rise to the probability distribution $\vec{q}$. For instance, we can imagine that Alice throws a dice and tells the result j to Bob. Then they both apply local channels corresponding to this result and hence obtain $\mathcal{E}_j^A \otimes \mathcal{E}_j^B$. In other words the classical communication helps to synchronize the local actions of Alice and Bob, so that the desired channel $\mathcal{E}$ is obtained. We conclude that any convex mixture of factorized channels can be implemented if classical communication between Alice and Bob is allowed.

We may also consider a somewhat more general protocol. Let us assume that Alice and Bob are performing local measurements and communicating their measurement outcomes. In this setting each can adjust his or her action according to the outcomes communicated by the second party. Moreover, communication and the subsequent conditioned local actions may be repeated many times. We will see that having this additional freedom allows us to implement some channels which are not mixtures of factorized channels. Could we even implement all channels by this method? Before answering this question, let us first formalize the idea in the following definition.

Definition 6.16 A channel $\mathcal{E} : \mathcal{T}(\mathcal{H}_A \otimes \mathcal{H}_B) \to \mathcal{T}(\mathcal{H}_A \otimes \mathcal{H}_B)$ is a local operations and classical communication (*LOCC*) channel if it can be implemented by a sequence of local actions (instruments and channels) and an exchange of classical communication. We denote by $\mathcal{O}_c^{\mathrm{LOCC}}$ the set of all LOCC channels.

To see established the mathematical form of an LOCC channel, let us assume that in the first round of the protocol Bob performs a measurement described by an instrument defined on a finite outcome space Ω_1 and consisting of operations $\mathcal{F}_{j_1}^B$. The sum $\sum_{j_1 \in \Omega_1} \mathcal{F}_{j_1}^B \equiv \mathcal{E}_1^B$ is a quantum channel. Bob communicates his observed measurement outcome j_1 to Alice. In the second round of the protocol Alice chooses a measurement described by an instrument defined on a finite outcome space Ω_2 and consisting of operations $\mathcal{F}_{j_2}^A$. Her choice may depend on the information j_1 received from Bob. Alice then communicates her measurement outcome j_2 back to Bob, who continues his action, and so on. After the last (nth) information exchange both Alice and Bob apply a local channel $\mathcal{E}_{n+1|j_n,\ldots,j_1}^A \otimes \mathcal{E}_{n+1|j_n,\ldots,j_1}^B$. The

realized *LOCC communication protocol* consists of n communication rounds, and
as a result Alice and Bob have jointly applied an LOCC channel:

$$\mathcal{E}_{\text{LOCC}} = \sum_{j_n,\dots,j_2,j_1} (\mathcal{E}^A_{n+1|j_n,\dots,j_1} \otimes \mathcal{E}^B_{n+1|j_n,\dots,j_1}) \cdots (\mathcal{F}^A_{j_2|j_1} \otimes \mathcal{F}^B_{j_1}), \qquad (6.9)$$

where $\sum_{j_1} \mathcal{F}^B_{j_1} = \mathcal{E}^B_1$, $\sum_{j_2} \mathcal{F}^A_{j_2|j_1} = \mathcal{E}^A_{2|j_1}$ etc. are channels.

In summary, we conclude that

$$\mathcal{O}^{\text{fac}}_{\text{c}} \subseteq co(\mathcal{O}^{\text{fac}}_{\text{c}}) \subseteq \mathcal{O}^{\text{LOCC}}_{\text{c}} \subseteq \mathcal{O}^{\text{sep}}_{\text{c}} \subseteq \mathcal{O}_{\text{c}}. \qquad (6.10)$$

The first two inclusions are obvious from the definitions. The third inclusion is also
clear since any LOCC channel is a sum of factorized operations.

Example 6.17 (*Unitary LOCC channels*)
A unitary channel σ_U on $\mathcal{H} = \mathcal{H}_A \otimes \mathcal{H}_B$ is separable if and only if it is factorized, in
which case the associated unitary operator is of the factorized form $U = U_A \otimes U_B$.
This follows directly from the 'trivial' ambiguity of the operator-sum decomposi-
tion of unitary channels. By Proposition 4.25, $\sigma_U(\cdot) = \sum_j A_j \cdot A_j^*$ with $A_j = c_j U$
and $\sum_j |c_j|^2 = 1$. That is, on the one hand, if $U \neq U_A \otimes U_B$ then the Kraus
operators $A_j = c_j U$ for σ_U are never factorized, which is in contradiction with the
property of being separable. On the other hand, if $U = U_A \otimes U_B$ then σ_U is separable. We conclude that for unitary channels the three classes, factorized, LOCC
and separable channels are the same.

An essential property of local unitary channels is that they map factorized states
into factorized states, i.e. $\xi_A \otimes \zeta_B \mapsto \xi'_A \otimes \zeta'_B$. This is, however, not a feature unique
to them: a unitary channel may have this property even if it is not factorized. An
innocent-looking SWAP gate V_{SWAP} (see Example 4.15) also maps factorized states
into factorized states since $V_{\text{SWAP}}(\xi_A \otimes \zeta_B) = \zeta_A \otimes \xi_B$. However, $V_{\text{SWAP}} \neq U_A \otimes V_B$
for any unitary operators U_A and V_B. Indeed, as we will see in Example 6.57 none
of the eigenvectors of V_{SWAP} associated with the eigenvalue -1 can be factorized.
One may ask whether there are other unitary channels mapping factorized states
into factorized states. It was shown by Busch [24] that unitary channels that do not
entangle at least some initially factorized states are either factorized unitaries, or
factorized unitaries composed with the SWAP gate.   △

It is instructive to look at a special class of LOCC channels. Suppose that communication between Alice and Bob is restricted in such a way that Alice can send
information to Bob, but not vice versa. In this case it is sufficient to consider
a single communication round from Alice to Bob. Alice makes a measurement
described by an instrument $j \mapsto \mathcal{F}^A_j$. After Bob receives the measurement outcome j, he applies a channel $\mathcal{E}^B_j$. In this way the resulting channel is a so-called
one-way LOCC channel, having the form

$$\mathcal{E}_{1-\mathrm{LOCC}} = \sum_j \mathcal{F}_j^A \otimes \mathcal{E}_j^B, \qquad (6.11)$$

where $\mathcal{E}_j^B$ are the channels and $\mathcal{F}_j^A$ the operations forming the instrument.

We can now use the explicit form of one-way LOCC channels to see that not all these channels are convex combinations of factorized channels. Let us make the counter-assumption that every one-way LOCC channel can be written as a convex sum of factorized channels, $\mathcal{E}_{1-\mathrm{LOCC}} = \sum_j p_j \tilde{\mathcal{E}}_j^A \otimes \tilde{\mathcal{E}}_j^B$. Then the action on the subsystems is described by the channels $\tilde{\mathcal{E}}^A = \sum_j p_j \tilde{\mathcal{E}}_j^A$ and $\tilde{\mathcal{E}}^B = \sum_j p_j \tilde{\mathcal{E}}_j^B$. Moreover, using the form (6.11) we see that the local action on the subsystem A is $\mathcal{E}^A = \sum_j \mathcal{F}_j^A$. These two descriptions should lead to the same result, and we should have $\mathcal{E}^A = \tilde{\mathcal{E}}^A$.

Consider a one-way LOCC channel such that $\mathcal{E}^A$ is an extremal channel but not an extremal operation. As an example we can choose the qubit amplitude-damping channel (Example 4.28). Extremality implies that $\tilde{\mathcal{E}}_j^A = \mathcal{E}^A$ and hence $\mathcal{E}_{1-\mathrm{LOCC}} = \mathcal{E}^A \otimes \mathcal{E}^B$ is necessarily factorized. However, the amplitude-damping channel under consideration, $\mathcal{E}^A$, can be expressed as a positive sum of simple operations (see Example 4.7) $\mathcal{A}_\theta^A(\cdot) = A_\theta \cdot A_\theta^*$ and $\mathcal{B}_\theta^A(\cdot) = B_\theta \cdot B_\theta^*$, where A_θ, B_θ are given in Example 4.28. We can write $\mathcal{E}_{1-\mathrm{LOCC}} = \mathcal{A}_\theta^A \otimes \mathcal{E}_0^B + \mathcal{B}_\theta^A \otimes \mathcal{E}_1^B$ for arbitrary local channels $\mathcal{E}_0^B, \mathcal{E}_1^B$. Suppose also that B is a qubit system, and set $\mathcal{E}_j^B(\cdot) = \sigma_j \cdot \sigma_j$ for $j = 0, 1$. As $A_\theta \varrho A_\theta^* \neq c(\varrho) B_\theta \varrho B_\theta^*$ for all ϱ, it follows that the operator $\mathcal{E}_{1-\mathrm{LOCC}}(\varrho \otimes \xi)$ is not factorized, which is in contradiction with the assumption that $\mathcal{E}_{1-\mathrm{LOCC}}$ is factorized. In conclusion, there is a one-way LOCC channel which cannot be expressed as a convex combination of factorized channels.

It follows that the set of LOCC channels, and thus also the set of separable channels, is strictly larger than the convex hull of all factorized channels. Our next question is, of course, whether all separable channels can be implemented by means of local operations and classical communication. The following example will be useful for the construction of a counter-example showing the existence of a separable but not LOCC channel.

Example 6.18 (*LOCC preparation of nonlocal factorized basis*)
Suppose that $\varphi_1, \varphi_2, \varphi_3$ form an orthonormal basis of $\mathcal{H}_3$. It is straightforward to verify that the following vectors form an orthonormal basis of $\mathcal{H}_3 \otimes \mathcal{H}_3$:

$$\begin{array}{cccccc} \varphi_1 \otimes \varphi_{1+2}, & \varphi_1 \otimes \varphi_{1-2}, & \varphi_3 \otimes \varphi_{2+3}, & \varphi_3 \otimes \varphi_{2-3}, & \varphi_2 \otimes \varphi_2, \\ \varphi_{2+3} \otimes \varphi_1, & \varphi_{2-3} \otimes \varphi_1, & \varphi_{1+2} \otimes \varphi_3, & \varphi_{1-2} \otimes \varphi_3, \end{array} \qquad (6.12)$$

where we have used the shorthand notation $\varphi_{j\pm k} = \frac{1}{\sqrt{2}}(\varphi_j \pm \varphi_k)$. A remarkable property of this orthonormal basis (also called a *nonlocal product basis*) is that all the basis vectors are factorized but the set of local vectors

$$\{\varphi_1, \varphi_2, \varphi_3, \varphi_{2\pm3}, \varphi_{1\pm2}\}$$

is overcomplete on $\mathcal{H}_3$, i.e. even if we drop a vector the set still generates $\mathcal{H}_3$.

A nonlocal factorized basis $\psi_{jk} \otimes \phi_{jk}$ specified in (6.12) defines a separable channel on $\mathcal{T}(\mathcal{H}_3 \otimes \mathcal{H}_3)$:

$$\mathcal{E}(\varrho) = \sum_{j,k} A_{jk}\varrho A_{jk}^*, \tag{6.13}$$

where $A_{jk} = |\psi_{jk} \otimes \phi_{jk}\rangle\langle\varphi_j \otimes \varphi_k|$. Specifically, this channel transforms a local factorized basis into a nonlocal one. Our question is whether it is LOCC.

Fortunately, the simplest idea works. Let Alice and Bob perform local orthogonal measurements associated with the orthonormal basis $\varphi_1, \varphi_2, \varphi_3$. Observing the outcomes j on Alice's side and k on Bob's side, the whole system transforms into

$$\varrho \mapsto (\mathcal{F}_j^A \otimes \mathcal{F}_k^B)(\varrho) = \varrho_{jk,jk}|\varphi_j \otimes \varphi_k\rangle\langle\varphi_j \otimes \varphi_k|, \tag{6.14}$$

where $\varrho_{jk,jk} = \langle\varphi_j \otimes \varphi_k|\varrho\varphi_j \otimes \varphi_k\rangle$. By exchanging information on their obtained measurement outcomes, Alice and Bob can apply local channels $\mathcal{E}_{jk}^A \otimes \mathcal{E}_{jk}^B$ depending on both outcomes j, k:

$$\mathcal{E}_{jk}^A : |\varphi_j\rangle\langle\varphi_j| \mapsto |\psi_{jk}\rangle\langle\psi_{jk}|, \qquad \mathcal{E}_{jk}^B : |\varphi_k\rangle\langle\varphi_k| \mapsto |\phi_{jk}\rangle\langle\phi_{jk}|.$$

As a result, they obtain the desired state

$$\varrho' = \sum_{j,k} \varrho_{jk,jk}|\psi_{jk} \otimes \phi_{jk}\rangle\langle\psi_{jk} \otimes \phi_{jk}|. \tag{6.15}$$

This proves that the separable channel defined in (6.13) is LOCC. △

The proof of the following statement is based on an example which was originally reported by Bennett *et al.* [18] and named by these authors *nonlocality without entanglement*.

Proposition 6.19 There exists a separable channel which is not LOCC.

Proof Consider the adjoint channel (but now understood as a channel in the Schrödinger picture) of that introduced in Example 6.18, i.e.

$$\mathcal{E}^*(\varrho) = \sum_{j,k} B_{jk}\varrho B_{jk}^*, \tag{6.16}$$

with $B_{jk} = A_{jk}^* = |\varphi_j \otimes \varphi_k\rangle\langle\psi_{jk} \otimes \phi_{jk}|$. It turns out that this adjoint channel $\mathcal{E}^*$ is not LOCC although it is separable. As we will see, the crucial difference between the channels $\mathcal{E}$ and $\mathcal{E}^*$ is that the Lüders measurement in the basis $\psi_{jk} \otimes \phi_{jk}$ cannot be implemented by LOCC. Therefore, the LOCC procedure exploited for the implementation of $\mathcal{E}$ fails for $\mathcal{E}^*$. Let us stress that if $\mathcal{E}^*$ were an LOCC channel then the vectors $\psi_{jk} \otimes \phi_{jk}$ could be discriminated perfectly by means of

LOCC actions, because the vectors $\varphi_j \otimes \varphi_k$ are locally orthogonal. Is it possible to discriminate between the vectors $\psi_{jk} \otimes \phi_{jk}$ in an LOCC manner?

Suppose we want to distinguish the vector states $\varphi_1 \otimes \varphi_{1+2}$ and $\varphi_1 \otimes \varphi_{1-2}$. Successful LOCC strategies must perform measurements of a unique (up to unitary freedom) sharp observable of subsystem B in the basis $\varphi_{1+2}, \varphi_{1-2}, \varphi_3$ and an arbitrary action on the subsystem A. Similarly, for vectors $\varphi_{1+2} \otimes \varphi_3, \varphi_{1-2} \otimes \varphi_3$ we need to perform measurements on a sharp observable on subsystem A in the basis $\varphi_{1+2}, \varphi_{1-2}, \varphi_3$ and we are free to choose an action on subsystem B. However, let us consider a pair of vectors $\varphi_2 \otimes \varphi_2, \varphi_1 \otimes \varphi_{1\pm2}$. Perfect distinguishability requires a sharp observable to be in the basis $\varphi_1, \varphi_2, \varphi_3$ on subsystem A, which is not compatible with the previous conditions, hence, we conclude that these five vectors cannot be LOCC perfectly discriminated. In conclusion, the separable channel $\mathcal{E}^*$ cannot be LOCC, which proves the proposition. $\qquad\square$

Example 6.18 and Proposition 6.19 illustrate a kind of *pointwise LOCC irreversibility*. While the preparation of vector states forming a nonlocal factorized basis from local factorized basis is LOCC, the reverse transformation is not. In a sense, if either Alice or Bob forgets which basis state he or she has actually prepared by an LOCC procedure, then he or she is no longer able to identify it perfectly by means of LOCC.

Let us note that in the classical theory all channels are LOCC. In fact, classical communication is nothing other than an exchange of classical systems. Suppose that we want to implement a classical transformation on a composite system, one part being in the possession of Alice and the second in the possession of Bob. Alice can send her system to Bob (by classical communication). Bob can apply an arbitrary desired operation locally on the joint system and send Alice's system, thus modified back to Alice. In this way any operation on a joint system can be accomplished. Thus, all classical operations are LOCC. Moreover, it is sufficient to use at most two communication rounds.

Summarizing the previous discussion we conclude that

$$co(\mathcal{O}_c^{\text{fac}}) \subsetneq \mathcal{O}_c^{\text{LOCC}} \subsetneq \mathcal{O}_c^{\text{sep}}. \qquad (6.17)$$

Perhaps the most interesting fact here is that $\mathcal{O}_c^{\text{LOCC}} \neq \mathcal{O}_c^{\text{sep}}$. In particular, we have learned that there are channels which are not LOCC, and this is already a quite intriguing observation. We end this section with a concrete example of an LOCC channel.

Example 6.20 (*Twirling channel*)
Suppose that both Alice and Bob apply the same unitary channel σ_U to their subsystems and that the choice of U is made in a random fashion. The randomness

is defined via the Haar measure dU on $\mathcal{U}(\mathcal{H})$, and the resulting map is therefore

$$S \mapsto T(S) = \int_{\mathcal{U}(\mathcal{H})} (U \otimes U)S(U^* \otimes U^*) \, dU, \tag{6.18}$$

which is a LOCC channel. The group invariance property of the Haar measure implies that $(U \otimes U)T(S)(U^* \otimes U^*) = T(S)$ for every $U \in \mathcal{U}(\mathcal{H})$ and hence the operator $T(S)$ commutes with all unitary operators of the type $U \otimes U$.

Since T is a linear map, it is completely determined by its action on selfadjoint operators (recall from the start of subsection 1.2.2 that every operator can be expressed as a linear combination of selfadjoint operators T_R and T_I). We also notice from (6.18) that $T(S)^* = T(S^*)$, hence $T(S)$ is selfadjoint whenever S is selfadjoint. If S is a selfadjoint operator on $\mathcal{H} \otimes \mathcal{H}$, we can write $T(S)$ in the spectral decomposition form $T(S) = \sum_j x_j P_j$, where the x_j are real numbers and the P_j are orthogonal projections. The commutativity of $T(S)$ with an operator $U \otimes U$ implies that each projection P_j commutes with $U \otimes U$ as well. That is, the subspaces $\mathcal{H}_j$ are invariant under the action of the operator $U \otimes U$.

Let us recall that the SWAP operator V_{SWAP} is defined as $V_{\text{SWAP}} \psi \otimes \varphi = \varphi \otimes \psi$. It is straightforward to verify that $V_{\text{SWAP}}(U \otimes U) = (U \otimes U)V_{\text{SWAP}}$ for all $U \in \mathcal{U}(\mathcal{H})$. It follows that the symmetric subspace $\mathcal{H}_+$ and the antisymmetric subspace $\mathcal{H}_-$ of $\mathcal{H} \otimes \mathcal{H}$, introduced in Example 6.4, are invariant under the action of the operators $U \otimes U$. These are the only subspaces of $\mathcal{H} \otimes \mathcal{H}$ which are invariant under the action of any unitary operators $U \otimes U$ (see the following exercise). We thus conclude that the only projections that commute with all unitary operators $U \otimes U$ are O, I and P_-, P_+. Let us note that P_+ and P_- are orthogonal projections and the $P_+ + P_- = I$.

As a result of the above discussion, the operator $T(S)$ is a linear combination of P_+ and P_-, i.e.

$$T(S) = a_+(S)P_+ + a_-(S)P_- \tag{6.19}$$

for some functionals a_+ and a_-. From formula (6.19) we obtain $\text{tr}[T(S)P_+] = d_+ a_+(S)$, but since P_+ commutes with all the operators $U \otimes U$ we also have $\text{tr}[T(S)P_+] = \text{tr}[T(SP_+)] = \text{tr}[SP_+]$. Therefore, $a_+(S) = \text{tr}[SP_+]/d_+$ and similarly $a_-(S) = \text{tr}[SP_-]/d_-$. In conclusion, the action of the twirling channel can be expressed by the formula

$$T(S) = \frac{\text{tr}[SP_+]}{d_+} P_+ + \frac{\text{tr}[SP_-]}{d_-} P_- . \tag{6.20}$$

We have derived this formula for selfadjoint operators, but by writing $S = S_R + i S_I$ we can see that it holds for all operators. △

Exercise 6.21 The basis vectors of the subspaces of the symmetric and antisymmetric vectors are given in Example 6.4. Find the explicit form of the unitary operators $U \otimes U$ transforming between the vectors in $\{\varphi_j \otimes \varphi_j\}_{j=1}^d$, $(\varphi_j \otimes \varphi_k + \frac{1}{\sqrt{2}}\varphi_k \otimes \varphi_j)_{j<k}$ and $\frac{1}{\sqrt{2}}(\varphi_j \otimes \varphi_k - \varphi_k \otimes \varphi_j)_{j<k}$, respectively. Moreover, verify that $\varphi_j \otimes \varphi_k + \varphi_k \otimes \varphi_j$ is a linear combination of $\varphi_{j+k} \otimes \varphi_{j+k}$ and $\varphi_{j-k} \otimes \varphi_{j-k}$, and where, as usual, $\varphi_{j\pm k} = \frac{1}{\sqrt{2}}(\varphi_j \pm \varphi_k)$. Conclude that there are no subsets of P_+, P_- that are invariant under the action of the unitary operators $U \otimes U$. [Hint: Observe that $\varphi_{j\pm k} \otimes \varphi_{j\pm k} = (U \otimes U)(\varphi_j \otimes \varphi_j)$.]

6.2 Entanglement and LOCC

Roughly speaking, entanglement is a property of quantum states of a composite quantum system that exhibits their 'quantumness', i.e. quantum features that are missing in classical composite systems.

In this section we will exploit the concept of local operations and classical communication (LOCC) channels to provide an operational intuition for entanglement. For any pair of quantum states, there exist channels transforming one of these states to the other, and vice versa. However, it could happen that for some pairs of states there is no LOCC channel relating them. Why are we interested in the existence of LOCC channels? The motivation is that LOCC channels can be implemented on spatially separated systems in the same manner as classical channels.

Interestingly, the existence of an LOCC channel in one direction does not guarantee the existence of an LOCC channel in the opposite direction. In this section we will use the semigroup structure of LOCC channels to define a partial ordering on the set of quantum states, and this will be interpreted as the entanglement ordering of states. This partial ordering allows us to say which states are more entangled than some others. Although not all states can be compared in this sense, the extreme cases of LOCC-smallest and LOCC-greatest states do exist.

6.2.1 LOCC ordering and separable states

Definition 6.22 A state ϱ is *LOCC smaller* than a state ξ, denoted $\varrho \leq_{\text{LOCC}} \xi$, if there exists an LOCC channel $\mathcal{E}$ mapping ξ into ϱ. We say that states ϱ and ξ are *LOCC equivalent* (denoted as $\varrho \sim_{\text{LOCC}} \xi$) if $\varrho \leq_{\text{LOCC}} \xi$ and $\xi \leq_{\text{LOCC}} \varrho$.

For each state $\varrho \in S(\mathcal{H}_A \otimes \mathcal{H}_B)$, we write

$$O_\varrho^{\text{LOCC}} = \{\varrho' \in S(\mathcal{H}_A \otimes \mathcal{H}_B) : \varrho' = \mathcal{E}(\varrho) \text{ for some } \mathcal{E} \in O_c^{\text{LOCC}}\}$$

and say that O_ϱ^{LOCC} is the *LOCC orbit* of ϱ. The ordering relation $\varrho \leq_{\text{LOCC}} \xi$ can than be restated as $O_\varrho^{\text{LOCC}} \subseteq O_\xi^{\text{LOCC}}$. Consequently, if ϱ and ξ are LOCC equivalent, this means that $O_\varrho^{\text{LOCC}} = O_\xi^{\text{LOCC}}$.

In the following proposition we formulate basic statements showing that LOCC ordering splits the state space into subsets of separable and entangled states.

Proposition 6.23 The following statements hold.

(a) The LOCC orbit of any separable state ω equals the set of separable states, i.e. $O_\omega^{\text{LOCC}} = S^{\text{sep}}(\mathcal{H}_A \otimes \mathcal{H}_B)$. Thus, all separable states are LOCC equivalent.

(b) $S^{\text{sep}}(\mathcal{H}_A \otimes \mathcal{H}_B) \subseteq O_\varrho^{\text{LOCC}}$ for any state ϱ.

(c) The set of separable states is the smallest LOCC equivalence class.

Proof (a): Consider a general separable state $\xi = \sum_j p_j |\psi_j \otimes \varphi_j\rangle\langle\psi_j \otimes \varphi_j|$. A factorized channel $\mathcal{A}_{\psi_j} \otimes \mathcal{A}_{\varphi_j}$, where $\mathcal{A}_\psi$ is the contraction into a pure state $|\psi\rangle\langle\psi|$, is LOCC and we have $(\mathcal{A}_{\psi_j} \otimes \mathcal{A}_{\varphi_j})(\omega) = |\psi_j \otimes \varphi_j\rangle\langle\psi_j \otimes \varphi_j|$. Since the set of LOCC channels is convex, it follows that $\mathcal{E}_{\text{LOCC}} = \sum_j p_j \mathcal{A}_{\psi_j} \otimes \mathcal{A}_{\varphi_j}$ is also a LOCC channel. Hence, for an arbitrary separable state ξ there exists an LOCC channel $\mathcal{E}_{\text{LOCC}}$ such that $\xi = \mathcal{E}_{\text{LOCC}}(\omega)$. Channels that are LOCC form a subset of separable channels and it is easy to verify that a separable channel applied to separable states can produce only separable states. Therefore, no entangled state belongs to the LOCC orbit of a separable state ω, and we have $O_\omega^{\text{LOCC}} = S^{\text{sep}}(\mathcal{H}_A \otimes \mathcal{H}_B)$.

(b): Starting with a general state $\varrho \in S(\mathcal{H}_A \otimes \mathcal{H}_B)$ we can apply the local channel $\mathcal{A}_\varphi \otimes \mathcal{A}_\varphi$, resulting in the transformation $\varrho \mapsto |\varphi \otimes \varphi\rangle\langle\varphi \otimes \varphi| \equiv \omega$, by which ϱ becomes a separable state. Now using (a) we obtain (b).

(c): This is a direct consequence of a combination of (a) and (b). $\qquad\square$

As a consequence of Proposition 6.23 we can formulate an operational definition of entangled and separable states.

Corollary 6.24 A state ϱ is entangled if and only if it cannot be prepared from any factorized state by means of an LOCC channel or, equivalently, if it does not belong to the LOCC equivalence class of separable states.

An interesting and surprising feature is that LOCC channels do not preserve LOCC ordering [87], [150], [151]. Even more surprisingly, no communication is needed to demonstrate this property.

Proposition 6.25 The LOCC ordering is not preserved under the action of local channels. In particular, there exist states $\varrho, \omega \in S(\mathcal{H}_A \otimes \mathcal{H}_B)$ and a channel $\mathcal{E}$ on $\mathcal{H}_A$ such that

$$\varrho \leq_{\text{LOCC}} \omega \quad \text{but} \quad (\mathcal{E} \otimes \mathcal{I})(\varrho) \npreceq_{\text{LOCC}} (\mathcal{E} \otimes \mathcal{I})(\omega). \tag{6.21}$$

Proof   Let $\mathcal{H}_3$ be a three-dimensional Hilbert space and $\{\varphi_1, \varphi_2, \varphi_3\}$ a correspond-ing orthonormal basis. We define two pure entangled states ψ_1, ψ_2 of a composite system $\mathcal{H}_3 \otimes \mathcal{H}_3$ using these basis vectors:

$$\psi_1 = \tfrac{1}{\sqrt{2}}(\varphi_2 \otimes \varphi_2 + \varphi_3 \otimes \varphi_3), \tag{6.22}$$

$$\psi_2 = \sqrt{q}\, \varphi_1 \otimes \varphi_1 + \sqrt{1-q}\, \varphi_2 \otimes \varphi_2 . \tag{6.23}$$

Consider a local channel $\mathcal{E}^A$ defined using the Kraus operators $M_1 = |\varphi_1\rangle\langle\varphi_1| + |\varphi_2\rangle\langle\varphi_2|$ and $M_2 = |\varphi_3\rangle\langle\varphi_3|$. Applying this local channel to the above pure states we obtain the states

$$\varrho_1^{\text{out}} = \tfrac{1}{\sqrt{2}} \left(|\varphi_2 \otimes \varphi_2\rangle\langle\varphi_2 \otimes \varphi_2| + |\varphi_3 \otimes \varphi_3\rangle\langle\varphi_3 \otimes \varphi_3| \right), \tag{6.24}$$

$$\varrho_2^{\text{out}} = |\psi_2\rangle\langle\psi_2| . \tag{6.25}$$

First, by a local unitary channel we can transform ψ_1 into $\tilde{\psi}_2 = \tfrac{1}{\sqrt{2}}(\varphi_1 \otimes \varphi_1 + \varphi_2 \otimes \varphi_2)$, i.e. they are LOCC equivalent. We can think of the vectors $\tilde{\psi}_2, \psi_2$ as elements of $\mathcal{H}_2 \otimes \mathcal{H}_2$ with $\mathcal{H}_2$ spanned by the vectors φ_1, φ_2. Then the vector $\tilde{\psi}_2$ is maximally entangled with respect to $\mathcal{H}_2 \otimes \mathcal{H}_2$ and, therefore, there exists an LOCC channel mapping $|\tilde{\psi}_2\rangle\langle\tilde{\psi}_2|$ into $|\psi_2\rangle\langle\psi_2|$; hence $|\psi_2\rangle\langle\psi_2| \leq_{\text{LOCC}} |\psi_1\rangle\langle\psi_1|$.

Second, if $|\psi_2\rangle\langle\psi_2| \mapsto |\psi_1\rangle\langle\psi_1|$ were implemented by some LOCC channel then $|\psi_2\rangle\langle\psi_2| \mapsto |\tilde{\psi}_2\rangle\langle\tilde{\psi}_2|$ would also be achieved by an LOCC channel, however, this would mean that $|\psi_2\rangle\langle\psi_2|$ is maximally entangled on $\mathcal{H}_2 \otimes \mathcal{H}_2$, which is not true unless $q = \tfrac{1}{\sqrt{2}}$. Therefore, the ordering relation is strict and $|\psi_2\rangle\langle\psi_2| <_{\text{LOCC}} |\psi_1\rangle\langle\psi_1|$.

Since ϱ_1^{out} is separable the transformation $\varrho_1^{\text{out}} \to \varrho_2^{\text{out}}$ cannot be an LOCC chan-nel, but the transformation $\varrho_2^{\text{out}} \to \varrho_1^{\text{out}}$ can still be implemented by an LOCC channel because the set of separable states is contained in the LOCC orbit of any state. That is, $\varrho_1^{\text{out}} <_{\text{LOCC}} \varrho_2^{\text{out}}$ although $|\psi_2\rangle\langle\psi_2| <_{\text{LOCC}} |\psi_1\rangle\langle\psi_1|$, which proves the proposition. $\qquad\square$

6.2.2 Maximally entangled states

We have seen that all separable states are LOCC equivalent and strictly LOCC smaller than any entangled state. In this subsection we investigate the other extreme – *maximal elements* of the LOCC ordered set of states. These are states that are LOCC-greater than any other state. As we will see, not only do there exist LOCC maximal elements but also an LOCC equivalence class of greatest elements. These states are LOCC greater than any other state. On the basis of this observation we may introduce the concept of *maximally entangled states*.

Definition 6.26 If a state ϱ belongs to the LOCC equivalence class of the LOCC-greatest element (if such an element exists!) then the state ϱ is called *maximally entangled*.

From this definition follows that the LOCC orbit of a maximally entangled state equals the whole state space: $O_\varrho^{\text{LOCC}} = \mathcal{S}(\mathcal{H}_A \otimes \mathcal{H}_B)$.

First we need to prove that maximally entangled states do exist. Let us define on $\mathcal{H}_A \otimes \mathcal{H}_B$ the vector state

$$\psi_+ = \frac{1}{\sqrt{d}} \sum_{j=0}^{d-1} \varphi_j \otimes \varphi_j', \tag{6.26}$$

where $\varphi_0, \ldots, \varphi_{d-1}$ is an orthonormal basis of $\mathcal{H}_A$ and $\varphi_0', \ldots, \varphi_{d-1}'$ is an orthonormal subset of $\mathcal{H}_B$. Here $d = \dim \mathcal{H}_A$, and the systems are ordered in such a way that $\dim \mathcal{H}_A \leq \dim \mathcal{H}_B$. We write

$$\mathcal{H}_{\text{max}} = \{\psi \in \mathcal{H}_A \otimes \mathcal{H}_B : \psi = (U \otimes V)\psi_+ \text{ for } U, V \text{ unitary}\}.$$

Proposition 6.27 If $\psi \in \mathcal{H}_{\text{max}}$ then P_ψ is a maximally entangled state.

Proof Since elements of $\mathcal{H}_{\text{max}}$ are locally unitarily equivalent, they are LOCC equivalent. Without loss of generality we can fix a state shared by two spatially separated partners, Alice and Bob to be $P_+ = |\psi_+\rangle\langle\psi_+|$. From the Choi isomorphism it follows directly that there exists an operation $\mathcal{A}_\omega$ such that $\omega = (\mathcal{I} \otimes \mathcal{A})[P_+]$ for all states $\omega \in \mathcal{S}(\mathcal{H}_A \otimes \mathcal{H}_B)$. Although this transformation $(\mathcal{I} \otimes \mathcal{A})$ is not physical, its action on P_+ can be simulated by an LOCC channel. Moreover, since any mixed quantum state can be written as a convex combination of pure states, it is sufficient to employ an LOCC channel transforming P_+ into $P_\phi = |\phi\rangle\langle\phi|$ for all vector states $\phi \in \mathcal{H}_A \otimes \mathcal{H}_B$. This LOCC channel transforming P_+ into $\omega = \sum_j q_j P_{\phi_j}$ is then implemented as a convex combination of LOCC channels transforming P_+ into P_{ϕ_j}.

According to the Schmidt decomposition, a general vector is locally unitarily equivalent (i.e. LOCC equivalent) to a state $\phi = \sum_j a_j \varphi_j \otimes \varphi_j'$. Assume that Alice and Bob share the state P_+. The Choi–Jamiolkowski isomorphism implies that $\phi = (I \otimes R_\phi)\psi_+$ with $R_\phi = \sqrt{d} \sum_j a_j |\varphi_j'\rangle\langle\varphi_j'|$. In the first step Alice adds a $(d_B = \dim\mathcal{H}_B)$-dimensional ancilla in a state $|\varphi_0'\rangle_{A'}\langle\varphi_0'|$ and applies an isometry transformation $G : \varphi_0' \otimes \varphi_j \to (U_j \otimes R_\phi)\psi_+$ to the systems A' and A, where the U_j are shift operators, i.e. $U_j \varphi_{j'}' = \varphi_{j' \oplus j}'$. Let us note that $j' = 0, \ldots, d_B$ and that we have added unit vectors $\varphi_d', \ldots, \varphi_{d_B-1}'$ to complete the basis on $\mathcal{H}_{A'} = \mathcal{H}_B$. After Alice's action the composite system is described by the vector state

$$\psi_{A'AB} = \sum_{j,j'} a_{j'} \varphi'_{j \oplus j'} \otimes \varphi_{j'} \otimes \varphi'_j = \sum_k \varphi'_k \otimes \left(\sum_{j,j':j \oplus j'=k} a_{j'} \varphi_{j'} \otimes \varphi'_j \right).$$

In the second step Alice performs a projective measurement, of the ancillary system A', described by the Lüder's instrument $k \mapsto \mathcal{I}_k(\varrho) = \langle \varphi'_k | \varrho | \varphi'_k \rangle |\varphi'_k\rangle \langle \varphi'_k|$; hence if she measures an outcome k the composite system ends up in the vector state

$$\psi^{(k)}_{A'AB} = \varphi'_k \otimes \sum_{j,j':j \oplus j'=k} a_{j'} \varphi_{j'} \otimes \varphi'_j. \tag{6.27}$$

In the third step Alice submits the outcome k that she found to Bob. Let us note that in this step the ancillary system becomes factorized from the systems A and B and does not play any further role. In the fourth step Bob applies a unitary channel given by the unitary operators $U^*_k : \varphi'_{j'} \mapsto \varphi'_{j' \ominus k}$ (the inverse of the shifts applied by Alice) to obtain the desired vector state

$$\psi^{(k)}_{AB} = \sum_{j,j':j \oplus j'=k} a_{j'} \varphi_{j'} \otimes \varphi'_{j \ominus k} = \sum_{j'} a_{j'} \varphi_{j'} \otimes \varphi'_{j'} = \phi_{AB}. \tag{6.28}$$

In summary, the state P_+ is LOCC-greater than any pure state and, consequently, LOCC-greater than any mixed state ω; thus $O^{\mathrm{LOCC}}_{P_+} = \mathcal{S}(\mathcal{H}_A \otimes \mathcal{H}_B)$, which proves the proposition. $\qquad\square$

Lemma 6.28 Suppose that $d \equiv \dim \mathcal{H}_A \leq \dim \mathcal{H}_B$. For an operator $A : \mathcal{H}_A \to \mathcal{H}_A$ define an operator $A^T : \mathcal{H}_B \to \mathcal{H}_B$ via the following identities:

$$\text{for } j < d, \qquad A^T \varphi'_j = \sum_{k=0}^{d-1} \langle \varphi_j | A \varphi_k \rangle \varphi'_k;$$

$$\text{for } j \geq d \qquad A^T \varphi'_j = \varphi'_j.$$

Then we have the identity

$$(A \otimes I)\psi_+ = (I \otimes A^T)\psi_+. \tag{6.29}$$

Proof A direct derivation gives

$$(A \otimes I)\psi_+ = \frac{1}{\sqrt{d}} \sum_{j=0}^{d-1} A\varphi_j \otimes \varphi'_j = \frac{1}{\sqrt{d}} \sum_{j,k=0}^{d-1} \langle \varphi_k | A\varphi_j \rangle \varphi_k \otimes \varphi'_j$$

$$= \frac{1}{\sqrt{d}} \sum_{k=0}^{d-1} \varphi_k \otimes A^T \varphi'_k.$$

$\qquad\square$

Using the above lemma we can simplify the definition of the set $\mathcal{H}_{\max}$. In particular, the lemma implies that $(U \otimes V)\psi_+ = (I \otimes U^T V)\psi_+$, where $U^T V$ defines a
unitary operator on $\mathcal{H}_B$. Consequently,

$$\mathcal{H}_{\max} = \{\psi \in \mathcal{H}_A \otimes \mathcal{H}_B : \psi = (I \otimes U)\psi_+ \text{ for } U \in \mathcal{U}(\mathcal{H}_B)\}.$$

The remaining question is whether there are maximally entangled states other than
the vector states belonging to $\mathcal{H}_{\max}$. Clearly, if a state ω is maximally entangled
then there must exist an LOCC channel transforming ω into ψ_+. In what follows
we will show that no other pure state is maximally entangled, but there exist mixed
maximally entangled states if the dimensions of $\mathcal{H}_A$ and $\mathcal{H}_B$ are different.

Proposition 6.29 If $\phi \notin \mathcal{H}_{\max}$ then P_ϕ is not maximally entangled.

Proof Consider a pure state $P_\phi = |\phi\rangle\langle\phi|$ for some $\phi \in \mathcal{H}_A \otimes \mathcal{H}_B$ for which
$\phi \notin \mathcal{H}_{\max}$; that is, in Schmidt form $\phi = \sum_j \sqrt{\lambda_k}\varphi_k \otimes \varphi'_k$, where the numbers
λ_k form the Schmidt vector $\vec{\lambda}_\phi$. A potential LOCC channel mapping ϕ into $\psi_+ =$
$\frac{1}{\sqrt{d}}\sum_k \varphi_k \otimes \varphi'_k$ is separable and thus can be written in the form

$$\sum_j |(A_j \otimes B_j)\phi\rangle\langle(A_j \otimes B_j)\phi| = |\psi_+\rangle\langle\psi_+|, \qquad (6.30)$$

with $\sum_j A_j^* A_j \otimes B_j^* B_j = I$. Since P_+ is a rank-1 operator, it follows that $(A_j \otimes
B_j)\phi = \alpha_j \psi_+$ for some complex numbers α_j, $|\alpha_j| \leq 1$. Writing the vector ϕ in
Schmidt form this condition reads

$$\sum_k \sqrt{\lambda_k} A_j \varphi_k \otimes B_j \varphi'_k = \frac{\alpha_j}{\sqrt{d}} \sum_k \varphi_k \otimes \varphi'_k. \qquad (6.31)$$

The uniqueness of the expression for a vector in a fixed basis requires that
$\sqrt{\lambda_k}(A_j \otimes B_j)(\varphi_k \otimes \varphi'_k) = \frac{1}{\sqrt{d}}\alpha_j(\varphi_k \otimes \varphi'_k)$, which implies that

$$\langle(A_j \otimes B_j)(\varphi_k \otimes \varphi'_k)|(A_j \otimes B_j)(\varphi_k \otimes \varphi'_k)\rangle = \frac{|\alpha_j|^2}{d\lambda_k}.$$

Using $\sum_j |\alpha_j|^2 = 1$ we obtain

$$\sum_j \langle(A_j \otimes B_j)(\varphi_k \otimes \varphi'_k)|(A_j \otimes B_j)(\varphi_k \otimes \varphi'_k)\rangle = \frac{1}{d\lambda_k}.$$

Owing to the normalization condition $\sum_k \lambda_k = 1$ there necessarily exist $\lambda_k < 1/d$,
and in such a case the right-hand side is strictly larger than 1. However, since
$\sum_j A_j^* A_j \otimes B_j^* B_j = I$ it follows that

$$\sum_j \langle(A_j \otimes B_j)(\varphi_k \otimes \varphi'_k)|(A_j \otimes B_j)(\varphi_k \otimes \varphi'_k)\rangle = 1,$$

independently of k, which contradicts the previous conclusion. Therefore, our assumption of the existence of a separable channel transforming P_ϕ into P_+ is wrong, which means that maximally entangled vector states from $\mathcal{H}_{\max}$ do not belong to the orbit of P_ϕ, and this proves the proposition. □

Exercise 6.30 Show that even if there exist LOCC channels $\mathcal{E}_1^{\text{LOCC}}$, $\mathcal{E}_2^{\text{LOCC}}$ mapping states ω_1, ω_2 into ϱ, respectively (which would mean that $\varrho \in O_{\omega_1}^{\text{LOCC}} \cap O_{\omega_2}^{\text{LOCC}}$), it may happen that there is no LOCC channel mapping their convex combination into ϱ, i.e. $\varrho \notin O_{q\omega_1+(1-q)\omega_2}^{\text{LOCC}}$ for some value q. [Hint: See Example 6.12.]

Lemma 6.31 If $P_\phi \leq_{\text{LOCC}} \omega$ for $\omega \in \mathcal{S}(\mathcal{H}_A \otimes \mathcal{H}_B)$, $\phi \in \mathcal{H}_A \otimes \mathcal{H}_B$ and $\omega = q P_\psi + (1-q)\omega'$ with $q > 0$ then $P_\phi \leq_{\text{LOCC}} P_\psi$.

Proof This statement follows from the following fact. If a channel $\mathcal{E}$ maps a mixture ω into a pure state P_ϕ then, owing to the convexity of the state space, this channel maps any pure state decomposing ω into the state P_ϕ. □

Combining this lemma with the previous proposition we can argue that, for equal-dimensional systems A and B the set $\mathcal{H}_{\max}$ contains all the maximally entangled states. No pure state outside $\mathcal{H}_{\max}$ is maximally entangled and therefore only mixtures of maximally entangled states remain as potential candidates for maximally entangled states. However, it must also hold that any given mixture of maximally entangled states cannot be written as a mixture that includes some nonmaximally entangled state. For a mixed state ω (a mixture of maximally entangled states) there are two cases. Either all its eigenvectors (corresponding to nonzero eigenvalues) are maximally entangled, or they are not. If they are not then by (an inversion of) Lemma 6.31 the state ω cannot be maximally entangled. In what follows we will provide an explicit example of a mixed maximally entangled state that is a convex combination of suitable maximally entangled vector states.

Example 6.32 (*Mixed maximally entangled state*)
Consider a system of three qubits. One qubit forms a system A and the remaining pair of qubits forms a system B, i.e. $\mathcal{H}_{AB} = \mathcal{H}_2 \otimes \mathcal{H}_4$. Consider a state ω that is a mixture of the orthogonal vector states

$$\psi_{AB_1} = \tfrac{1}{\sqrt{2}}(\varphi \otimes \varphi \otimes \varphi + \varphi_\perp \otimes \varphi_\perp \otimes \varphi);$$
$$\psi_{AB_2} = \tfrac{1}{\sqrt{2}}(\varphi_\perp \otimes \varphi \otimes \varphi_\perp + \varphi \otimes \varphi_\perp \otimes \varphi_\perp),$$

i.e.

$$\omega = q|\psi_{AB_1}\rangle\langle\psi_{AB_1}| + (1-q)|\psi_{AB_2}\rangle\langle\psi_{AB_2}|. \tag{6.32}$$

Both vectors are expressed in Schmidt form and are entangled with respect to the splitting $A|B$. Moreover, they belong to $\mathcal{H}_{\max}$ for $\mathcal{H}_2 \otimes \mathcal{H}_4$ and hence are maximally entangled. In the first step Bob implements a Lüders measurement

$$1 \mapsto (I \otimes Q_1)\omega(I \otimes Q_1); \quad 2 \mapsto (I \otimes Q_2)\omega(I \otimes Q_2), \qquad (6.33)$$

where $Q_1 = |\varphi \otimes \varphi\rangle\langle\varphi \otimes \varphi| + |\varphi_\perp \otimes \varphi\rangle\varphi_\perp \otimes \varphi|$ and $Q_2 = |\varphi \otimes \varphi_\perp\rangle\langle\varphi \otimes \varphi_\perp| + |\varphi_\perp \otimes \varphi_\perp\rangle\langle\varphi_\perp \otimes \varphi_\perp|$ are projections. If the state ω is inserted into this instrument, the joint system transforms into conditional states:

$$\omega \mapsto \omega_1 = |\psi_{AB_1}\rangle\langle\psi_{AB_1}|; \quad \omega \mapsto \omega_2 = |\psi_{AB_2}\rangle\langle\psi_{AB_2}|.$$

On observing outcome 1 (see (6.33)) Bob will do nothing, but if he records outcome 2 then he performs a unitary transformation $I \otimes U$ transforming ψ_{AB_2} into ψ_{AB_1}. In conclusion, Bob can locally implement a channel transforming ω into the maximally entangled state $|\psi_{AB_1}\rangle\langle\psi_{AB_1}|$; hence, the mixed state ω is maximally entangled as well. However, due to its mixedness it is obvious that $\omega \notin \mathcal{H}_{\max}$. △

This example shows that if $\dim \mathcal{H}_B \geq 2 \dim \mathcal{H}_A$, then there is enough room in the state space to accommodate *mixed maximally entangled states*. What about the case $\dim \mathcal{H}_B < 2 \dim \mathcal{H}_A$? Can we classify all maximally entangled states?

Proposition 6.33 A state $\omega \in \mathcal{S}(\mathcal{H}_A \otimes \mathcal{H}_B)$ is maximally entangled if and only if $\omega = \sum_{k=1}^{K} q_k P_k$, where $P_k = (I \otimes U_k)P_+(I \otimes U_k)$, $\sum_{k=1}^{K} q_k = 1$ and $\mathrm{tr}_A[P_k] \perp \mathrm{tr}_A[P_l]$ for $k \neq l$.

Proof Since $\frac{1}{d}Q_k = \mathrm{tr}_A[P_k] = U_k\mathrm{tr}_A[P_+]U_k^* = \frac{1}{d}U_k Q_0 U_k^*$, where $Q_0 = \sum_{j=0}^{d}|\varphi_j'\rangle\langle\varphi_j'|$ is a projection onto a d-dimensional subspace of $\mathcal{H}_B$, it follows that the Q_k are also d-dimensional projections onto subspaces of $\mathcal{H}_B$. The orthogonality condition $Q_k \perp Q_l$ for $k \neq l$ implies that $\dim \mathcal{H}_B \geq K \dim \mathcal{H}_A = Kd$. Using the same idea as in the LOCC protocol described in Example 6.32, we can transform an ω of the form given above into the state P_+. Thus, such mixed states are indeed maximally entangled; this proves the sufficiency of the form ω. The proof of necessity is formulated as the following exercise. □

Exercise 6.34 Complete the proof of Proposition 6.33. [Hint: Show that there are nonmaximally entangled eigenvectors in the spectral decomposition of ω with nonorthogonal $\mathrm{tr}_A[P_k]$. Then apply Lemma 6.31.]

The following property is a direct consequence of Proposition 6.33. If a state ω is maximally entangled then

$$\mathrm{tr}_B[\omega] = \frac{1}{d}I; \quad \text{and} \quad \mathrm{tr}_A[\omega] = \sum_j p_j Q_j,$$

where the Q_j are mutually orthogonal projections onto $\mathcal{H}_B$ of rank (at least) $\dim \mathcal{H}_A$. This proposition also implies that if $\dim \mathcal{H}_A \leq \dim \mathcal{H}_B < 2 \dim \mathcal{H}_A$ then the only maximally entangled states are elements of $\mathcal{H}_{\max}$; hence, in this case all the maximally entangled states are pure.

6.2.3 Majorization criterion for LOCC

Thus far, we have employed the concept of LOCC channels to define a partial ordering on the state space of a bipartite system, and we have identified classes of separable states and maximally entangled states. In general, the question on the existence of an LOCC channel connecting a given pair of states ω_1, ω_2 may be difficult to solve. A major achievement in this direction was due to Nielsen [102], who showed that for pure states the LOCC ordering problem can be resolved by majorization. We will state this result without a proof; the latter can be found either in the original paper [102] or in the textbook by Nielsen and Chuang [104].

Theorem 6.35 (*Nielsen's majorization theorem*)
Let $\psi, \phi \in \mathcal{H}_A \otimes \mathcal{H}_B$ be two unit vectors and let $\vec{\lambda}_\psi, \vec{\lambda}_\phi$ be the corresponding Schmidt vectors. Then $P_\phi \leq_{\mathrm{LOCC}} P_\psi$ if and only if $\vec{\lambda}_\phi$ majorizes $\vec{\lambda}_\psi$, which means that their components $\lambda_\psi(j), \lambda_\phi(j)$ satisfy

$$\sum_{j=1}^{n} \lambda_\psi(j) \leq \sum_{j=1}^{n} \lambda_\phi(j)$$

for all $n \leq d = \min\{d_A, d_B\}$.

This theorem gives a general method for deciding whether a pure state is LOCC-smaller than another pure state. A general characterization of an LOCC ordering that includes mixed states is still open.

Exercise 6.36 Let $\{\varphi_1, \varphi_2, \varphi_3, \varphi_4\}$ be an orthonormal basis for $\mathcal{H}_4$. Show that vector states $\psi, \phi \in \mathcal{H}_4 \otimes \mathcal{H}_4$ defined by

$$\psi = \sqrt{\tfrac{2}{5}}(\varphi_1 \otimes \varphi_1 + \varphi_2 \otimes \varphi_2) + \sqrt{\tfrac{1}{10}}(\varphi_3 \otimes \varphi_3 + \varphi_4 \otimes \varphi_4),$$

$$\phi = \sqrt{\tfrac{1}{2}}\varphi_1 \otimes \varphi_1 + \tfrac{1}{2}(\varphi_2 \otimes \varphi_2 + \varphi_3 \otimes \varphi_3),$$

are LOCC incomparable (this term means that neither $P_\psi \leq_{\mathrm{LOCC}} P_\phi$ nor $P_\phi \leq_{\mathrm{LOCC}} P_\psi$ holds).

The majorization theorem leads to some interesting observations on the nature of LOCC ordering. In what follows we ask how additional resources would change this relation.

Let ψ, ϕ be as in Exercise 6.36. It follows from their incomparability that Alice and Bob cannot transform the state P_ψ into P_ϕ just by means of LOCC channels. Suppose that they are given an additional resource in the form of a two-qubit entangled state P_η, where

$$\eta = \sqrt{\tfrac{3}{5}}\varphi \otimes \varphi + \sqrt{\tfrac{2}{5}}\varphi_\perp \otimes \varphi_\perp$$

and $\varphi, \varphi_\perp$ are orthogonal unit vectors. The Schmidt coefficients of the pure states $P_\psi \otimes P_\eta$ and $P_\phi \otimes P_\eta$ are

$$\vec{\lambda}_{\psi \otimes \eta} = \left(\tfrac{6}{25}, \tfrac{6}{25}, \tfrac{4}{25}, \tfrac{4}{25}, \tfrac{3}{50}, \tfrac{3}{50}, \tfrac{1}{25}, \tfrac{1}{25} \right),$$
$$\vec{\lambda}_{\phi \otimes \eta} = \left(\tfrac{3}{10}, \tfrac{2}{10}, \tfrac{3}{20}, \tfrac{3}{20}, \tfrac{1}{10}, \tfrac{1}{10}, \tfrac{0}{10}, \tfrac{0}{10} \right).$$

It is straightforward to verify that $\vec{\lambda}_{\phi \otimes \eta}$ majorizes $\vec{\lambda}_{\psi \otimes \eta}$. Therefore, by Nielsen's majorization theorem there exists an LOCC channel transforming $P_\psi \otimes P_\eta$ into $P_\phi \otimes P_\eta$, although P_ψ and P_ϕ are LOCC incomparable. In conclusion, the shared two-qubit state P_η plays the role of an LOCC *catalyst* and enables one to transform a specific pair of LOCC-incomparable states. This phenomenon is known as *entanglement catalysis* [86]. Notice that the crucial thing here is that the catalyst state P_η is not consumed but can be given back to its original owner after the LOCC channel has been implemented.

Exercise 6.37 One might expect that a maximally entangled pair would be the best catalyst. This is, however, not true. On the contrary, a maximally entangled pair cannot catalyze any transformation. Namely, prove that if two pure states P_ψ and P_ϕ are LOCC incomparable then $P_\psi \otimes P_\eta$ and $P_\phi \otimes P_\eta$ are also LOCC incomparable whenever P_η is a maximally entangled state. [Hint: The Schmidt vector of a maximally entangled state is uniform.]

6.3 Entanglement detection

The paradigm of LOCC channels divides the quantum state space into two disjoint subsets containing separable and entangled states, respectively. But how can one find out whether a given state is entangled or separable? That is the question addressed in this section.

Even if a complete mathematical description of a state is given, entanglement detection is not an easy task. The point is that it is not at all clear which parameters of the density operator are relevant in resolving this problem. Two main goals of the entanglement theory are on the one hand to find out how to capture the fingerprints of entanglement and on the other hand how to design suitable experimental tests.

6.3.1 Entanglement witnesses

Suppose that we have an unknown state ϱ of a composite system and that we want to know whether ϱ is entangled. To discover this property of ϱ, we perform a measurement of a real observable A. As discussed in subsection 3.1.4, we can then calculate the mean value $\langle A \rangle_\varrho$, and this number contains at least part of the required information.

The key idea in detecting entanglement via the mean value $\langle A \rangle_\varrho$ is very simple. Suppose that the set of mean values has minimal and maximal values, i.e.

$$a_{\min} \leq \langle A \rangle_\varrho \leq a_{\max}$$

as ϱ run over all states in the set. The state space is a convex set, hence we thus obtain all the values between $a_{\min}$ and $a_{\max}$. For the subset of separable states we have different extremal values:

$$a_{\min} \leq a_{\min}^{\text{sep}} \leq \langle A \rangle_\varrho \leq a_{\max}^{\text{sep}} \leq a_{\max}$$

for all separable states ϱ. The set of separable states is a closed and convex subset of the state space, and therefore each value from the interval $[a_{\min}^{\text{sep}}, a_{\max}^{\text{sep}}]$ is attained by some separable state. Whenever the interval $[a_{\min}^{\text{sep}}, a_{\max}^{\text{sep}}]$ is strictly contained in the main interval $[a_{\min}, a_{\max}]$, the mean value $\langle A \rangle_\varrho$ can potentially lead to entanglement detection. Namely, if $\langle A \rangle_\varrho \notin [a_{\min}^{\text{sep}}, a_{\max}^{\text{sep}}]$ then we can draw the conclusion that the state ϱ is entangled. Notice, however, that $\langle A \rangle_\varrho \in [a_{\min}^{\text{sep}}, a_{\max}^{\text{sep}}]$ does *not* imply that ϱ is separable.

Let us now make some observations to simplify the discussion. First, we are free to label the outcomes of A as we wish, and our choice can be used to shift the mean values $\langle A \rangle_\varrho$. We can therefore always assume that $a_{\min}^{\text{sep}} = 0$, meaning that $\langle A \rangle_\varrho \geq 0$ for all separable states ϱ.

The second point is that the mean-value mapping $\varrho \mapsto \langle A \rangle_\varrho$ determines a selfadjoint operator A by the formula $\langle A \rangle_\varrho = \text{tr}[\varrho A]$, and any selfadjoint operator can be obtained by thus calculating the mean value of some observable (see subsection 3.2.2). We can therefore switch the discussion completely to selfadjoint operators.

Definition 6.38 A selfadjoint operator $W \in \mathcal{L}_s(\mathcal{H}_A \otimes \mathcal{H}_B)$ is an *entanglement witness* if W is not a positive operator but $\langle \psi \otimes \varphi | W \psi \otimes \varphi \rangle \geq 0$ for all factorized vectors $\psi \otimes \varphi \in \mathcal{H}_A \otimes \mathcal{H}_B$. We say that an entangled state ϱ is *detected by* W if $\text{tr}[\varrho W] < 0$.

To explain this terminology, consider first the factorized state $\varrho = \varrho^A \otimes \varrho^B$. By writing canonical convex decompositions for ϱ^A and ϱ^B we obtain

$$\mathrm{tr}[\varrho^A \otimes \varrho^B W] = \sum_{i,j} \lambda_i^A \lambda_j^B \mathrm{tr}[|\psi_i\rangle\langle\psi_i| \otimes |\varphi_j\rangle\langle\varphi_j|W]$$

$$= \sum_{i,j} \lambda_i^A \lambda_j^B \langle\psi_i \otimes \varphi_j|W\psi_i \otimes \varphi_j\rangle \geq 0.$$

By definition, any separable state is a convex combination of factorized states and we thus conclude that $\mathrm{tr}[\varrho W] \geq 0$ for all separable states ϱ. In particular, the condition $\mathrm{tr}[\varrho W] < 0$ implies that ϱ is entangled. The following observation is due to Horodecki *et al.* [77]; see Figure 6.2.

Theorem 6.39 A state ϱ is separable if and only if $\mathrm{tr}[\varrho W] \geq 0$ for all entanglement witnesses W.

Proof If ϱ is separable then $\mathrm{tr}[\varrho W] \geq 0$ for all entanglement witnesses W, as previously discussed.

In order to prove the '*only if*' part, let us assume that ϱ is entangled. We need to find an entanglement witness W such that $\mathrm{tr}[\varrho W] < 0$. The set $\mathcal{S}^{\mathrm{sep}}(\mathcal{H}_A \otimes \mathcal{H}_B)$ of separable states is a closed and convex subset of the real Banach space $\mathcal{T}_s(\mathcal{H}_A \otimes \mathcal{H}_B)$. It follows from the Hahn–Banach theorem (see e.g. [123, Theorem 3.4]) that there exist a linear functional f on $\mathcal{T}_s(\mathcal{H}_A \otimes \mathcal{H}_B)$ and $r\epsilon\mathcal{R}$ such that

$$f(\varrho) < r < f(\eta) \tag{6.34}$$

for all $\eta \in \mathcal{S}^{\mathrm{sep}}(\mathcal{H}_A \otimes \mathcal{H}_B)$. However, the dual space of $\mathcal{T}_s(\mathcal{H}_A \otimes \mathcal{H}_B)$ is $\mathcal{L}_s(\mathcal{H}_A \otimes \mathcal{H}_B)$, and hence there is an operator $F \in \mathcal{L}_s(\mathcal{H}_A \otimes \mathcal{H}_B)$ such that $f(\varrho) = \mathrm{tr}[\varrho F]$ for every $\varrho \in \mathcal{T}_s(\mathcal{H}_A \otimes \mathcal{H}_B)$ (see subsection 1.3.4). Defining $W = F - r\mathcal{I}$ we obtain the required entanglement witness, for which $\mathrm{tr}[\varrho W] = f(\varrho) - r < 0$. $\square$

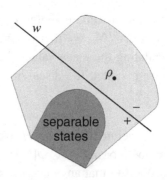

Figure 6.2 The Hahn–Banach theorem for entanglement witnesses.

Corollary 6.40 A state ϱ is entangled if and only if there exists an entanglement witness W such that $\mathrm{tr}[\varrho W] < 0$.

It would be convenient if there were a single entanglement witness detecting all entangled states. Unfortunately, this is not the case.

Proposition 6.41 There is no entanglement witness that detects all entangled states.

Proof Let us suppose, on the contrary, that W is an entanglement witness detecting all entangled states. Let $\varphi_1, \varphi_2 \in \mathcal{H}$ be two orthogonal unit vectors and write $\eta_\pm = \frac{1}{\sqrt{2}}(\varphi_1 \otimes \varphi \pm \varphi_2 \otimes \varphi_2)$. Both $\eta_\pm$ are entangled vectors, as shown in Example 6.2, and therefore we have $\langle \eta_+ | W \eta_+ \rangle < 0$ and $\langle \eta_- | W \eta_- \rangle < 0$. This implies that

$$0 > \langle \eta_+ | W \eta_+ \rangle + \langle \eta_- | W \eta_- \rangle$$
$$= \langle \varphi_1 \otimes \varphi_1 | W \varphi_1 \otimes \varphi_1 \rangle + \langle \varphi_2 \otimes \varphi_2 | W \varphi_2 \otimes \varphi_2 \rangle,$$

which contradicts the fact that W is positive for factorized states. $\square$

The previous statements characterize the power of entanglement witnesses in two extreme cases. There is no entanglement witness detecting all entangled states but the set of all entanglement witnesses is sufficient to detect every entangled state.

Example 6.42 (*CHSH entanglement witness*)
Historically, an inequality based on the CHSH entanglement witness was one of the first tests for quantum entanglement. The acronym CHSH comes from the authors Clauser, Horne, Shimony and Holt [44].

Consider the following selfadjoint operator defined on $\mathcal{H}_2 \otimes \mathcal{H}_2$:

$$X_{\mathrm{CHSH}} = \vec{a} \cdot \vec{\sigma} \otimes (\vec{b} \cdot \vec{\sigma} + \vec{b}' \cdot \vec{\sigma}) + \vec{a}' \cdot \vec{\sigma} \otimes (\vec{b} \cdot \vec{\sigma} - \vec{b}' \cdot \vec{\sigma}), \qquad (6.35)$$

where $\vec{a}, \vec{a}', \vec{b}, \vec{b}'$ are unit vectors in $\mathbb{R}^3$. We will use the notation $A = \vec{a} \cdot \vec{\sigma}$, $A' = \vec{a}' \cdot \vec{\sigma}$, $B = \vec{b} \cdot \vec{\sigma}$, $B' = \vec{b}' \cdot \vec{\sigma}$. For product states $\varrho_A \otimes \varrho_B$ we have

$$\mathrm{tr}[(A \otimes B)(\varrho_A \otimes \varrho_B)] = \mathrm{tr}[A\varrho_A] \, \mathrm{tr}[B\varrho_B]$$

and $-1 \le \mathrm{tr}[\varrho(\vec{x} \cdot \vec{\sigma})] \le 1$ if $\| \vec{x} \| = 1$. Using these relations we see that

$$\mathrm{tr}[X_{\mathrm{CHSH}}\varrho_A \otimes \varrho_B] = \mathrm{tr}[A\varrho_A]\mathrm{tr}[(B + B')\varrho_B] + \mathrm{tr}[A'\varrho_A]\mathrm{tr}[(B - B')\varrho_B]$$
$$\le \mathrm{tr}[(B + B' + B - B')\varrho_B] \le 2$$

and

$$\mathrm{tr}[X_{\mathrm{CHSH}}\varrho_A \otimes \varrho_B] = \mathrm{tr}[A\varrho_A]\mathrm{tr}[(B + B')\varrho_B] + \mathrm{tr}[A'\varrho_A]\mathrm{tr}[(B - B')\varrho_B]$$
$$\ge -\mathrm{tr}[(B + B' + B - B')\varrho_B] \ge -2 .$$

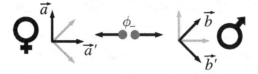

Figure 6.3 CHSH experiment.

Owing to the linearity of the mean value $(\mathrm{tr}[X(\omega_1 + \omega_2)] = \mathrm{tr}[X\omega_1] + \mathrm{tr}[X\omega_2])$ and the convexity of the set of separable states, the above inequalities imply that

$$- 2 \leq \mathrm{tr}[X_{\mathrm{CHSH}}\omega] \leq 2 \tag{6.36}$$

for all $\omega \in \mathcal{S}^{\mathrm{sep}}$; thus, $f_{\pm}^{\mathrm{sep}} = \pm 2$. We can define witness operators

$$W_+ = 2I - X_{\mathrm{CHSH}}, \qquad W_- = X_{\mathrm{CHSH}} - 2I. \tag{6.37}$$

These witnesses are of interest because the mean value of X_{CHSH} can be deduced from classically coordinated local measurements only. Namely, it consists of all bipartite combinations of measurements A, A' Alice's side and B, B' on Bob's side.

Setting the local measurements (see Figure 6.3) as

$$\vec{a} = (1, 0, 0), \quad \vec{a}' = (0, 1, 0), \quad \vec{b} = \tfrac{1}{\sqrt{2}}(1, 1, 0), \quad \vec{b}' = \tfrac{1}{\sqrt{2}}(1, -1, 0),$$

and inserting them into the formula for X_{CHSH} we obtain

$$X_{\mathrm{CHSH}} = \sqrt{2}(\sigma_x \otimes \sigma_x + \sigma_y \otimes \sigma_y) = 2\sqrt{2}(|\phi_+\rangle\langle\phi_+| - |\phi_-\rangle\langle\phi_-|),$$

where $\phi_\pm = \tfrac{1}{\sqrt{2}}(\varphi \otimes \varphi_\perp \pm \varphi_\perp \otimes \varphi)$ are the eigenvectors of X_{CHSH}. It follows that, for a general state $\varrho \in S(\mathcal{H}_2 \otimes \mathcal{H}_2)$,

$$- 2\sqrt{2} \leq \mathrm{tr}[X_{\mathrm{CHSH}}\varrho] \leq 2\sqrt{2}. \tag{6.38}$$

Thus, in this case both the associated witnesses $W_\pm$ are nontrivial and detect some entanglement. In particular, $W_\pm$ achieve their minimal (negative) mean values for the vector states $\phi_\pm$. The entanglement of these vector states can be certified by measuring of X_{CHSH}.   △

Definition 6.43 Let W and W' be two entanglement witnesses. We say that W is a *better witness* than W' (denoted $W \succ W'$) if all the states detected by W' are detected also by W, i.e. if for all states ϱ the following implication holds:

$$\mathrm{tr}[W'\varrho] < 0 \qquad \Rightarrow \qquad \mathrm{tr}[W\varrho] < 0.$$

The above relation $\succ$ defines a partial ordering on the set of entanglement witnesses. An entanglement witness W_{opt} is called *optimal* if it is a maximal element with respect to this partial ordering, i.e. for all witnesses W' either $W_{opt} \succ W'$ or W_{opt} and W' are incomparable.

Exercise 6.44 Show that the CHSH witnesses $W_\pm$ defined in the previous example are optimal. [Hint: It is sufficient to show that $\text{tr}[W_\pm \omega] = 0$ for some separable state.]

There are several ways of determining the average value of a witness operator W. The canonical way is to measure the sharp positive operator-valued measure (POVM) associated with the selfadjoint operator W via its spectral decomposition. Providing that the witness operator W and the POVM effects F_j form a linearly dependent set, we may associate values x_j with effects F_j such that $W = \sum_j x_j F_j$. Then the average value of an observable $x_j \mapsto F_j$ equals $\langle F \rangle_\varrho = \sum_j x_j p_j = \sum_j x_j \text{tr}[\varrho F_j] = \text{tr}[\varrho W]$, which determines the average value of the witness operator W. Further, we will now generalize this concept to include nonlinear entanglement witnesses.

Nonlinear entanglement witnesses

Let us consider the measurement of an observable F on n copies of an unknown bipartite state ϱ. After the experiment our knowledge is represented by a probability distribution $p_\varrho(k) = \text{tr}[F_k(\varrho \otimes \cdots \otimes \varrho)]$, where the F_k are joint quantum effects forming a POVM on n copies of the system. As before, we may define the average value $\langle W \rangle_\varrho = \text{tr}[\varrho W^{\otimes n}]$ for an operator $W = \sum_j x_j F_j$. Let f_{min}, f_{max} stand for the minimal and maximal values of $\langle W \rangle_\varrho$; then f_{min}^{sep}, f_{max}^{sep} are the minimal and maximal values of $\langle W \rangle_\varrho$ when we restrict ourselves to separable states. If either $f_{min} \leq f_{min}^{sep}$ or $f_{max}^{sep} \leq f_{max}$ then we say the operator W is a *nonlinear entanglement witness*. In such case, if we observe a value $\text{tr}[W \varrho^{\otimes n}] \notin [f_{min}^{sep}, f_{max}^{sep}]$ then we conclude that ϱ is entangled.

The nonlinearity becomes clear if we express the states as Bloch vectors (see subsection 2.1.3), i.e. we write $\varrho = \frac{1}{d}(I + \vec{r} \cdot \vec{E})$, where $E_1, \ldots, E_{D^2-1}$ are traceless mutually orthogonal operators on $\mathcal{H}_A \otimes \mathcal{H}_B$ and $D = (\dim \mathcal{H}_A)(\dim \mathcal{H}_B)$. The multiple-copy state $\varrho^{\otimes n}$ is then given by

$$\varrho^{\otimes n} = \frac{1}{D^n}(I^{\otimes n} + \vec{r} \cdot \vec{E} \otimes I \otimes \cdots \otimes I + \cdots + \vec{r} \cdot \vec{E} \otimes \cdots \otimes \vec{r} \cdot \vec{E}).$$

Similarly, for an entanglement witness $W \in \mathcal{L}_s((\mathcal{H}_A \otimes \mathcal{H}_B)^{\otimes n})$ we have

$$W = w_0 I^{\otimes n} + \vec{w}_1 \cdot \vec{E} \otimes I \otimes \cdots \otimes I + \cdots + I \otimes \cdots \otimes I \otimes \vec{w}_n \cdot \vec{E}$$
$$+ \vec{w}_{12} \vec{E} \otimes \vec{E} \otimes I \otimes \cdots \otimes I + \cdots + \vec{w}_{1 \cdots n} \vec{E} \otimes \cdots \otimes \vec{E},$$

where $\vec{w}_{j_1 \cdots j_K}$ are real $K(D^2 - 1)$-dimensional vectors (for $K = 1, \ldots, n$) and the labels $j_1, \ldots, j_K$ list K places in the tensor product of the operator basis at which the operators are traceless. Evaluation of the mean gives us the formula

$$\mathrm{tr}[\varrho^{\otimes n} W] = w_0 + \vec{r} \cdot (\vec{w}_1 + \cdots + \vec{w}_n) + \cdots + \sum_{j_1,\dots,j_n} r_{j_1} \cdots r_{j_n} w^{1\cdots n}_{j_1 \cdots j_n},$$

which determines (for each entanglement witness) a polynomial of the nth degree in ϱ expressed in terms of the Bloch vector $\vec{r}$. For a single-copy witness this expression is linear,

$$\mathrm{tr}[\varrho W] = w_0 + \vec{w} \cdot \vec{r}, \tag{6.39}$$

and the condition $\mathrm{tr}[\varrho W] = 0$ defines a hyperplane in the Bloch-sphere representation of the state space.

By evaluating the average value of some observable we reduce significantly the information provided by the whole probability distribution, which combines the mean values $\mathrm{tr}[\omega F_j]$ of a set of operators F_j. Even if the individual operators F_j are not themselves entanglement witnesses, it may happen that the observed probability distributions may be used for entanglement detection. An example that is in some senses trivial is provided by an informationally complete observable, for which each probability distribution specifies a unique quantum state. Unfortunately, the derivation of the conditions under which the probability vector is induced by separable states is exactly the problem of perfect entanglement detection. Moreover, the implementation of a complete tomography measurement is also quite a complex experimental task. Nevertheless, one may hope that density operators with only a few parameters are relevant for entanglement detection, and this motivates the investigation of simpler experimental tests and entanglement measures. The discussion of these topics is beyond the scope of this book, however.

6.3.2 Quantum nonlocal realism

Before proceeding further with the entanglement-detection problem we will expand the discussion of Example 6.42. Let us have a taster in the form of quantum nonlocal realism – the phenomenon that brought the concept of entanglement into our 'everyday' life.

For the general setting of measurements A, A', B, B', the inequality

$$|\mathrm{tr}[X_{\mathrm{CHSH}}\varrho]| \leq 2 \tag{6.40}$$

derived in Example 6.42 is known as the *CHSH inequality* and was originally proposed as a testbed for the paradigm of *local realism*. The concept of local realism can be traced back to the famous article 'Can the quantum-mechanical description of physical reality be considered complete?' by Albert Einstein, Boris Podolsky and Nathan Rosen [58].

We have seen that quantum systems can violate the CHSH inequality, and this leads to the conclusion that quantum theory is not in accordance with the idea of local realism. In this section we will briefly describe the background of the EPR paradox and of its experimental tests, provided by the Bell inequalities.

So, what is this local realism that was so important for Einstein, Podolsky and Rosen? Consider two spatially separated laboratories. On the one hand, by 'local' we understand that the choice of measurement (A or A', B or B') in one laboratory does not instantaneously affect the measured outcome in the second. The term 'realism', on the other hand, means that if for some measurement the outcome can be predicted with certainty without affecting the system, then this has to be a property of the system before the experiment was performed. Let us note provocatively that separately neither condition seems to be in direct contradiction to quantum theory. It is their combination which turns out to be problematic.

Example 6.45 (*EPR paradox*)
Consider a pair of qubits, each qubit being located in a spatially separate laboratory. Their composite state is $P_- = |\psi_-\rangle\langle\psi_-|$, where $\psi_- = \frac{1}{\sqrt{2}}(\varphi \otimes \varphi_\perp - \varphi_\perp \otimes \varphi)$. Since ψ_- is the only element of the antisymmetric subspace of $\mathcal{H}_2 \otimes \mathcal{H}_2$, it takes the same form in any basis $\varphi, \varphi_\perp$ of $\mathcal{H}_2$. Therefore, if Alice and Bob run the same ideal Stern–Gerlach experiment, each on one member of the qubit pair (in both cases aligned with the same axis, determined by the basis $\varphi, \varphi_\perp$), then they will always record mutually opposite outcomes.

After Alice learns the outcome of her measurement she knows with certainty what will be the outcome of Bob's measurement. Local realism implies that Bob's system must have possessed this feature before the experiment. Such a situation is usual also in the classical case and simply reflects the presence of correlations between the measurements A and B. However, the problem with quantum systems concerning local realism is that Alice has the freedom to choose between noncommuting measurements and yet perfect predictability still holds. Local realism then requires that Bob's qubit must have definite values for all spin-direction measurements, which is not compatible with the intrinsic uncertainty in quantum theory. In fact, we saw in Chapter 3 that in general it is impossible to predict with certainty the individual outcome of a Stern–Gerlach experiment in any given direction. △

Einstein, Podolsky and Rosen were perplexed by the probabilistic nature of quantum theory, hence, they were happy to find a contradiction between the quantum formalism and the idea of local realism. On the basis of this inconsistency they argued that the quantum description of physical reality is incomplete.

The EPR paradox suggests that there could be a local hidden variable model containing some (potentially) unmeasurable quantities. However, once we know their values we are able to predict at least the probabilities of the individual outcomes in any measurement. Before we proceed let us formalize this idea.

Let us introduce a hidden variable λ taking values in some measurable space Λ. The statistics of any measurement is described by a distribution $\chi_A(a, \lambda)$ determining the probability of recording a given that the hidden parameter equals λ. We meet the reality condition in cases when the value a is uniquely determined by the value of λ, i.e. $\chi_A(a, \lambda) = 1$ for some value of a. In our setting the system consists of two spatially separated subsystems and, by the locality assumption, the measurements A on Alice's side and B on Bob's side are independent. Hence, for a fixed value of λ characterizing the state of the composite system, the product $\chi_A(a, \lambda)\chi_B(b, \lambda)$ defines the probability of observing value a in experiment A and value b in experiment B. Not having access to the actual values of λ, we will assume that they are distributed according to some probability distribution μ on Λ; thus

$$P(a, b|A, B) = \int \chi_A(a, \lambda)\chi_B(b, \lambda)\mu(d\lambda) \qquad (6.41)$$

is the observed joint probability of the measurable quantities A and B. The question is how to relate the probability measures μ to the density operators $\varrho \in \mathcal{S}(\mathcal{H}_A \otimes \mathcal{H}_B)$. But, is it even possible? Is there a suitable probability measure μ for any quantum state?

Clearly, for a fixed density operator and a fixed pair of measurements A and B, one can find μ, χ_A, χ_B compatible with the associated distribution $P(a, b|A, B) = \mathrm{tr}[\varrho E_A(a) \otimes F_B(b)]$, where E_A, F_B are POVMs associated with A and B, respectively. However, for a given state ϱ it turns out to be difficult to find a probability measure μ that is simultaneously compatible with all pairs of measurements represented by χ_A, χ_B. If it is possible then we say the state ϱ admits a *local hidden variable model* represented by the probability measure μ. In the following example we will derive an inequality that is valid for any local hidden variable model.

Example 6.46 (*CHSH inequality*)
Suppose that both A and B are binary observables, with outcomes ± 1. Let us use quantum notation to denote the mean value of a product of the individual outcomes of A and B, i.e.

$$\langle A \otimes B \rangle = \sum_{a,b \in \{\pm 1\}} ab P(a, b|A, B) = \int \mu(d\lambda)\hat{a}(\lambda)\hat{b}(\lambda), \qquad (6.42)$$

where $\hat{a}(\lambda) = (\chi_A(1, \lambda) - \chi_A(-1, \lambda))$ and $\hat{b}(\lambda) = (\chi_B(1, \lambda) - \chi_B(-1, \lambda))$. Let us stress that if λ gives the outcomes a and b deterministically then $\hat{a}(\lambda), \hat{b}(\lambda) \in \{\pm 1\}$. Otherwise $\hat{a}(\lambda), \hat{b}(\lambda) \in [-1, 1]$.

Let us consider the following combination of mean values of different pairs of measurements,

$$\mathcal{B} = |\langle A \otimes B \rangle + \langle A \otimes B' \rangle + \langle A' \otimes B \rangle - \langle A' \otimes B' \rangle|, \qquad (6.43)$$

where A, A' and B, B' are measurements performed on Alice's and Bob's sides, respectively. Then

$$\mathcal{B} = |\hat{a}(\lambda)(\hat{b}(\lambda) + \hat{b}'(\lambda)) + \hat{a}'(\lambda)(\hat{b}(\lambda) - \hat{b}'(\lambda))| \le 2, \qquad (6.44)$$

because $\hat{a}(\lambda), \hat{a}'(\lambda), \hat{b}(\lambda), \hat{b}'(\lambda) \in [-1, 1]$. In Example 6.42 we showed that in quantum settings (where $\mathcal{B} = |\text{tr}[X_{\text{CHSH}}\varrho]|$) the inequality (6.40) holds for all separable states; however, there are entangled states that violate it. $\triangle$

In the above example we derived the CHSH inequality, which must be valid for any locally realistic theory. However, some entangled states of quantum theory violate this inequality (see Example 6.42); hence either these states do not occur in nature or quantum theory is not compatible with the paradigm of local realism. This feature of entanglement, the violation of the consequences of local hidden variable models, is called *quantum nonlocal realism* or *quantum nonlocality*.

What is the relation between entanglement and quantum nonlocality or, equivalently, between entanglement and local hidden-variable models? Any separable state admits a local hidden variable model and hence quantum nonlocality requires entanglement. However, it was demonstrated by Werner [141] that there are entangled states admitting local hidden-variable models, implying that entanglement does not necessarily imply quantum nonlocality.

6.3.3 Positive but not completely positive maps

In this subsection we will see that the complete-positivity constraint on physical transformations is intimately related to entanglement. Consider a linear map $\mathcal{F} = \mathcal{F}_A \otimes \mathcal{F}_B$ from $T(\mathcal{H}_A \otimes \mathcal{H}_B)$ to $T(\mathcal{H}_{A'} \otimes \mathcal{H}_{B'})$, where $\mathcal{F}_A$, $\mathcal{F}_B$ are positive linear maps acting on subsystems A, B, respectively. We say that $\mathcal{F}$ is a *factorized locally positive map*.

If operators $T \in T(\mathcal{H}_A)$ and $S \in T(\mathcal{H}_B)$ are positive then $\mathcal{F}_A(T)$ and $\mathcal{F}_B(S)$ are positive as well. Consequently, if $\mathcal{F}$ is applied to a separable state $\varrho_{AB} = \sum_j p_j \varrho_A^{(j)} \otimes \varrho_B^{(j)}$ of a composite system $A + B$ then the operator

$$\mathcal{F}(\varrho_{AB}) = (\mathcal{F}_A \otimes \mathcal{F}_B)(\varrho_{AB}) = \sum_j p_j \mathcal{F}_A(\varrho_A^{(j)}) \otimes \mathcal{F}_B(\varrho_B^{(j)})$$

is positive. However, if ϱ_{AB} is entangled then the operator $\mathcal{F}(\varrho_{AB})$ need not be positive.

As an aside, we recall from Chapter 4 that if both $\mathcal{F}_A$ and $\mathcal{F}_B$ are not only positive but completely positive then $\mathcal{F} = \mathcal{F}_A \otimes \mathcal{F}_B$ is a positive map. In this case the operator $\mathcal{F}(\varrho_{AB})$ is positive whenever ϱ_{AB} is positive. This was exactly the motivation for including complete positivity in the definition of an operation.

This observation leads us to an important application of maps $\mathcal{F}_A$, $\mathcal{F}_B$ which are positive but not completely positive. Namely, if the operator $(\mathcal{F}_A \otimes \mathcal{F}_B)(\varrho_{AB})$ is not positive then the state ϱ_{AB} is entangled. As in the case of entanglement witnesses, every factorized locally positive map $\mathcal{F} = \mathcal{F}_A \otimes \mathcal{F}_B$ divides the state space $\mathcal{S}(\mathcal{H}_A \otimes \mathcal{H}_B)$ into two subsets,

$$S_+^{\mathcal{F}} = \{\varrho : \mathcal{F}(\varrho) \geq 0\}, \qquad S_-^{\mathcal{F}} = \{\varrho : \mathcal{F}(\varrho) \ngeq 0\}.$$

As before, the question is how strong are the entanglement tests that we obtain from factorized locally positive maps? In particular, is there a universal factorized locally positive map $\mathcal{F}$ deciding the entanglement of any state?

The paradigmatic example of positive, but not completely positive, linear maps is the transposition map $\tau : \varrho \to \varrho^T$. In Example 4.3 we showed that the induced linear maps $\mathcal{I} \otimes \tau$ and $\tau \otimes \mathcal{I}$, called partial transpositions, are not positive. We will use the shorthand notation $\tau_A(\varrho) = (\tau \otimes \mathcal{I})(\varrho)$ and $\tau_B(\varrho) = (\mathcal{I} \otimes \tau)(\varrho)$ and formulate a useful lemma concerning partial transpositions and partial traces.

Lemma 6.47 For operators $X, Y \in \mathcal{L}(\mathcal{H}_A \otimes \mathcal{H}_B)$, the following identity holds:

$$\mathrm{tr}_B[\tau_B(X)Y] = \mathrm{tr}_B[X\tau_B(Y)]. \tag{6.45}$$

Proof Consider an orthonormal basis $\{\psi_j\}$ of $\mathcal{H}_B$ and let $E_{jk} = |\psi_j\rangle\langle\psi_k|$ be the associated operator basis of $\mathcal{L}(\mathcal{H}_B)$. We can write $X = \sum_{j,k} X_{jk} \otimes E_{jk}$, $Y = \sum_{j,k} Y_{jk} \otimes E_{jk}$, where X_{jk}, Y_{jk} are suitable operators from $\mathcal{L}(\mathcal{H}_A)$. Then

$$\mathrm{tr}_B[\tau_B(X)Y] = \sum X_{kl} Y_{k'l'} \langle\psi_j|E_{kl}^T E_{k'l'}\psi_j\rangle = \sum X_{kj} Y_{kj}$$

and, simultaneously,

$$\mathrm{tr}_B[X\tau_B(Y)] = \sum X_{kl} Y_{k'l'} \langle\psi_j|E_{kl} E_{k'l'}^T \psi_j\rangle = \sum X_{jl} Y_{jl}$$

which proves the lemma. $\qquad\qquad\qquad\square$

It is an important result [77] due to Horodecki *et al.* that for the purposes of entanglement detection it is sufficient to restrict one's attention to a special family of one-sided local channels, by setting $\mathcal{F}_A = \mathcal{I}$ and $A' = B' = A$.

Proposition 6.48 A state $\varrho \in \mathcal{S}(\mathcal{H}_A \otimes \mathcal{H}_B)$ is separable if and only if for all positive maps $\mathcal{F} : \mathcal{L}(\mathcal{H}_B) \to \mathcal{L}(\mathcal{H}_A)$ the operator $(\mathcal{I} \otimes \mathcal{F})(\varrho)$ is positive.

Proof We will prove that under the Choi–Jamiolkowski isomorphism there is a one-to-one correspondence between linear witnesses $W : \mathcal{H}_A \otimes \mathcal{H}_B \to \mathcal{H}_A \otimes \mathcal{H}_B$ and positive maps $\mathcal{F} : \mathcal{L}(\mathcal{H}_B) \to \mathcal{L}(\mathcal{H}_A)$. Consequently, the results achieved for entanglement witnesses can be (to some extent) translated into the language of positive maps and vice versa. In particular, this correspondence makes the proposition equivalent to Theorem 6.39.

Assume that $d = \dim \mathcal{H}_A \leq \dim \mathcal{H}_B = d'$. Suppose that $\mathcal{F} : \mathcal{L}(\mathcal{H}_B) \to \mathcal{L}(\mathcal{H}_A)$ is a positive linear map. Then the associated Choi–Jamiolkowski operator acting on $\mathcal{H}_A \otimes \mathcal{H}_B$ is

$$W_{\mathcal{F}} = (\mathcal{F} \otimes \mathcal{I})(P_+), \tag{6.46}$$

where $P_+ = |\psi_+\rangle\langle\psi_+|$, $\psi_+ = \frac{1}{\sqrt{d'}} \sum_j \psi_j \otimes \psi_j$ and the vectors $\{\psi_j\}$ form an orthonormal basis of $\mathcal{H}_B$. For all $\varphi \in \mathcal{H}_A$ and $\psi \in \mathcal{H}_B$,

$$\langle\varphi \otimes \psi | W_{\mathcal{F}} | \varphi \otimes \psi\rangle = \frac{1}{d'} \sum_{j,k} \langle\varphi \otimes \psi | (\mathcal{F} \otimes \mathcal{I})(E_{jk} \otimes E_{jk}) | (\varphi \otimes \psi)\rangle$$

$$= \frac{1}{d'} \langle\varphi | \mathcal{F}(\sum_{jk} \mathrm{tr}[E_{jk} P_\psi] E_{jk}) | \varphi\rangle$$

$$= \frac{1}{d'} \langle\varphi | \mathcal{F}(P_\psi^T) | \varphi\rangle \geq 0 \,,$$

where we have used the notation $E_{jk} = |\psi_j\rangle\langle\psi_k|$ and $P_\psi = |\psi\rangle\langle\psi|$, and P_ψ^T stands for transposition with respect to the basis $\{\psi_j\}$ used in the definition of ψ_+. Let us note that the inequality in the last line of the right-hand side holds owing to the positivity of the transposition map and of $\mathcal{F}$ itself. If an operator is positive on factorized vector states then, by convexity, its mean value is positive on all separable states; hence $W_{\mathcal{F}}$ defines an entanglement witness on $\mathcal{H}_A \otimes \mathcal{H}_B$.

Suppose that $W \in \mathcal{L}(\mathcal{H}_A \otimes \mathcal{H}_B)$ is an entanglement witness. Then the induced linear map $\mathcal{F}_W : \mathcal{L}(\mathcal{H}_B) \to \mathcal{L}(\mathcal{H}_A)$ is defined via the inverse formula

$$\mathcal{F}_W(X) = d' \, \mathrm{tr}_B[W(I \otimes X^T)]. \tag{6.47}$$

Setting $X = P_\psi^T = (|\psi\rangle\langle\psi|)^T$ we obtain

$$\langle\varphi | \mathcal{F}_W[P_\psi^T] | \varphi\rangle = d' \, \langle\varphi | \mathrm{tr}_B[W(I \otimes P_\psi^{TT})] | \varphi\rangle$$

$$= d' \, \langle\varphi \otimes \psi | W(I \otimes P_\psi)(\varphi \otimes \psi)\rangle$$

$$= d' \, \langle\varphi \otimes \psi | W(\varphi \otimes \psi)\rangle \geq 0,$$

where the last inequality follows from the property of the entanglement witness. Since this derivation holds for all vectors ψ and since a general positive operator is a positive sum of one-dimensional projectors it follows that $\mathcal{F}_W[X] \geq O$ for all $X \in \mathcal{L}_+(\mathcal{H}_B)$. In summary, operators that are positive on product vectors (i.e. linear entanglement witnesses) and positive linear maps are related by the Choi–Jamiolkowski isomorphism.

According to Theorem 6.39, for an entangled state ϱ there exists an entanglement witness W such that $\mathrm{tr}[W\varrho] < 0$. Consider a linear map $\mathcal{F}$ induced by the operator W^T defined by the formula

$$(\mathcal{I}_A \otimes \mathcal{F}_B)(\varrho_{AB}) = d' \,\mathrm{tr}_B[(I_A \otimes (\tau_{A'} \otimes \tau_B)(W_{A'B}))((\tau_B(\varrho_{AB}) \otimes I_{A'})],$$

where the systems A, A' are identical and $(\tau_{A'} \otimes \tau_B)(W_{A'B}) = W_{A'B}^T$ defines the complete transposition of $W_{A'B}$. Let us note that W is an entanglement witness if and only if W^T is an entanglement witness; thus the considered linear map $\mathcal{F}$ is positive. Using Lemma 6.47 and the identity $\tau_B\tau_B = \mathcal{I}$ we obtain

$$(\mathcal{I}_A \otimes \mathcal{F}_B)(\varrho_{AB}) = d' \,\mathrm{tr}_B[(I_A \otimes \tau_{A'}(W_{A'B}))(\varrho_{AB} \otimes I_{A'})]. \qquad (6.48)$$

In order to show the nonpositivity of $(\mathcal{I}_A \otimes \mathcal{F}_B)(\varrho_{AB})$ let us evaluate its mean value,

$$\Delta = \langle \psi_+|(\mathcal{I}_A \otimes \mathcal{F}_B)(\varrho_{AB})\psi_+\rangle = \mathrm{tr}[(\mathcal{I}_A \otimes \mathcal{F}_B)(\varrho_{AB})P_+], \qquad (6.49)$$

where $P_+ = |\psi_+\rangle\langle\psi_+|$, $\psi_+ = \frac{1}{\sqrt{d}}\sum_j \varphi_j \otimes \varphi_j$ and the unit vectors $\{\varphi_j\}$ form an orthonormal basis of $\mathcal{H}_A = \mathcal{H}_{A'}$. The operators $G_{jk} = |\varphi_j\rangle\langle\varphi_k|$ form an operator basis of $\mathcal{L}(\mathcal{H}_A)$ and $\mathcal{L}(\mathcal{H}_{A'})$. The operators $E_{xy} = |\psi_x\rangle\langle\psi_y|$ form an operator basis of $\mathcal{L}(\mathcal{H}_B)$. Writing $W = \sum w_{ab,xy} G_{ab} \otimes E_{xy}$ and $\varrho = \sum \varrho_{ab,xy} G_{ab} \otimes E_{xy}$ we obtain

$$\Delta = \frac{d'}{d} \sum w_{ab,xy}\varrho_{a'b',x'y'} \langle \varphi_k|G_{a'b'}\varphi_j\rangle\langle \varphi_k|G_{ab}^T\varphi_j\rangle\langle \psi_l|E_{xy}E_{x'y'}\psi_l\rangle$$

$$= \frac{d'}{d} \sum w_{ab,xy}\varrho_{a'b',x'y'} \delta_{ka'}\delta_{b'j}\delta_{kb}\delta_{aj}\delta_{lx}\delta_{yx'}\delta_{y'l}$$

$$= \frac{d'}{d} \sum w_{jk,ly}\varrho_{kj,yl} = \frac{d'}{d} \,\mathrm{tr}[W_{AB}\varrho_{AB}].$$

Since we have assumed that $\mathrm{tr}[W\varrho] < 0$ it follows that $\Delta < 0$ and, consequently, the operator $(\mathcal{I}_A \otimes \mathcal{F}_B)(\varrho_{AB})$ is not positive, which proves the proposition. $\qquad\square$

Exercise 6.49 (*Reduction map*)
Define a linear map $\mathcal{R}$ on $\mathcal{L}(\mathcal{H})$ by $\mathcal{R}(T) = \mathrm{tr}[T]I - T$. Show that this map is positive and formulate the corresponding entanglement criterion. Show that either of the inequalities $S(\varrho_A) > S(\varrho_{AB})$ and $S(\varrho_B) > S(\varrho_{AB})$, where S is the von Neuman entropy, implies that the state ϱ_{AB} is entangled. Using the corresponding entanglement witness show that $\mathrm{tr}[\varrho P_+] > \frac{1}{d}$ implies that ϱ is entangled.

6.3.4 Negative partial transpose (NPT) criterion

We have shown that for each entangled state on $\mathcal{H}_A \otimes \mathcal{H}_B$ there is a positive map $\mathcal{F} : \mathcal{T}(\mathcal{H}_B) \rightarrow \mathcal{T}(\mathcal{H}_A)$ such that the operator $(\mathcal{I} \otimes \mathcal{F})(\varrho)$ is not positive. Although Proposition 6.48 is closely related to Theorem 6.39, the 'power' of entanglement tests based on positive maps is not equivalent to the 'power' of entanglement witnesses. In fact, as illustrated in the following example, the conditions $(\mathcal{I} \otimes \mathcal{F})(\varrho) \not\geq 0$ and $\text{tr}[W_{\mathcal{F}}\varrho] < 0$ are satisfied for different states ϱ.

Example 6.50 (*Entanglement witness for transposition*)
The entanglement witness corresponding to the transposition map $\tau : \mathcal{T}(\mathcal{H}_A) \rightarrow \mathcal{T}(\mathcal{H}_A)$ reads

$$W_\tau = \tau_B(P_+) = \frac{1}{d} \sum_{jk} |\varphi_j\rangle\langle\varphi_k| \otimes |\varphi_k\rangle\langle\varphi_j| = \frac{1}{d} W_{\text{SWAP}},$$

where W_{SWAP} is the SWAP operator, which was defined in Example 4.15. We have $\text{tr}[W_{\text{SWAP}} P_+] > 0$ but the operator $(I \otimes \tau)[P_+]$ is not positive (see Example 4.3). In other words the entanglement witness W_{SWAP} does not detect the entanglement of a maximally entangled state P_+, but the transposition map τ is capable of detecting the entanglement of any pure maximally entangled state. △

Exercise 6.51 For a two-qubit system show that $\tau_B(\varrho) \not\geq O$ if ϱ is an entangled pure state. Note that the spectrum of the partially transposed operators is independent of the basis with respect to which the transposition is accomplished. [Hint: First just do the calculation for a general pure state. Then use Schmidt decomposition and make a partial transposition in the Schmidt basis.]

While very simple, the transposition map provides a surprisingly strong entanglement test. The relation between positive maps and entanglement was recognized for the first time by Peres [115]. He asked whether the nonpositivity of a density operator under partial transposition is not in fact a necessary condition for the state to be entangled. Using old results of Stormer [134] and of Woronowicz [148], it was shown by Horodecki *et al.* [77] that this is the case for density operators defined on $\mathbb{C}^2 \otimes \mathbb{C}^2$ and $\mathbb{C}^2 \otimes \mathbb{C}^3$.

Definition 6.52 A state $\varrho \in \mathcal{S}(\mathcal{H}_A \otimes \mathcal{H}_B)$ is called a *positive partial transpose (PPT) state* if it remains positive under partial transpositions. We denote the convex set of PPT states by $\mathcal{S}^{\text{PPT}}$. If $\varrho \notin \mathcal{S}^{\text{PPT}}$ then we say it is a *negative partial transpose (NPT) state*.

Theorem 6.53 (*Stromer–Woronowicz* [148])
All positive linear maps $\mathcal{F} : \mathcal{L}(\mathbb{C}^2) \rightarrow \mathcal{L}(\mathbb{C}^2)$ and $\mathcal{F} : \mathcal{L}(\mathbb{C}^3) \rightarrow \mathcal{L}(\mathbb{C}^2)$ are *decomposable* in the sense that

$$\mathcal{F} = \mathcal{F}_{\text{cp}}^1 + \mathcal{F}_{\text{cp}}^2 \circ \tau, \tag{6.50}$$

where $\mathcal{F}_{\text{cp}}^1, \mathcal{F}_{\text{cp}}^2$ are completely positive maps.

Theorem 6.54 (*NPT/Peres–Horodecki criterion*)
If a state $\varrho_{AB} \in \mathcal{S}(\mathcal{H}_A \otimes \mathcal{H}_B)$ is NPT then it is entangled. Moreover, if $\mathcal{H}_A = \mathbb{C}^2$ and $\mathcal{H}_B \in \{\mathbb{C}^2, \mathbb{C}^3\}$ then ϱ_{AB} is NPT if and only if it is entangled.

Proof The first part of the theorem is a direct consequence of Proposition 6.48 applied to the transposition map τ. According to Theorem 6.53 all positive maps on systems $\mathcal{H}_2 \otimes \mathcal{H}_2$ and $\mathcal{H}_2 \otimes \mathcal{H}_3$ are decomposable and thus related to the transposition map. By Proposition 6.48, for any entangled state there exists a linear map $\mathcal{F}$ such that $(\mathcal{I} \otimes \mathcal{F})(\varrho)$ is nonpositive. However, owing to the decomposability of $\mathcal{F}$ it follows that $(\mathcal{I} \otimes \tau)(\varrho)$ is also nonpositive. From the other side, the positivity of $(\mathcal{I} \otimes \tau)(\varrho)$ implies that $(\mathcal{I} \otimes \mathcal{F})(\varrho)$ is positive for all positive maps $\mathcal{F}$, as well. In other words all entangled states for the considered systems are NPT. $\qquad\square$

Exercise 6.55 Show that the states X_λ from Example 6.12 are entangled. [Hint: Try the partial transposition test.]

6.3.5 Range criterion

In order to show that for larger-dimensional systems the set of entangled PPT states is not empty, we will employ the so-called *range criterion* for separability [79]. By definition, if a state ϱ is separable then there exists a family of product states $\{\psi_j \otimes \phi_j\}$ spanning the range of ϱ. If we find that there is no such family of product vectors $\{\psi_j \otimes \phi_j\}$ spanning the range of ϱ we can conclude that the state is not separable and hence entangled.

Proposition 6.56 (*Range criterion*)
If the range of $\varrho \in \mathcal{S}(\mathcal{H}_A \otimes \mathcal{H}_B)$ is not spanned by a family of product vectors $\{\psi_j \otimes \phi_j\}$ then ϱ is entangled.

This criterion does not look very practical. However, there are some interesting special cases where it turns out to be useful.

Example 6.57 (*Symmetric and antisymmetric states*)
Consider a composite Hilbert space $\mathcal{H} \otimes \mathcal{H}$ of two systems of the same type. In Example 6.4 we introduced symmetric and antisymmetric vectors. We can also define symmetric and antisymmetric states. If the range of ϱ is contained in the symmetric (antisymmetric) subspace we say the state is symmetric (antisymmetric). An important property of antisymmetric vectors is that they cannot be of product form. In particular, applying the swap operator to a product vector $\psi \otimes \phi$

we obtain $\phi \otimes \psi \neq -\psi \otimes \phi$. Thus, antisymmetric states have no product vectors in their range, and according to the range criterion, they are entangled. For symmetric states the situation is different and they can be either separable or entangled. $\quad \triangle$

In what follows we will show that the range criterion and the Peres–Horodecki criterion are different in a sense that there exist entangled states that pass the PPT test but are still identifiable by the range criterion. Recall from Proposition 1.52 the complex conjugate operator J related to an orthonormal basis $\{\varphi_j\}$. We assume that the orthonormal basis is fixed and write $\varphi^* = J\varphi$. Then $\tau[|\varphi\rangle\langle\varphi|] = |\varphi^*\rangle\langle\varphi^*|$. Therefore, if a product vector $\psi \otimes \varphi$ belongs to the range of ϱ then the product vector $\psi \otimes \varphi^*$ belongs to the range of $\tau_B[\varrho]$. We will exploit this observation in order to design a PPT entangled state.

The following key concept was introduced in [19].

Definition 6.58 A set of orthogonal product vectors $\{\psi_1 \otimes \varphi_1, \ldots, \psi_m \otimes \varphi_m\}$ with $m < \dim \mathcal{H}_A \dim \mathcal{H}_B$ is an *unextendible product basis* if there is no factorized vector $\phi \in \mathcal{H}_A \otimes \mathcal{H}_B$ such that $\phi \perp \psi_j \otimes \varphi_j$ for all j.

Proposition 6.59 (*PPT entangled states*)
Consider an unextendible product basis (upb) of $\mathcal{H}_A \otimes \mathcal{H}_B$ consisting of $m < d_A d_B$ vectors $\omega_j = \psi_j \otimes \varphi_j$. (Here $d_A = \dim \mathcal{H}_A$ and $d_B = \dim \mathcal{H}_B$.) Define a projection $\Pi_{\text{upb}} = \sum_j |\omega_j\rangle\langle\omega_j|$. Then the state $\varrho_{\text{upb}} = \frac{1}{d_A d_B - m}(I - \Pi_{\text{upb}})$ is a PPT entangled state.

Proof By definition the range of ϱ_{upb} is orthogonal to the vectors ω_j; thus it does not contain any product state. Consequently, the range criterion implies that ϱ_{upb} is entangled. The vectors $\psi_j \otimes \varphi_j^*$ are also mutually orthogonal, hence, $\tau_B[\Pi_{\text{upb}}] = \sum_j |\psi_j \otimes \varphi_j^*\rangle\langle\psi_j \otimes \varphi_j^*|$ and $I - \tau_B[\Pi_{\text{upb}}]$ are also projections. It follows that, although ϱ_{upb} is entangled after the partial transposition, the operator $\tau_B[\varrho_{\text{upb}}] = \frac{1}{d_A d_B - m}(I - \tau_B[\Pi_{\text{upb}}])$ is positive. In summary, ϱ_{upb} is a PPT entangled state. To conclude the proof it remains to show the existence of an unextendible product basis; this is left for the reader (see the following exercise). $\quad \square$

Exercise 6.60 Consider a composite system $\mathcal{H}_3 \otimes \mathcal{H}_3$. Show that the vectors $\varphi_0 \otimes \varphi_{0-1}$, $\varphi_2 \otimes \varphi_{1-2}$, $\varphi_{0-1} \otimes \varphi_2$, $\varphi_{1-2} \otimes \varphi_0$ and $\varphi_{0+1+2} \otimes \varphi_{0+1+2}$ form an unextendible product basis and verify that the associated state ϱ_{upb} is indeed a PPT entangled state; the shorthand subscript notation denotes a superposition of orthonormal states $\varphi_0, \varphi_1, \varphi_2$ with equal absolute values of their amplitudes but possibly different signs. For example, φ_{0-1} stands for $\frac{1}{\sqrt{2}}(\varphi_0 - \varphi_1)$.

Example 6.61 (*Entanglement witnesses for ϱ_{upb}*)
An unextendible product basis can be used to design nondecomposable positive maps determining new entanglement tests that are not equivalent to the partial

transposition criterion. In this example we will design a linear entanglement witness for the entanglement of ϱ_{upb}, as defined in Proposition 6.59. Let us start with
the observation that the projection Π_{upb} is strictly positive on all factorized vectors,
i.e. $\min_{\psi \otimes \varphi} \langle \psi \otimes \varphi | \Pi_{\mathrm{upb}} | \psi \otimes \varphi \rangle = \epsilon > 0$, because of the unextendibility of the
product basis. Let us note that $\epsilon = 0$ would imply that there exists a factorized
vector orthogonal to all vectors in the unextendible product basis, which is a contradiction. Define $P_\phi = |\phi\rangle\langle\phi|$ for an entangled vector ϕ satisfying $\Pi_{\mathrm{upb}}\phi = 0$.
Since for the operator $W_{\mathrm{upb}}^\phi = \Pi_{\mathrm{upb}} - \epsilon P_\phi$ we have

$$\langle \psi \otimes \varphi | W_{\mathrm{upb}}^\phi | \psi \otimes \varphi \rangle \geq \epsilon \left(1 - |\langle \phi | \psi \otimes \varphi \rangle|^2 \right) \geq 0 \tag{6.51}$$

for all product vectors $\psi \otimes \varphi$, it follows that W_{upb}^ϕ is positive on all separable states
and thus is an entanglement witness. Moreover, the inequality

$$\mathrm{tr}\left[\varrho_{\mathrm{upb}} W_{\mathrm{upb}}^\phi\right] = -\frac{\epsilon}{d_1 d_2 - m} < 0 \tag{6.52}$$

implies that it detects the PPT entangled state ϱ_{upb}. Using the Choi–Jamiolkowski
isomorphism we can define an indecomposable positive map providing an entanglement criterion qualitatively different from the PPT criterion.   △

6.4 Additional topics in entanglement theory

Over last 20 years, approximately the age of entanglement theory, there have
been many interesting and important discoveries; so far, we have covered only the
absolute basics of the phenomenon of entanglement. In this section we will introduce three additional topics from entanglement theory, extending the paradigm of
LOCC channels to LOCC operations, bipartite systems to multipartite systems and
entanglement kinematics to entanglement dynamics.

6.4.1 Entanglement teleportation and distillation

Most papers on quantum entanglement glorify its importance in quantum information theory and highlight its conceptual position within quantum theory itself.
Indeed, particular entangled states (especially maximally entangled ones) exhibit
very interesting and puzzling features and they deserve the special attention of
mathematicians, physicists and computer scientists. However, it is also fair to say
that we lack an operationally clear understanding of the difference between separable and entangled states. In our approach we have exploited the paradigm of LOCC
channels to introduce the concepts of separable and entangled states and have
identified a particularly interesting equivalence class, that of maximally entangled
states.

It turns out that, in the field of quantum information processing, maximally entangled states constitute a rich resource allowing us to perform tasks that are either impossible or have no counterpart in the classical information domain. Perhaps the most prominent example is the quantum communication protocol for *quantum state teleportation*. Using the weird properties of maximally entangled states, it provides a purely quantum alternative for the perfect (and secure) transmission of a quantum state from side A to side B without transmitting any quantum systems.

Example 6.62 (*Quantum state teleportation* [18])
Let us start with the remark that quantum state teleportation can be viewed as yet another quantum generalization of a classical cryptographic protocol, the one-time pad (see subsection 5.2.4). In this case the goal is to communicate quantum information (identified with the concept of a quantum state) using a quantum key (represented by a pair of shared maximally entangled quantum systems) and a classical public line (used to transfer a classical message). In what follows we will describe the quantum state teleportation protocol of arbitrary (finite-dimensional) quantum systems.

Consider a composite system composed of three subsystems $\mathcal{H}_{A'} \otimes \mathcal{H}_A \otimes \mathcal{H}_B$ such that $\mathcal{H}_{A'} = \mathcal{H}_A = \mathcal{H}_B$. Two of these (A', A) are in the possession of Alice and one (B) is in the possession of Bob. We assume that (prior to communication) Alice and Bob share a pure maximally entangled state P_+ playing the role of the quantum key. The goal is to transfer a state A' of the subsystem (not the subsystem itself!) from Alice to Bob.

The joint state of all three subsystems is $\Omega_{A'AB} = \varrho_{A'} \otimes P_+^{AB}$, where $\varrho_{A'}$ is a state that is going to be transmitted from Alice to Bob. The 'magic' of quantum state teleportation is based on the following identity. For all unitary operators $U : \mathcal{H} \rightarrow \mathcal{H}$ and any operator $X \in \mathcal{L}(\mathcal{H})$,

$$\mathrm{tr}_{A'A}\left[(P_U^{A'A} \otimes I)(X_{A'} \otimes P_+^{AB})(P_U^{A'A} \otimes I)\right] = \frac{1}{d^2}U^* X_B U, \qquad (6.53)$$

where $P_U = (U \otimes I)P_+(U^* \otimes I)$, $P_+ = \frac{1}{d}\sum_{jk}|\varphi_j \otimes \varphi_j\rangle\langle\varphi_k \otimes \varphi_k|$ is the projection onto the maximally entangled state and $X_{A'} = X_B$. Indeed, denoting by Δ the left-hand side of (6.53) we get

$$d^2\Delta = d^2\mathrm{tr}_{A'A}[(P_+^{A'A} \otimes I)(U^*XU \otimes P_+^{AB})]$$

$$= \sum_{j,k,m,n} \mathrm{tr}_{A'}[|\varphi_j\rangle\langle\varphi_k|U^*XU]\,\mathrm{tr}_A[|\varphi_j\rangle\langle\varphi_k|\varphi_m\rangle\langle\varphi_n|]|\varphi_m\rangle_B\langle\varphi_n|$$

$$= \sum_{j,k,m,n} \delta_{km}\delta_{jn}\langle\varphi_k|U^*XU\varphi_j\rangle|\varphi_m\rangle_B\langle\varphi_n| = U^*X_B U.$$

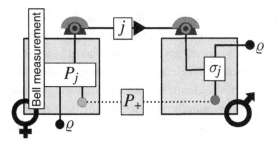

Figure 6.4 Quantum state teleportation; see Example 6.62.

Moreover, in a similar way one can prove that

$$d^2 \operatorname{tr}_B \left[(P_U^{A'A} \otimes I)(X_{A'} \otimes P_+^{AB})(P_U^{A'A} \otimes I) \right] = \operatorname{tr}[X] \, P_U^{A'A}.$$

Equation (6.53) is interpreted as a postselective state transformation of system B in the measurement process described by the Lüders state transformer associated with the effect $P_U^{A'A} \otimes I$. This effect is observed with probability $p = \operatorname{tr}[\Omega_{A'AB}(P_U^{A'A} \otimes I)] = \frac{1}{d^2}$. In other words, if Alice performs a local measurement and observes the effect P_U then Bob's subsystem is described by the unitarily transformed state ϱ. This fact is hidden to Bob unless this information is communicated from Alice. After receiving the information that Alice has observed the particular outcome P_U, Bob can reverse the unwanted unitary transformation and retrieve the original state ϱ on his subsystem (see Figure 6.4).

Does there exist a measurement for which all effects are projections onto maximally entangled vector states? Luckily, the Bell basis (see Example 6.5) is of exactly this type; thus Alice's measurement can be composed of d^2 projections $P_{U_1}, \ldots, P_{U_{d^2}}$ such that $\sum_j P_{U_j} = I \otimes I$ and $\operatorname{tr}[U_j^\dagger U_k] = d\delta_{jk}$. Such a measurement is known as a Bell measurement and is described in Example 6.5. It follows from the identity in (6.53) that, for each outcome associated with the projection P_{U_j} Bob can perform a unitary transformation U_j to retrieve the original state ϱ on his side while the subsystems A' and A on Alice's side end up in the state P_{U_j}. Applying a conditioned unitary channel $\sigma_{U_j \otimes I}$ on Alice's side, this state can be transformed into $P_+^{A'A}$.

In summary, the procedure described implements an LOCC channel that effectively swaps the subsystems A' and B, i.e.

$$\varrho_{A'} \otimes P_+^{AB} \quad \mapsto \quad P_+^{A'A} \otimes \varrho_B.$$

It consists of the following steps.

1. The maximally entangled state P_+ is distributed between Alice and Bob.
2. Alice performs a Bell measurement on systems A, A' and submits the recorded outcome j to Bob.
3. Both Alice and Bob apply the conditional correcting unitary channel σ_{U_j} to systems A and B, respectively.

State teleportation achieves the deterministic and perfect transmission of an unknown quantum state from Alice to Bob (just using classical communication). In each run Alice sends only $\log d^2 = 2 \log d$ bits of classical information that in themselves contain no information about the state; that is, the sequence of recorded outcomes (if the teleportation is repeated) is completely random and independent of ϱ; hence, security is achieved.   △

A natural question is whether the teleportation protocol described is unique. In particular, is it necessary to have a maximally entangled state? The answer was given by Werner [142], who showed that the family of teleportation schemes described in Example 6.62 is essentially unique. In particular, a general teleportation protocol consists of an observable $x \mapsto F_x^{A'A}$ and a set of 'undo' channels $\{\mathcal{E}_x^B\}$ satisfying the identity

$$\mathrm{tr}[\varrho E] = \sum_x \mathrm{tr}\left[(I^{A'A} \otimes E^B)(\mathcal{I}^{A'A} \otimes \mathcal{E}_x^B)[(F_x^{A'A} \otimes I^B)(\varrho_{A'} \otimes \omega^{AB})]\right],$$

for all states ϱ and all effects E. This guarantees that the final state of the system B and the initial state of the system A' coincide. Using the dual channel $\mathcal{E}^*$ we obtain a more compact form:

$$\mathrm{tr}[\varrho E] = \sum_x \mathrm{tr}\left[(F_x^{A'A} \otimes \mathcal{E}_x^{B*}(E))(\varrho_{A'} \otimes \omega^{AB})\right]. \qquad (6.54)$$

It was shown in [142] that, in order to achieve the perfect quantum-state teleportation of a d-dimensional quantum system, it is crucial that the state ω^{AB} initially shared by Alice and Bob is maximally entangled, i.e. $F_x^{A'A} = (U_x \otimes I) P_+^{A'A} (U_x^* \otimes I)$ and $\mathcal{E}_x(\cdot) = U_x \cdot U_x^*$ where $\{U_1, \ldots, U_{d^2}\}$ are orthogonal unitary operators $(\mathrm{tr}[U_x^* U_{x'}] = d\delta_{xx'})$.

As maximally entangled states are very fragile in the presence of any type of noise, it is of interest (not only from the perspective of teleportation) to investigate to what extent the maximally entangled state can be 'distilled' out of general mixed states by means of LOCC actions. Suppose that we have access to n copies of a shared state ω. The goal of the *entanglement distillation protocol* is to transform asymptotically these n copies of ω into $m < n$ (approximate) copies of P_+. The limiting fraction m/n is called the *entanglement distillation rate*.

Example 6.63 (*Entanglement distillation protocol*)

The goal of this example is to illustrate that distillation is indeed possible. Suppose Alice and Bob share two pairs of two-qubit states $\varrho = q P_{00} + (1-q)Q_+$, where $P_{00} = |\varphi_0 \otimes \varphi_0\rangle\langle\varphi_0 \otimes \varphi_0|$ and $Q_+ = |\phi_+\rangle\langle\phi_+|$ with $\phi_+ = \frac{1}{\sqrt{2}}(\varphi_0 \otimes \varphi_1 + \varphi_1 \otimes \varphi_0)$. Suppose that both Alice and Bob apply to the qubits in their possession the controlled-NOT gate such that the qubits forming the first pair play the role of the control and the remaining qubits represent the target. Since $U \equiv V_{\mathrm{CNOT}}^{A_1 A_2} \otimes V_{\mathrm{CNOT}}^{B_1 B_2} = \sum_{j,k=0,1} P_j^{A_1} \otimes P_k^{B_1} \otimes \sigma_j^{A_2} \otimes \sigma_k^{B_2}$, with $P_{0/1} = |\varphi_{0/1}\rangle\langle\varphi_{0/1}|$ and $\sigma_0 = I$, $\sigma_1 = \sigma_x$, it follows that

$$U(\varphi_{00}^{A_1 B_1} \otimes \varphi_{00}^{A_2 B_2}) = \varphi_{0000}^{A_1 B_1 A_2 B_2},$$
$$U(\varphi_{00}^{A_1 B_1} \otimes \phi_+^{A_2 B_2}) = \varphi_{00}^{A_1 B_1} \otimes \phi_+^{A_2 B_2},$$
$$U(\phi_+^{A_1 B_1} \otimes \varphi_{00}^{A_2 B_2}) = \tfrac{1}{\sqrt{2}}(\varphi_{0101}^{A_1 B_1 A_2 B_2} + \varphi_{1010}^{A_1 B_1 A_2 B_2}),$$
$$U(\phi_+^{A_1 B_1} \otimes \phi_+^{A_2 B_2}) = \phi_+^{A_1 B_1} \otimes \psi_+^{A_2 B_2}.$$

Using these expressions we obtain, setting $\omega' = U(\varrho \otimes \varrho)U^*$,

$$\omega' = q^2 P_{0000} + q(1-q)(P_{00} \otimes Q_+ + P_{0101+1010}) + (1-q)^2 Q_+ \otimes P_+.$$

Further, let Alice and Bob measure locally the target qubits in the basis $\{\varphi_0, \varphi_1\}$. When they both record the effect P_1 then the first pair of qubits ends up in the maximally entangled vector state $\phi_+^{A_1 B_1}$; thus with probability $p = \langle\omega'|(I \otimes I \otimes P_0 \otimes P_0)\omega'\rangle = \frac{1}{2}(1-q)^2$ Alice and Bob have prepared a single copy of the maximally entangled state. Otherwise the remaining pair of (control) qubits is discarded and hence the distillation algorithm fails. Using this (nonoptimal) distillation procedure on n pairs of ϱ we can (on average) distil $\frac{1}{4}(1-q)^2 n$ pairs of maximally entangled states. △

Exercise 6.64 Show that the two-qubit state ϱ from the previous example is entangled for all $q > 0$. [Hint: It is possible to use the NPT criterion but the above example also provides an evidence that separable states cannot be distilled.]

Entanglement distillation procedures are very complex and we will omit any further details here. Instead, we will formulate a surprising connection between the Peres–Horodecki criterion and the existence of an entanglement distillation procedure. For a reader interested in more details on entanglement distillation we recommend the review articles [80] and [118], mentioned earlier.

Proposition 6.65 If a state is positive under partial transposition, i.e. $\tau_B[\varrho] \geq O$, then it cannot be distilled, i.e. the entanglement distillation rate is zero.

Proof First, the distillation process is an LOCC channel and LOCC channels cannot change a PPT state into an NPT state. Second, if ω is PPT then $\omega^{\otimes n}$ is also PPT, because $\tau_B(\omega^{\otimes n}) = [\tau_B(\omega)]^{\otimes n}$. Third, maximally entangled states are NPT. In summary, these three statements imply that no maximally entangled state can be distilled from PPT states. In order to show that PPT is LOCC invariant, we will employ the following identity. It is quite simple to check that, for operators $A, C \in \mathcal{L}(\mathcal{H}_A^{\otimes n})$, $B, D \in \mathcal{L}(\mathcal{H}_B^{\otimes n})$ and $X \in \mathcal{L}(\mathcal{H}_A^{\otimes n} \otimes \mathcal{H}_B^{\otimes n})$,

$$\tau_B[(A \otimes B)X(C \otimes D)] = (A \otimes D^T)\tau_B[X](C \otimes B^T). \tag{6.55}$$

As any LOCC channel takes the separable form $\mathcal{F}_n(X) = \sum_j A_j \otimes B_j X A_j^* \otimes B_j^*$, it follows that

$$\tau_B(\mathcal{F}_n(\omega^{\otimes n})) = \sum_j (A_j \otimes B_j^{*T})\tau_B(\omega^{\otimes n})(A_j^* \otimes B_j^T). \tag{6.56}$$

Since $B_j^{*T} = B_j^{T*}$ we see that $\tau_B(\mathcal{F}_n(\omega^{\otimes n})) = \tilde{\mathcal{F}}_n(\tau_B(\omega^{\otimes n}))$ and $\tilde{\mathcal{F}}_n$ is a completely positive map. That is, $\tau_B(\mathcal{F}_n(\omega^{\otimes n}))$ is positive providing that $\tau_B(\omega^{\otimes n})$ is positive. Equivalently, we could prove that if ω is PPT then $\mathcal{F}_n(\omega^{\otimes n})$ is PPT, i.e. 'being PPT' is LOCC invariant and the proposition follows. $\qquad\square$

This result shows that there are entangled states that cannot be distilled; hence, they are useless for teleportation and have a special name.

Definition 6.66 An entangled state $\omega \in \mathcal{S}(\mathcal{H} \otimes \mathcal{H})$ is called *bound entangled* if it cannot be distilled.

The existence of bound entangled states is a strange feature of quantum entanglement and has attracted much attention. Examples can be found in subsection 6.3.5, where PPT entangled states were identified. A main unsolved problem of entanglement theory is the existence or nonexistence of so-called *NPT bound entangled states*, i.e. entangled states that cannot be distilled but can be identified by means of the Peres–Horodecki entanglement criterion.

6.4.2 Multipartite entanglement

Our discussion so far concerns only bipartite systems. In this subsection we briefly sketch some elementary features of the entanglement of multipartite composite systems.

We consider a composite system consisting of N subsystems, which we now label by numbers $1, \ldots, N$ instead of letters. The associated Hilbert space $\mathcal{H} = \mathcal{H}_1 \otimes \cdots \otimes \mathcal{H}_N$ has dimension $D = d_1 \cdots d_N$, where $d_j = \dim \mathcal{H}_j$ is the dimension of the jth subsystem.

Assuming that each subsystem is in the possession of a different experimenter we say that a multipartite channel $\mathcal{E}$ is LOCC if it can be implemented by the joint action of all experimenters coordinated by the unlimited exchange of classical information. In analogy with bipartite systems the related multipartite LOCC ordering relation identifies the set of multipartite separable states as the equivalence class of the LOCC-smallest states. In all other cases we say that a state is entangled. A multipartite state $\varrho \in S(\mathcal{H}_1 \otimes \cdots \otimes \mathcal{H}_N)$ is factorized if $\varrho = \varrho_1 \otimes \cdots \otimes \varrho_N$, where $\varrho_j = \text{tr}_{\bar{j}}[\varrho]$ and by $\text{tr}_{\bar{j}}$ we denote the partial trace over all subsystems except the jth.

Proposition 6.67 A state $\varrho \in S(\mathcal{H}_1 \otimes \cdots \otimes \mathcal{H}_N)$ is separable if and only if it can be written as a convex sum of factorized states, i.e.

$$\varrho = \sum_j q_j \varrho_1^{(j)} \otimes \cdots \otimes \varrho_N^{(j)}, \tag{6.57}$$

where $0 \le q_j \le 1$, $\sum_j q_j = 1$.

Proof The proof is analogous to that of Proposition 6.23. There is no need to go through the details again; we will just repeat the basic line of argument. First, a single-point contraction $S(\mathcal{H}_1 \otimes \cdots \mathcal{H}_N) \mapsto \omega_0 = P_1 \otimes \cdots \otimes P_N$, where the P_j are one-dimensional projections on $\mathcal{H}_j$, is an LOCC channel. Second, there is an LOCC channel mapping ω_0 into an arbitrary state of the form (6.57). Finally, there is no separable channel mapping ω_0 into an entangled state. This proves the proposition that the states expressed in (6.57) are the only LOCC-smallest, i.e. separable, states of a multipartite system. $\square$

Although the composite system consists of N subsystems we can always split them into two subsets, forming a pair of (composite) subsystems A and B each composed of N_A and N_B subsystems, respectively ($N_A + N_B = N$). Let us stress that such a *bipartite cut* is not unique; however, once it is fixed we can employ the results derived for bipartite systems and investigate the properties of multipartite systems from a bipartite perspective. In the following example we illustrate the rather surprising fact that separability with respect to any bipartite cut is not sufficient to guarantee separability in the multipartite sense. This result was first presented by Bennett *et al.* [19].

Example 6.68 Suppose that an unextendible product basis of three qubits labelled A, B, C consists of vectors $\psi_1 = \varphi \otimes \varphi_\perp \otimes \varphi_+ \equiv \varphi \otimes \zeta_1$, $\psi_2 = \varphi_\perp \otimes \varphi_+ \otimes \varphi \equiv \varphi_\perp \otimes \zeta_2$, $\psi_3 = \varphi_+ \otimes \varphi \otimes \varphi_\perp \equiv \varphi_+ \otimes \kappa_1$ and $\psi_4 = \varphi_- \otimes \varphi_- \otimes \varphi_- \equiv \varphi_- \otimes \kappa_2$, where $\varphi_\pm = \frac{1}{\sqrt{2}}(\varphi \pm \varphi_\perp)$. The furthest right equalities determine vectors $\zeta_1, \zeta_2, \kappa_1, \kappa_2$ on $\mathcal{H}_4 = \mathcal{H}_B \otimes \mathcal{H}_C$ satisfying $\langle \zeta_j | \kappa_k \rangle = 0$. For simplicity we omit the use of the indices A, B, C. The state $\omega = \frac{1}{4}(I - \sum_{j=1}^4 |\psi_j\rangle\langle\psi_j|)$ is an example of a multipartite entangled state (see Proposition 6.59), because it does not contain any three-partite product vector in its support. Choose $\zeta_j^\perp$ such that $\langle \zeta_j^\perp \zeta_j \rangle = 0$ and such that the $\zeta_j^\perp$ are from the subspace spanned by the vectors ζ_1, ζ_2, and set $\psi_5 = \varphi \otimes \zeta_1^\perp$, $\psi_6 = \varphi_\perp \otimes \zeta_2^\perp$. In the same manner we define unit vectors $\kappa_j^\perp$ and set $\psi_7 = \varphi_+ \otimes \kappa_1^\perp$, $\psi_8 = \varphi_- \otimes \kappa_2^\perp$. The vectors $\{\psi_j\}$ form an orthonormal factorized basis of the bipartite cut $\mathcal{H}_2 \otimes \mathcal{H}_4$; however, let us note that the vectors $\zeta_j^\perp, \kappa_j^\perp$ are entangled on $\mathcal{H}_4 = \mathcal{H}_B \otimes \mathcal{H}_C$. Using the notation $\omega = \frac{1}{4} \sum_{j=5}^{8} |\psi_j\rangle\langle\psi_j|$, we can see that ω is separable with respect to a bipartite cut $\mathcal{H}_2 \otimes \mathcal{H}_4$ separating qubit A from qubits B and C. Because of the symmetry of the vectors $\{\psi_j\}$ a similar construction works for any other bipartite cuts; thus the entangled state ω is separable with respect to any bipartite cut. $\triangle$

Which multipartite states are maximally entangled? One can show that even for three qubits there is no unique equivalence class of maximally entangled states. In the following we introduce two paradigmatic examples of pure three-qubit states, for which an even stronger inequivalence holds (details can be found in [55]).

Example 6.69 (*GHZ state and paradox*)
Let $\{\varphi, \varphi_\perp\}$ be an orthonormal basis for $\mathcal{H}_2$. We write

$$\psi_{\text{GHZ}} = \tfrac{1}{\sqrt{2}}(\varphi \otimes \varphi \otimes \varphi + \varphi_\perp \otimes \varphi_\perp \otimes \varphi_\perp) \in \mathcal{H}_2 \otimes \mathcal{H}_2 \otimes \mathcal{H}_2,$$

and the related pure state $P_{\text{GHZ}} = |\psi_{\text{GHZ}}\rangle\langle\psi_{\text{GHZ}}|$ is called the *Greenberger–Horn–Zeilinger (GHZ) state*. There are three possible bipartite cuts, and in any of them the state is maximally entangled on $\mathcal{H}_2 \otimes \mathcal{H}_4$. We thus conclude that the GHZ state is an entangled multipartite state.

However, on discarding any qubit the state of the remaining pair of qubits is $\varrho_{\text{GHZ}} = \frac{1}{2}(P_\varphi \otimes P_\varphi + P_\varphi^\perp \otimes P_\varphi^\perp)$, and this state is separable. The remarkable feature of the GHZ state is that, although no pair of subsystems is entangled, the joint state of all three qubits is entangled.

This state is exploited in the so-called GHZ paradox, in which three experimenters (A, B and C) are asked to take part in the following game. They are individually asked one of two questions (labelled by X and Y), the answers to which are either $+1$ or -1. Their answers are required to satisfy the relations

$$X(A)X(B)X(C) = 1,$$
$$X(A)Y(B)Y(C) = Y(A)X(B)Y(C) = Y(A)Y(B)X(C) = -1. \quad (6.58)$$

Multiplying together the left-hand sides and also the right-hand sides of all four equations we find the contradiction $1 = -1$, which means that it is impossible to satisfy the rules of this game. However, if the experimenters share a GHZ state and questions X, Y are replaced by measurements associated with operators σ_x, σ_y (with outcomes labelled ± 1) then the conditions in (6.58) are met whenever the challenge is given. This paradox illustrates a (deterministic) violation of local realism by quantum systems. It was originally proposed by Greenberger, Horne and Zeilinger [67] and is known as the GHZ paradox.

In particular, let us consider the XXX case. In the eigenbasis of σ_x the GHZ vector state takes the form $\psi_{\text{GHZ}} = \frac{1}{2}(\varphi_+ \otimes \varphi_+ \otimes \varphi_+ - \varphi_+ \otimes \varphi_- \otimes \varphi_- - \varphi_- \otimes \varphi_+ \otimes \varphi_- - \varphi_- \otimes \varphi_- \otimes \varphi_+)$, where $\varphi_\pm = \frac{1}{\sqrt{2}}(\varphi \pm \varphi_\perp)$ are the eigenvectors of σ_x. That is, by measuring $\sigma_x \otimes \sigma_x \otimes \sigma_x$ the experimenters will record with equal probabilities one of the outcomes $+++$, $+--$, $-+-$, $--+$, all of which satisfy the constraint $X(A)X(B)X(C) = 1$. The validity of the remaining conditions can be verified in a similar way by expressing the GHZ state in the corresponding measurement bases.   △

Example 6.70 (*W state*)
Again let $\{\varphi, \varphi_\perp\}$ denote an orthonormal basis for $\mathcal{H}_2$. We write

$$\psi_{\text{W}} = \frac{1}{\sqrt{3}}(\varphi \otimes \varphi \otimes \varphi_\perp + \varphi \otimes \varphi_\perp \otimes \varphi + \varphi_\perp \otimes \varphi \otimes \varphi), \qquad (6.59)$$

and the related pure state P_{W} is called the *W state*. In any of the three possible bipartite cuts, the state of the $\mathcal{H}_2 \otimes \mathcal{H}_4$ system is not maximally entangled. Thus the existence of an LOCC channel mapping P_{W} into P_{GHZ} is forbidden from the bipartite entanglement perspective. Let us notice that multipartite LOCC channels are included in the set of LOCC channels with respect to any bipartite cut. In contrast with the GHZ state, for the W state each pair of qubits is in an entangled state (see Example 6.12 and Exercise 6.55):

$$\varrho_W = \frac{1}{3}|\varphi_\perp \otimes \varphi_\perp\rangle\langle\varphi \otimes \varphi| + \frac{2}{3}Q_+, \qquad (6.60)$$

where Q_+ is the projection onto a unit vector $\phi_+ = \frac{1}{\sqrt{2}}(\varphi \otimes \varphi_\perp + \varphi_\perp \otimes \varphi)$. That is, while ϱ_{GHZ} is separable, the state ϱ_W is entangled. Since no LOCC channel can transform separable states into entangled ones we conclude that neither can P_W be (multipartite) LOCC transformed into P_{GHZ}. The states P_{GHZ} and P_W are not comparable from the multipartite entanglement point of view.   △

6.4.3 Evolution of quantum entanglement

It follows from our discussion on entanglement that two spatially separated experimenters cannot create a desired entangled state unless entangled states are already involved in the process or they are able to exchange quantum systems. The reason is that separable states can be turned into entangled ones only via a nonlocal operation (in the sense of a tensor product of Hilbert spaces), and no (spatially) long-range physical interaction has been discovered yet. As a result, in order to entangle a pair of quantum systems produced by independent (uncorrelated) sources, they must be put in contact with some entangling device. It could be either a measurement acting via a quantum operation or a quantum gate implementing an interaction between the systems by means of some quantum channel. Theoretically, there are many different ways to prepare a maximally entangled state from a fixed factorized state. In the following example we describe one that is canonical in the area of quantum computing.

Example 6.71 (*Creation of maximally entangled state*)
Consider a pair of qubits described by the vector state $\varphi \otimes \varphi$. Our goal is to prepare the vector state $\frac{1}{\sqrt{2}}(\varphi \otimes \varphi + \varphi_\perp \otimes \varphi_\perp)$ for some fixed basis $\varphi, \varphi_\perp$. On the basis of the quantum circuit model the experimenters are, typically, trying to implement this process in two steps. First, they prepare a superposition of states φ and $\varphi_\perp$ of the second qubit, i.e. a state transformation

$$\varphi \otimes \varphi \mapsto \varphi \otimes \tfrac{1}{\sqrt{2}}(\varphi + \varphi_\perp). \tag{6.61}$$

This can be achieved by application of the so-called *Hadamard gate* implementing a unitary channel induced by the unitary operator

$$V_{\mathrm{H}}(\varphi) = \tfrac{1}{\sqrt{2}}(\varphi + \varphi_\perp), \quad V_{\mathrm{H}}(\varphi_\perp) = \tfrac{1}{\sqrt{2}}(\varphi - \varphi_\perp). \tag{6.62}$$

In the second step, a so-called *controlled-NOT gate*,

$$V_{\mathrm{CNOT}} = |\varphi\rangle\langle\varphi| \otimes I + |\varphi_\perp\rangle\langle\varphi_\perp| \otimes \sigma_x, \tag{6.63}$$

is applied to both systems, which results in the desired transformation:

$$\varphi \otimes \tfrac{1}{\sqrt{2}}(\varphi + \varphi_\perp) \mapsto \tfrac{1}{\sqrt{2}}(\varphi \otimes \varphi + \varphi_\perp \otimes \varphi_\perp). \tag{6.64}$$

In summary, the sequence of unitary operators $V_{\mathrm{CNOT}}(I \otimes V_{\mathrm{H}})$ transforms the vector state $\varphi \otimes \varphi$ into a maximally entangled vector state ψ_+. The remaining question is how to realize the above mentioned (universal) quantum gates for a particular physical system. This is exactly the type of question that is not addressed in this book. We refer interested readers to textbooks on quantum information theory. △

The following example shows that entanglement creation between systems can be mediated by another system, which means that entangled systems do not have to have met in the past.

Example 6.72 (*Entanglement swapping*)

Quantum teleportation enables us to transfer an unknown state from Alice to Bob just by using classical communication channels and a maximally entangled state P_+^{AB} shared by both sides. We assume that the systems A, A' are both in the possession of Alice. Suppose now that the system A' is in fact entangled with a third system C. Let us denote by $\varrho_{A'C}$ their joint state. By running a teleportation protocol between Alice and Bob the transformation

$$\varrho_{A'C} \otimes P_+^{AB} \mapsto \varrho_{BC} \otimes P_+^{AA'}, \tag{6.65}$$

is implemented (verify!); thus systems B and C exchange roles. In conclusion, although never being close to each other (at any time in the past), the systems B and C become entangled. Simply, the entanglement of the systems A' and C is transformed into the entanglement of B and C.   △

We have shown that, in order to entangle a given system with another, some nonlocal quantum operation (with respect to the tensor product) must take place. In what follows we will briefly discuss the opposite case. In particular, what is needed to disentangle a given system from another? Suppose that we do not know the exact nature of this other system, which is entangled with the given system. In such a case we are restricted to actions on this system. Is it possible, by applying a channel $\mathcal{E}$, to disentangle the system $\mathcal{H}$ from any other system $\mathcal{H}_{\text{other}}$ with which it was originally entangled? It turns out that this is possible, and the class of such channels forms as a very interesting subset of the channels acting on $\mathcal{H}$.

Definition 6.73 A channel $\mathcal{E} : T(\mathcal{H}) \to T(\mathcal{H})$ is called *entanglement-breaking* if, for all $\mathcal{H}_{\text{other}}$, the state $\omega' = \mathcal{E} \otimes \mathcal{I}_{\text{other}}(\omega)$ is separable for all $\omega \in \mathcal{S}(\mathcal{H} \otimes \mathcal{H}_{\text{other}})$.

The following proposition, which was originally reported by Horodecki, Shor and Ruskai [78], identifies entanglement-breaking channels exactly with *measure-and-prepare* procedures, consisting of a measurement of some POVM and the subsequent conditional state preparation.

Proposition 6.74 A channel $\mathcal{E}$ is entanglement-breaking if and only if $\mathcal{E}$ is of the form $\mathcal{E}(\varrho) = \sum_j \xi_j \text{tr}[\varrho F_j]$, where the ξ_j are states and the positive operators F_j determines a discrete POVM, i.e. $\sum_j F_j = I$.

Proof By definition, if $\mathcal{E}$ is entanglement-breaking then the state $\omega = (\mathcal{E} \otimes \mathcal{I})[P_+]$ is separable, i.e. $\omega_{\mathcal{E}} = \sum_n p_n |\varphi_n\rangle\langle\varphi_n| \otimes |\phi_n\rangle\langle\phi_n|$. Consider a map $\mathcal{E}'$ of the form $\mathcal{E}'(\varrho) = \sum_n \xi_n \text{tr}[\varrho F_n]$ with $\xi_n = |\phi_n\rangle\langle\phi_n|$ and $F_n = d p_n |\varphi_n\rangle\langle\varphi_n|$, where d is the dimension of the Hilbert space on which the channel $\mathcal{E}$ acts. Let us note that $\text{tr}_A[\omega_{\mathcal{E}}] = \frac{1}{d} I = \sum_n p_n |\phi_n\rangle\langle\phi_n|$; thus the positive operators F_n are properly normalized ($\sum_n F_n = I$) and form a POVM. Owing to the Choi–Jamiolkowski isomorphism, the state $\omega_\mathcal{E}$ determines the channel $\mathcal{E}$ uniquely. Therefore, since $\omega_{\mathcal{E}'} = (\mathcal{E}' \otimes \mathcal{I})[P_+] = \sum_n p_n |\varphi_n\rangle\langle\varphi_n| \otimes |\phi_n\rangle\langle\phi_n| = \omega_\mathcal{E}$, we can conclude that $\mathcal{E} = \mathcal{E}'$; hence, if $\mathcal{E}$ is entanglement-breaking then it has the desired form.

Conversely, if $\mathcal{E}$ is defined as $\mathcal{E}(\varrho) = \sum_j \xi_j \, \mathrm{tr}[\varrho F_j]$ then

$$(\mathcal{E} \otimes \mathcal{I})(\omega) = \sum_j (\xi_j \otimes I) \, \mathrm{tr}_A[(F_j \otimes I)\Omega] = \sum_j q_j \xi_j \otimes Q_j,$$

where $q_j = \mathrm{tr}[(F_j \otimes I)\Omega]$ and $Q_k = q_j^{-1} \mathrm{tr}_A[\sqrt{F_j \otimes I}\Omega\sqrt{F_j \otimes I}]$ are positive operators with $\mathrm{tr}[Q_j] = q_j^{-1}\mathrm{tr}[(F_j \otimes I)\Omega] = 1$, i.e. the Q_j are density operators. That is, the state $(\mathcal{E} \otimes \mathcal{I})(\omega)$ is separable. □

By definition, entanglement-breaking channels disentangle any system from the rest of the world (excluding the local environment used to implement the channel). The Choi–Jamiolkowski operator of such a channel is necessarily separable. It turns out that the inverse statement is also true: if the Choi–Jamiolkowski operator associated with a channel $\mathcal{E}$ is separable then $\mathcal{E}$ is entanglement-breaking. In fact, in the above proof we assigned a measure-and-prepare channel for any separable Choi–Jamiolkowski state. Consequently, the decision whether a given channel is entanglement-breaking is as difficult as testing the separability of its Choi–Jamiolkowski representation.

Exercise 6.75 Consider the class of Pauli channels, introduced in Section 4.6, i.e. $\mathcal{E}_{\mathrm{Pauli}}(\cdot) = \sum_j p_j \sigma_j \cdot \sigma_j$. Find n such that the concatenation $\mathcal{E}_{\mathrm{Pauli}}^n$ is an entanglement-breaking channel.

6.5 Example: Werner states

In this section we introduce a specific family of states called *Werner states*, originally used by Werner in his seminal article [141] to show that there are entangled states admitting local hidden-variable models. Since then this class of states has been employed in many different case studies and provides important insights into the properties of entangled states.

Definition 6.76 A state $\varrho \in \mathcal{S}(\mathcal{H}_d \otimes \mathcal{H}_d)$ is called a *Werner state* if it is invariant under unitary channels of the form $\sigma_{U\otimes U}$, i.e. if $[\varrho, U \otimes U] = 0$ for all unitary operators U on $\mathcal{H}_d$.

Let us note that by definition a Werner state ϱ is invariant under the twirling channel introduced in Example 6.20, i.e. $\mathcal{T}(\varrho) = \varrho$. It follows that all Werner states can be written as linear combinations of projections P_+ and P_- onto symmetric and antisymmetric subspaces $\mathcal{H}_\pm$, respectively. Taking account of positivity and normalization, Werner states are of the form

$$\varrho_w = w\frac{1}{d_+}P_+ + (1 - w)\frac{1}{d_-}P_-, \qquad (6.66)$$

where $w \in [0, 1]$ and $d_\pm = \mathrm{tr}[P_\pm] = \frac{1}{2}d(d \pm 1)$ are the dimensions of $\mathcal{H}_\pm$.

We recall that $V_{\mathrm{SWAP}} = P_+ - P_-$ and $I = P_+ + P_-$; hence

$$P_\pm = \tfrac{1}{2}(I \pm V_{\mathrm{SWAP}}). \qquad (6.67)$$

Therefore, we can express Werner states alternatively as linear combinations of the identity and SWAP operators:

$$\begin{aligned}
\varrho_w &= \frac{w}{d(d + 1)}(I + V_{\mathrm{SWAP}}) + \frac{1 - w}{d(d - 1)}(I - V_{\mathrm{SWAP}}) \\
&= \frac{1}{d(d^2 - 1)}\Big((d + 1 - 2w)I + (2dw - d - 1)V_{\mathrm{SWAP}}\Big).
\end{aligned} \qquad (6.68)$$

Proposition 6.77 A Werner state ϱ_w is entangled if and only if $w < \frac{1}{2}$.

Proof Let us start with the observation that the functional $\omega \mapsto \mathrm{tr}[V_{\mathrm{SWAP}}\omega]$ is invariant under the action of the twirling channel $\mathcal{T}$, i.e. $\mathrm{tr}[V_{\mathrm{SWAP}}\mathcal{T}(\omega)] = \mathrm{tr}[V_{\mathrm{SWAP}}\omega]$ for all states ω. In fact, as a result of the twirling channel, an arbitrary state ω is mapped into some Werner state ϱ_μ with some specific value of w. Therefore, as the twirling channel is LOCC, the Werner state $\varrho_w = \mathcal{T}(\omega)$ is separable whenever ω is separable. Consequently, since for pure product states $\mathrm{tr}[V_{\mathrm{SWAP}}(|\varphi \otimes \phi\rangle\langle\varphi \otimes \phi|)] = |\langle\varphi|\phi\rangle|^2 \in [0, 1]$, it follows that for each Werner state ϱ_w with $\mathrm{tr}[V_{\mathrm{SWAP}}\varrho_w] \geq 0$ there exists a separable state $\omega_0 = |\varphi \otimes \phi\rangle\langle\varphi \otimes \phi|$ such that $\varrho_w = \mathcal{T}[|\varphi \otimes \phi\rangle\langle\varphi \otimes \phi|]$. Therefore, if $\mathrm{tr}[V_{\mathrm{SWAP}}\varrho_w] \geq 0$ then the Werner state ϱ_w must be separable.

Let us recall that the selfadjoint operator V_{SWAP} is proportional to the entanglement witness associated with the partial transposition criterion (see Example 6.50). That is, if $\mathrm{tr}[V_{\mathrm{SWAP}}\omega] < 0$ then the state ω is entangled. Since $\mathrm{tr}[V_{\mathrm{SWAP}}P_\pm] = \pm d_\pm$, we obtain

$$\mathrm{tr}[V_{\mathrm{SWAP}}\varrho_w] = w\frac{\mathrm{tr}[V_{\mathrm{SWAP}}P_+]}{d_+} + (1 - w)\frac{\mathrm{tr}[V_{\mathrm{SWAP}}P_-]}{d_-} = 2w - 1.$$

In summary, for values $w \geq \frac{1}{2}$ Werner states are separable and for $w < \frac{1}{2}$ they are entangled.   □

Exercise 6.78 For the two-qubit case show that $\varrho_w \leq_{\mathrm{LOCC}} \varrho_{w'}$ if $w' < w$, i.e. the larger is the fraction P_- the more entangled is the Werner state. [Hint: The depolarizing channels introduced in Section 4.6 are useful.]

Figure 6.5 Werner states: the projections $P_\pm$ represent unnormalized states.

Proposition 6.79 A Werner state ϱ_w is PPT if and only if $w \geq \frac{1}{2}$.

Proof It follows from the previous proposition that for $w \geq \frac{1}{2}$ the state is separable, and hence positive under partial transposition. It remains to show that for $w < \frac{1}{2}$ the operator $\tau_A(\varrho_w)$ has a negative eigenvalue. For the purposes of this proof we will use the parametrization of Werner states from (6.68). The identity operator is unaffected by the partial transposition and

$$\tau_A(V_{\text{SWAP}}) = \tau_A \left(\sum |\varphi_j \otimes \varphi_k\rangle\langle\varphi_k \otimes \varphi_j| \right) = \sum |\varphi_k \otimes \varphi_k\rangle\langle\varphi_j \otimes \varphi_j|$$
$$= d|\psi_+\rangle\langle\psi_+|,$$

where $\psi_+ = \frac{1}{\sqrt{d}} \sum_j \varphi_j \otimes \varphi_j$. Clearly, ψ_+ is an eigenvector of $\tau_A(\varrho_w)$ and

$$\tau_A(\varrho_w)\psi_+ = \frac{1}{d(d^2-1)}((2w-1)(d^2-1))\psi_+ = \frac{2w-1}{d}\psi_+ . \qquad (6.69)$$

The eigenvalue is negative for all $w < \frac{1}{2}$. $\qquad \square$

Exercise 6.80 Show that the reduction criterion formulated in Exercise 6.49 does not detect the entanglement of Werner states unless $d = 2$.

Combining the previous propositions we can conclude that the PPT criterion is necessary and sufficient to identify the entanglement of Werner states in any dimension.

Example 6.81 (*Mixed states distillation via Werner states*)
Starting from an arbitrary two-qubit state ω, the twirling channel (see Example 6.20) transforms this state into a Werner state. Namely,

$$T(\omega) = \frac{1}{3} \text{tr}[\omega P_+]P_+ + \text{tr}[\omega P_-]P_-; \qquad (6.70)$$

thus $T(\omega) = \varrho_w$ with $w = \text{tr}[\omega P_+]$. This fact is employed to reduce the distillation of a general bipartite state into the distillation of Werner states. This is not the most general, or the most optimal strategy; however, it possesses the sign of universality. Let us recall that twirling is a LOCC channel and thus represents a valid distillation step.

Suppose that Alice and Bob share two copies of the Werner state ϱ_w, i.e. $\varrho_w^{AB} \otimes \varrho_w^{A'B'}$. In the first step Alice implements the local unitary transformation

$\sigma_y \otimes \sigma_y$ on A, A', thus transforming the Werner states into a mixture of the canonical maximally entangled state ψ_+ and the mixture $\frac{1}{3}(P_1 + P_2 + P_3)$, i.e.

$$\varrho_w \mapsto \varrho'_w = \tfrac{1}{3}w(P_1 + P_2 + P_3) + (1 - w)P_0, \qquad (6.71)$$

where we use the notation $P_0 = |\psi_+\rangle\langle\psi_+|$ and $P_j = (\sigma_j \otimes I)(P_0)$. In the second step both Alice and Bob apply locally CNOT gates on their qubits, regarding A, B as the pair of control qubits and A', B' as the pair of target qubits. Under this action,

$$
\begin{aligned}
P_0 \otimes P_0 &\mapsto & P_0 \otimes P_0, \quad \tfrac{1}{4}I \otimes \tfrac{1}{4}I &\mapsto & \tfrac{1}{4}I \otimes \tfrac{1}{4}I, \\
\tfrac{1}{4}I \otimes P_0 &\mapsto & \tfrac{1}{4}((P_0 + P_3) \otimes P_0 + (P_1 + P_2) \otimes P_1); \\
P_0 \otimes \tfrac{1}{4}I &\mapsto & \tfrac{1}{4}(P_0 \otimes (P_0 + P_2) + P_3 \otimes (P_1 + P_3)).
\end{aligned}
$$

Let us note that $\varrho'_w = x P_0 + (1 - x)\tfrac{1}{4}I$, with $x = 1 - \tfrac{4}{3}w$. If, on performing σ_z measurements (see Chapter 3) on the target qubits A', B', they record the same outcomes then the unnormalized state of systems A and B can be post-selected:

$$
\begin{aligned}
\tilde\omega_1 &= \text{tr}_{A'B'}\left[(I^{AB} \otimes (P_0^{A'B'} + P_3^{A'B'}))(\sigma_{\text{CNOT}}^{AA'} \otimes \sigma_{\text{CNOT}}^{BB'})(\varrho'_w \otimes \varrho'_w)\right] \\
&= x^2 P_0 + \tfrac{1}{2}(1 - x)^2\tfrac{1}{4}I + \tfrac{1}{2}x(1 - x)(P_0 + P_3),
\end{aligned}
$$

where $\sigma_{\text{CNOT}}(\cdot) = V_{\text{CNOT}} \cdot V_{\text{CNOT}}$. Normalizing this state and applying $\sigma_y^A \otimes I^B$, we obtain the state

$$\omega_1 = \tfrac{1}{\text{tr}[\tilde\omega_1]}\left(x^2 P_2 + \tfrac{1}{2}(1 - x)^2\tfrac{1}{4}I + \tfrac{1}{2}x(1 - x)(P_1 + P_2)\right),$$

where $\text{tr}[\tilde\omega_1] = \tfrac{1}{2}(1 + x^2)$ is the probability that Alice and Bob record the same outcome. It equals unity when the Werner state is maximally entangled, i.e. if $x = 1$ and $w = 0$. By twirling the state ω_1 we obtain a new Werner state with $w_1 = 1 - \text{tr}[\omega_1 P_-] = 1 - \text{tr}[\omega_1 P_2]$; thus

$$x_1 = 1 - \tfrac{4}{3}w_1 = x\frac{4x + 2}{3(x^2 + 1)}. \qquad (6.72)$$

Since $(4x+2)/(3x^2+3) > 1$ for $\tfrac{1}{3} < x < 1$, or, equivalently, for $w = \text{tr}[\omega P_+] < \tfrac{1}{2}$, it follows that $x_1 > x$. As a result we find that the Werner state associated with ω_1 is more entangled (it contains a larger fraction of the singlet state, see Exercise 6.78) than the Werner state associated with the original state ω. One can apply this procedure recursively, consuming two pairs of ω_1 to produce ω_2, etc. Finally, whenever $\text{tr}[\omega P_-] > \tfrac{1}{2}$ the protocol that we have described enables Alice and Bob to distil some pairs of maximally entangled states, i.e. $\omega_\infty = P_-$ with $x \to 1$ and $w \to 0$. $\qquad \triangle$

Symbols

Sets of Hilbert space operators

- $\mathcal{L}(\mathcal{H})$ bounded linear operators
- $\mathcal{L}_s(\mathcal{H})$ bounded self-adjoint operators ($T \in \mathcal{L}(\mathcal{H})$ and $T = T^*$)
- $\mathcal{E}(\mathcal{H})$ effects ($T \in \mathcal{L}_s(\mathcal{H})$ and $O \leq T \leq I$)
- $\mathcal{P}(\mathcal{H})$ projections ($T \in \mathcal{L}_s(\mathcal{H})$ and $T = T^2$)
- $\mathcal{U}(\mathcal{H})$ unitary operators ($T \in \mathcal{L}(\mathcal{H})$ and $TT^* = T^*T = I$)
- $\mathcal{T}(\mathcal{H})$ trace class operators
- $\mathcal{T}_s(\mathcal{H})$ self-adjoint trace class operators ($T \in \mathcal{L}_s(\mathcal{H}) \cap \mathcal{T}(\mathcal{H})$)

Sets of states

- $\mathcal{S}(\mathcal{H})$ states ($T \in \mathcal{T}(\mathcal{H})$, $T \geq O$ and $\mathrm{tr}[T] = 1$)
- $\mathcal{S}^{\mathrm{ext}}(\mathcal{H})$ pure states ($T \in \mathcal{S}(\mathcal{H})$, $\mathrm{tr}[T^2] = 1$)
- $\mathcal{S}^{\mathrm{us}}(\mathcal{H})$ unnormalized states ($T \in \mathcal{T}(\mathcal{H})$, $T \geq O$ and $\mathrm{tr}[T] \leq 1$)
- $\mathcal{S}^{\mathrm{fac}}(\mathcal{H}_A \otimes \mathcal{H}_B)$ factorized states
- $\mathcal{S}^{\mathrm{sep}}(\mathcal{H}_A \otimes \mathcal{H}_B)$ separable states

Sets of channels and of operations

- $\mathcal{M}_{cp}$ completely positive linear mappings
- $\mathcal{O}$ quantum operations ($\mathcal{E} \in \mathcal{M}_{cp}$ and $\mathcal{E}$ is trace nonincreasing)
- $\mathcal{O}_c$ quantum channels ($\mathcal{E} \in \mathcal{M}_{cp}$ and $\mathcal{E}$ is trace preserving)
- $\mathcal{O}_c^{\mathrm{fac}}$ all local channels
- $\mathcal{O}_c^{\mathrm{sep}}$ separable channels
- $\mathcal{O}_c^{\mathrm{LOCC}}$ all LOCC channels

References

[1] A. Acin. Statistical distinguishability between unitary operations. *Phys. Rev. Lett.*, 87: 177 901, 2001.

[2] S.T. Ali, J.-P. Antoine and J.-P. Gazeau. *Coherent states, Wavelets and Their Generalizations*. Springer-Verlag, New York, 2000.

[3] S.T. Ali, C. Carmeli, T. Heinosaari and A. Toigo. Commutative povms and fuzzy observables. *Found. Phys.*, 39: 593–612, 2009.

[4] S.T. Ali and G.G. Emch. Fuzzy observables in quantum mechanics. *J. Math, Phys.* 15: 176–182, 1974

[5] A. Ambainis, M. Mosca, A. Tapp and R. de Wolf. Private quantum channels. In *FOCS, Proc. 41st Annual Symp. on Foundations of Computer Science*, pp. 547–553. IEEE Computer Society, 2000.

[6] H. Araki and M.M. Yanase. Measurement of quantum mechanical operators. *Phys. Rev. 2*, 120: 622–626, 1960.

[7] A. Arias, A. Gheondea and S. Gudder. Fixed points of quantum operations. *J. Math. Phys.*, 43: 5872–5881, 2002.

[8] K. Audenaert and S. Scheel. On random unitary channels. *New J. Phys.*, 10: 023 011, 2008.

[9] A. Barenco, C.H. Bennett, R. Cleve *et al.* Elementary gates for quantum computation. *Phys. Rev. A*, 52: 3457–3467, 1995.

[10] V. Bargmann. Note on Wigner's theorem on symmetry operations. *J. Math. Phys.*, 5; 862–868, 1964

[11] J.S. Bell. On the Einstein Podolsky Rosen paradox. *Physics*, 1: 195–200, 1964.

[12] J.S. Bell. On the problem of hidden variables in quantum mechanics. *Rev. Mod. Phys.*, 38: 447–452, 1966.

[13] E. Beltrametti, G. Cassinelli and P. Lahti. Unitary measurements of discrete quantities in quantum mechanics. *J. Math. Phys.*, 31: 91–98, 1990.

[14] I. Bengtsson and K. Życzkowski. *Geometry of quantum states*. Cambridge University Press, 2006.

[15] C.H. Bennett. Quantum cryptography using any two nonorthogonal states. *Phys. Rev. Lett.*, 68: 3121–3124, 1992.

[16] C.H. Bennett and G. Brassard. Quantum cryptography: public-key distribution and coin-tossing. In *Proc. IEEE Int. Conf. on Computers, Systems and Signal Processing*, pp. 175–179. IEEE, New York, 1984.

[17] C.H. Bennett and S.J. Wiesner. Communication via one- and two-particle operators on Einstein-Podolsky-Rosen states. *Phys. Rev. Lett.*, 69: 2881–2884, 1992.

[18] C.H. Bennett, G. Brassard, C. Crepeau, R. Jozsa, A. Peres, and W.K. Wootters. Teleporting an unknown quantum state via dual classical and Einstein–Podolsky–Rosen channels. *Phys. Rev. Lett.*, 70: 1895–1899, 1993.

[19] C.H. Bennett, D.P. DiVincenzo, C.A. Fuchs, *et al.* Quantum nonlocality without entanglement. *Phys. Rev. A*, 59: 1070–1091, 1999.

[20] C.H. Bennett, D.P. DiVincenzo, T. Mor, P.W. Shor, J.A. Smolin, and B.M. Terhal. Unextendible product bases and bound entanglement. *Phys. Rev. Lett.*, 82: 5385, 1999.

[21] J.A. Bergou, U. Herzog and M. Hillery. Discrimination of quantum states. In *Quantum State Estimation*, Vol. 649 of *Lecture Notes in Physics*, pp. 417–465. Springer, Berlin, 2004.

[22] P. Busch. Unsharp reality and joint measurements for spin observables. *Phys. Rev. D*, 33: 2253–2261, 1986.

[23] P. Busch. Quantum states and generalized observables: a simple proof of Gleason's theorem. *Phys. Rev. Lett.*, 91: 120403/1–4, 2003.

[24] P. Busch. The role of entanglement in quantum measurement and information processing. *Int. J. Theor. Phys.*, 42: 937–941, 2003.

[25] P. Busch. 'No information without disturbance': quantum limitations of measurement. In J. Christian and W. Myrvold, eds., *Quantum reality, relativistic causality, and closing the epistemic circle*. Springer-Verlag, Berlin, 2008.

[26] P. Busch and P. Lahti. The determination of the past and the future of a physical system in quantum mechanics. *Found. Phys.*, 19: 633–678, 1989.

[27] P. Busch and P Lahti. Some remarks on unsharp quantum measurements, quantum nondemolition, and all that. *Ann. Physik 7*, 47: 369–382, 1990.

[28] P. Busch and P. Lahti. The complementarity of quantum observables: Theory and experiments. *Riv. Nuovo Cimento*, 18: 1–27, 1995.

[29] P. Busch and H.-J. Schmidt. Coexistence of qubit effects. *Quantum Inf. Process.*, 9: 143–169, 2010.

[30] P. Busch and C. Shilladay. Complementarity and uncertainty in Mach–Zehnder interferometry and beyond. *Phys. Rep.*, 435: 1–31, 2006.

[31] P. Busch and J. Singh. Lüders theorem for unsharp quantum measurements. *Phys. Lett. A*, 249: 10–12, 1998.

[32] P. Busch, G. Cassinelli and P. Lahti. Probability structures for quantum state spaces. *Rev. Math. Phys.*, 7: 1105–1121, 1995.

[33] P. Busch, M. Grabowski and P. Lahti. Repeatable measurements in quantum theory: their role and feasibility. *Found. Phys.*, 25: 1239–1266, 1995.

[34] P. Busch, M. Grabowski and P. Lahti. *Operational Quantum Physics*. Springer-Verlag, Berlin, 1997; second, corrected, printing.

[35] P. Busch, P. Lahti and P. Mittelstaedt. *The Quantum Theory of Measurement*. Springer-Verlag, Berlin, second edition, 1996.

[36] A. Aspuru-Guzik. C.A. Rodríguez Rosario and K. Modi. Linear assignment maps for correlated system–environment states. *Phys. Rev. A*, 81: 012 313, 2010.

[37] C. Carmeli, T. Heinonen and A. Toigo. Intrinsic unsharpness and approximate repeatability of quantum measurements. *J. Phys. A*, 40: 1303–1323, 2007.

[38] G. Cassinelli, E. De Vito, P. Lahti and A. Levrero. *The Theory of Symmetry Actions in Quantum Mechanics*. Springer, Berlin, 2004.

[39] G. Cassinelli, E. De Vito and A. Levrero. On the decompositions of a quantum state. *J. Math. Anal. Appl.*, 210: 472–483, 1997.

[40] C.M. Caves, C.A.. Fuchs and R. Schack. Unknown quantum states: the quantum de Finetti representation. *J. Math. Phys.*, 43: 4537–4559, 2002.

[41] N.J. Cerf and J. Fiurášek. Optimal quantum cloning – a review. In *Progress in Optics*, Vol. 49. Elsevier, 2006.

[42] A. Chefles. Quantum state discrimination. *Contemp. Physics*, 41: 401–424, 2000.

[43] M. Choi. Completely positive linear maps on complex matrices. *Linear Alg. Appl.*, 10: 285–290, 1975.

[44] J.F. Clauser, M.A. Horne, A. Shimony and R.A. Holt. Proposed experiment to test local hidden-variable theories. *Phys. Rev. Lett.*, 23: 880–884, 1969.

[45] J.B. Conway. *A Course in Functional Analysis*, second edition. Springer, 1990.

[46] J.B. Conway. *A Course in Operator Theory*. American Mathematical Society, Providence, Rhode Island, 2000.

[47] R. Cooke, M. Keane and W. Moran. An elementary proof of Gleason's theorem. *Math. Proc. Cambridge Philos. Soc.*, 98: 117–128, 1985.

[48] J. Corbett. The Pauli problem, state reconstruction and quantum-real numbers. *Rep. Math. Phys.*, 57: 53–68, 2006.

[49] J. Corbett and C. Hurst. Are wave functions uniquely determined by their position and momentum distributions? *J. Austral. Math. Soc. Ser. B*, 20: 182–201, 1978.

[50] G.M. D'Ariano, P. Lo Presti and P. Perinotti. Classical randomness in quantum measurements. *J. Phys. A*, 38: 5979–5991, 2005.

[51] E.B. Davies. *Quantum Theory of Open Systems*. Academic Press, London, 1976.

[52] E.B. Davies and J.T. Lewis. An operational approach to quantum probability. *Comm. Math. Phys.*, 17: 239–260, 1970.

[53] W.M. de Muynck. *Foundations of Quantum Mechanics, An Empiricist Approach*. Kluwer, Dordrecht, 2002.

[54] P.A.M. Dirac. *The Principles of Quantum Mechanics*. Oxford University Press, 1930.

[55] W. Dür, G. Vidal and J.I. Cirac. Three qubits can be entangled in two inequivalent ways. *Phys. Rev. A*, 62: 062 314, 2000.

[56] T. Durt, B.-G. Englert, I. Bengtsson and K. Zyczkowski. On mutually unbiased bases. *Int. J. Quant. Inf.*, 8: 535–640, 2010.

[57] A. Dvurečenskij. *Gleason's Theorem and Its Applications*. Kluwer, Dordrecht, 1993.

[58] A. Einstein, B. Podolsky and N. Rosen. Can quantum-mechanical description of physical reality be considered complete? *Phys. Rev.*, 47: 777–780, 1935.

[59] Y. Feng, R. Duan and M. Ying. Unambiguous discrimination between mixed quantum states. *Phys. Rev. A*, 70: 012 308, 2004.

[60] G.B. Folland. *A Course in Abstract Harmonic Analysis*. CRC Press, Boca Raton, Florida, 1995.

[61] G.B. Folland. *Real Analysis*, second edition. John Wiley & Sons, New York, 1999.

[62] A. Friedman. *Foundations of Modern Analysis*. Dover, New York, 1982. Reprint of the 1970 original.

[63] C. Friedman. Some remarks on Pauli data in quantum mechanics. *J. Austral. Math. Soc. Ser. B*, 30: 298–303, 1989.

[64] C. Gerry and P. Knight. *Introductory Quantum Optics*. Cambridge University Press, 2005.

[65] N. Gisin, G. Ribordy, W. Tittel and H. Zbinden. Quantum cryptography. *Rev. Mod. Phys.*, 74: 145–195, 2002.

[66] A.M. Gleason. Measures on the closed subspaces of a Hilbert space. *J. Math. Mech.*, 6: 885–893, 1957.

[67] D.M. Greenberger, M.A. Horne and A. Zeilinger. Going beyond Bell's theorem. In M. Kafatos, ed., *Bell's Theorem, Quantum Theory, and Conceptions of the Universe*, pp. 69–72. Kluwer Academic Publishers, Dordrecht, 1989.

[68] S. Gudder. Lattice properties of quantum effects. *J. Math. Phys.*, 37: 2637, 1996.

[69] S. Gudder and R. Greechie. Effect algebra counterexamples. *Math. Slovaca*, 46: 317–325, 1996.

[70] R. Haag and D. Kastler. An algebraic approach to quantum field theory. *J. Math. Phys.*, 5: 848–861, 1964.

[71] N. Hadjisavvas. Properties of mixtures on non-orthogonal states. *Lett. Math. Phys.*, 5: 327–332, 1981.

[72] M. Hayashi. *Quantum Information*. Springer-Verlag, Berlin, 2006. Translated from the 2003 Japanese original.

[73] T. Heinonen. Optimal measurements in quantum mechanics. *Phys. Lett. A*, 346: 77–86, 2005.

[74] T. Heinosaari and M.M. Wolf. Nondisturbing quantum measurements. *J. Math. Phys.*, 51: 092201, 2010.

[75] C.W. Helstrom. *Quantum Detection and Estimation Theory*. Academic Press, New York, 1976.

[76] A.S. Holevo. *Probabilistic and Statistical Aspects of Quantum Theory*. North-Holland, Amsterdam, 1982.

[77] M. Horodecki, P. Horodecki and R. Horodecki. Separability of mixed states: necessary and sufficient conditions. *Phys. Lett. A*, 223: 1–8, 1996.

[78] M. Horodecki, P.W. Shor and M.B. Ruskai. Entanglement breaking channels. *Rev. Math. Phys.*, 15: 629–641, 2003.

[79] P. Horodecki. Separability criterion and inseparable mixed states with positive partial transposition. *Phys. Lett. A*, 232, 1997.

[80] R. Horodecki, P. Horodecki, M. Horodecki and K. Horodecki. Quantum entanglement. *Rev. Mod. Phys.* 81: 865–942, 209.

[81] L.P. Hughston, R. Jozsa and W.K. Wootters. A complete classification of quantum ensembles having a given density matrix. *Phys. Lett. A*, 183: 14–18, 1993.

[82] I.D. Ivanović. Geometrical description of quantal state determination. *J. Phys. A: Math. Gen.*, 14: 3241–3245, 1981.

[83] I.D. Ivanović. How to differentiate between nonorthogonal states. *Phys. Lett. A*, 123: 257–259, 1987.

[84] E.T. Jaynes. Information theory and statistical mechanics. II. *Phys. Rev.*, 108: 171–190, 1957.

[85] A. Jenčová and S. Pulmannová. Characterizations of commutative POV measures. *Found. Phys.*, 39: 613–624, 2009.

[86] D. Jonathan and M.B. Plenio. Entanglement-assisted local manipulation of pure quantum states. *Phys. Rev. Lett.*, 83: 3566–3569, 1999.

[87] Y. Kinoshita, R. Namiki, T. Yamamoto, M. Koashi and N. Imoto. Selective entanglement breaking. *Phys. Rev. A*, 75: 032 307, 2007.

[88] K. Kraus. General state changes in quantum theory. *Ann. Physics*, 64: 311–335, 1971.

[89] K. Kraus. *States, Effects, and Operations*. Springer-Verlag, Berlin, 1983.

[90] S. Kullback and R.A. Leibler. On information and sufficiency. *Ann. Math. Statistics*, 22: 79–86, 1951.

[91] P.J. Lahti and M. Mączynski. Partial order of quantum effects. *J. Math. Phys.*, 36: 1673–1680, 1995.

[92] L.J. Landau and R.F. Streater. On Birkhoff's theorem for doubly stochastic completely positive maps of matrix algebras. *Linear Algebra Appl.*, 193: 107–127, 1993.

[93] G. Lindblad. Completely positive maps and entropy inequalities. *Comm. Math. Phys.*, 40: 147–151, 1975.

[94] L. Loveridge and P. Busch. 'Measurement of quantum mechanical operators' revisited. *Eur. Phys. J. D*, 2011.

[95] G. Lüders. Über die Zustandsänderung durch den Messprozess. *Ann. Physik (6)*, 8: 322–328, 1951.

[96] G. Ludwig. *Foundations of Quantum Mechanics I*. Springer-Verlag, New York, 1983.

[97] D. Mayers. Unconditional security in quantum cryptography. *J. Assoc. Comp. Mach.*, 48: 351, 2001.

[98] C.B. Mendl and M.M. Wolf. Unital quantum channels – convex structure and revivals of Birkhoff's theorem. *Comm. Math. Phys.*, 289: 1057–1086, 2009.

[99] D.N. Mermin. *Quantum Computer Science*. Cambridge University Press, 2007.

[100] M. Mičuda, M. Ježek, M. Dušek and J. Fiurášek. Experimental realization of programmable quantum gate. *Phys. Rev. A*, 78: 062 311, 2008.

[101] L. Molnár. *Selected Preserver Problems on Algebraic Structures of Linear Operators and on Function Spaces*, Vol. 1895 of *Lecture Notes in Mathematics*. Springer-Verlag, Berlin, 2007.

[102] M.A. Nielsen. Conditions for a class of entanglement transformations. *Phys. Rev. Lett.*, 83: 436–439, 1999.

[103] M.A. Nielsen and I.L. Chuang. Programmable quantum gate arrays. *Phys. Rev. Lett.*, 79: 321–324, 1997.

[104] M.A. Nielsen and I.L. Chuang. *Quantum Computation and Quantum Information*. Cambridge University Press, 2000.

[105] M. Ohya and D. Petz. *Quantum Entropy and Its Use*. Springer-Verlag, Berlin. Second, corrected printing, 2004.

[106] M. Ozawa. Conditional expectation and repeated measurements of continuous quantum observables. In *Probability Theory and Mathematical Statistics (Tbilisi, 1982)*, Vol. 1021 of *Lecture Notes in Mathematics*, pp. 518–525. Springer, Berlin, 1983.

[107] M. Ozawa. Quantum measuring processes of continuous observables. *J. Math. Phys.*, 25: 79–87, 1984.

[108] M. Ozawa. Conditional probability and a posteriori states in quantum mechanics. *Publ. RIMS, Kyoto Univ.*, 21: 279–295, 1985.

[109] M. Ozawa. Operations, disturbance, and simultaneous measurability. *Phys. Rev. A*, 63: 032 109, 2001.

[110] M. Paris and J. Řeháček, eds. *Quantum State Estimation*, Vol. 649 of *Lecture Notes in Physics*. Springer-Verlag, Berlin, 2004.

[111] K.R. Parthasarathy. Extremal decision rules in quantum hypothesis testing. *Infin. Dimens. Anal. Quantum Probab. Relat. Top.*, 2(4): 557–568, 1999.

[112] V. Paulsen. *Completely Bounded Maps and Operator Algebras*. Cambridge University Press, Cambridge, 2003.

[113] G. Pedersen. *Analysis Now*. Springer-Verlag, New York, 1989.

[114] J.-P. Pellonpää. Complete characterization of extreme quantum observables in infinite dimensions. *J. Phys. A: Math. Theor.*, 44: 085 304, 2011.

[115] A. Peres. Separability criterion for density matrices. *Phys. Rev. Lett.*, 77: 1413–1415, 1996.

[116] D. Petz. *Quantum Information Theory and Quantum Statistics*. Springer, Berlin, 2008.

[117] D. Petz. Algebraic complementarity in quantum theory. *J. Math. Phys.*, 51: 015 215, 2010.

[118] M.B. Plenio and S. Virmani. An introduction to entanglement measures. *Quant. Inf. Comp.*, 7: 1, 2007.

[119] E. Prugovečki. Information-theoretical aspects of quantum measurements. *Int. J. Theor. Phys.*, 16: 321–331, 1977.

[120] M. Raginsky. Strictly contractive quantum channels and physically realizable quantum computers. *Phys. Rev. A*, 65: 032 306, 2002.

[121] M. Reed and B. Simon. *Methods of Modern Mathematical Physics, Vol. I: Functional Analysis*, revised and enlarged ed. Academic Press, London, 1980.

[122] W. Rudin. *Real and Complex Analysis*, third ed. McGraw-Hill, New York, 1987.

[123] W. Rudin. *Functional Analysis*, second ed., McGraw-Hill, New York, 1991.

[124] M.B. Ruskai, S. Szarek and E. Werner. An analysis of completely positive trace-preserving maps on $\mathcal{M}_2$. *Linear Algebra Appl.*, 347: 159–187, 2002.

[125] B. Russo and H.A. Dye. A note on unitary operators in C^*-algebras. *Duke Math. J.*, 33: 413–416, 1966.

[126] V. Scarani, S. Iblisdir, N. Gisin and A. Acin. Quantum cloning. *Rev. Mod. Phys.*, 77: 1225–1256, 2005.

[127] E. Schrödinger. Discussion of probability relations between separated systems. *Math. Proc. Camb. Phil. Soc.*, 31: 555–563, 1935.

[128] A.J. Scott and M. Grassl. Symmetric informationally complete positive-operator-valued measures: a new computer study. *J. Math. Phys.*, 51: 042 203, 2010.

[129] P. Shor. Polynomial-time algorithms for prime factorization and discrete logarithms on a quantum computer. *SIAM J. Comput.*, 26: 1484–1509, 1997.

[130] P.W. Shor and J. Preskill. Simple proof of security of the BB84 quantum key distribution protocol. *Phys. Rev. Lett.*, 85: 441–444, 2000.

[131] M. Singer and W. Stulpe. Phase-space representations of general statistical physical theories. *J. Math. Phys.*, 33: 131–142, 1992.

[132] P. Stano, D. Reitzner and T. Heinosaari. Coexistence of qubit effects. *Phys. Rev. A*, 78: 012 315, 2008.

[133] W.F. Stinespring. Positive functions on C^*-algebras. *Proc. Amer. Math. Soc.*, 6: 211–216, 1955.

[134] E. Størmer. Positive linear maps of operator algebras. *Acta Math.*, 110: 233–278, 1963.

[135] W. Thirring. Atoms, molecules and large systems, in *Quantum Mathematical Physics*, second ed. Springer-Verlag, Berlin, 2002. Translated from the 1979 and 1980 German originals by Evans M. Harrell II.

[136] L. Vaidman, Y. Aharonov and D. Albert. How to ascertain the values of σ_x, σ_y, and σ_z of a spin-1/2 particle. *Phys. Rev. Lett.*, 58: 1385–1387, 1987.

[137] G. Vernam. Cipher printing telegraph system for secret wire and radio telegraphic communications. *J. Am. Inst. Electr. Eng.*, 45: 109–115, 1926.

[138] J. von Neumann. *Mathematical Foundations of Quantum Mechanics*. Princeton University Press, 1955. Translated by R.T. Beyer from *Mathematische Grundlagen der Quantenmechanik*, Springer, Berlin, 1932.

[139] D.F. Walls and G.J. Milburn. *Quantum Optics*, second ed. Springer-Verlag, 2008.

[140] T.-C. Wei, K. Nemoto, P.M. Goldbart, P.G. Kwiat, W.J. Munro and F. Verstraete. Maximal entanglement versus entropy for mixed quantum states. *Phys. Rev. A*, 67: 022 110, 2003.

[141] R.F. Werner. Quantum states with Einstein–Podolsky–Rosen correlations admitting a hidden-variable model. *Phys. Rev. A*, 40: 4277–4281, 1989.

[142] R.F. Werner. All teleportation and dense coding schemes. *J. Phys. A*, 34: 7081, 2001.

[143] E.P. Wigner. Die Messung quantenmechanischer Operatoren. *Z. Physik*, 133: 101–108, 1952.

[144] E.P. Wigner. *Group Theory and Its Application to the Quantum Mechanics of Atomic Spectra*. Expanded and improved edition, translated from the German by J. J. Griffin. Academic Press, New York, 1959.

[145] M.M. Wolf and J.I. Cirac. Dividing quantum channels. *Commun. Math. Phys.*, 279: 147–168, 2008.

[146] M.M. Wolf, D. Perez-Garcia and C. Fernandez. Measurements incompatible in quantum theory cannot be measured jointly in any other no-signaling theory. *Phys. Rev. Lett.*, 103: 230 402, 2009.

[147] W.K. Wootters and B.D. Fields. Optimal state-determination by mutually unbiased measurements. *Ann. Physics*, 191: 363–381, 1989.

[148] S. L. Woronowicz. Positive maps of low dimensional matrix algebras. *Rep. Math. Phys.*, 10: 165–183, 1976.

[149] S. Yu, N. Liu, L. Li and C.H. Oh. Joint measurement of two unsharp observables of a qubit. *Phys. Rev. A.* 81: 062 116, 2010.

[150] M. Ziman and V. Bužek. Entanglement-induced state ordering under local operations. *Phys. Rev. A*, 73: 012 312, 2006.

[151] M. Ziman and V. Bužek. Entanglement measures: state ordering vs. local operations. In M. Zukowski, ed., *Quantum Communication and Security*, pp. 196–204. IOS Press, 2007.

Index